RETURNING MATERIALS:
Place in book drop to remove this checkout from your record. FINES will be charged if book is returned after the date stamped

Genetic Engineering: Applications to Agriculture

PREVIOUS SYMPOSIA IN THIS SERIES:

[1] Virology in Agriculture
May 10–12, 1976
Published, 1977

[2] Biosystematics in Agriculture
May 8–11, 1977
Published, 1978

[3] Animal Reproduction
May 14–17, 1978
Published, 1979

[4] Human Nutrition Research
May 6–9, 1979
Published, 1981

[5] Biological Control in Crop Production
May 18–21, 1980
Published, 1981

[6] Strategies of Plant Reproduction
May 17–20, 1981
Published, 1983

FORTHCOMING:

[8] Agricultural Chemicals of the Future

Beltsville Symposia in Agricultural Research

[7] Genetic Engineering: Applications to Agriculture

Lowell D. Owens, Editor

Invited papers presented at a symposium held May 16–19, 1982, at the Beltsville Agricultural Research Center (BARC), Beltsville, Maryland 20705

Organized by THE BARC SYMPOSIUM VII COMMITTEE
Lowell D. Owens, Chairman

Sponsored by
THE BELTSVILLE AGRICULTURAL RESEARCH CENTER
Northeastern Region, Agricultural Research Service
United States Department of Agriculture

ROWMAN & ALLANHELD, Publishers

GRANADA London Toronto Sydney

Published in the United States of America in 1983
by Rowman & Allanheld, Publishers
(A division of Littlefield, Adams & Company)
81 Adams Drive, Totowa, New Jersey 07512

First published in Great Britain 1983 by
Granada Publishing
Granada Publishing Limited–Technical Books Division
Frogmore, St Albans, Herts AL2 2NF
and
36 Golden Square, London W1R 4AH
117 York Street, Sydney, NSW 2000, Australia
61 Beach Road, Auckland, New Zealand

ISBN 0 246 11947 0

Library of Congress Cataloging in Publication Data
Main entry under title:

Genetic engineering, applications to agriculture.

(Beltsville symposia in agricultural research; 7)
Includes bibliographical references and index.
1. Genetic engineering—Congresses. 2. Genetic engineering—Industrial applications—Congresses. 3. Agriculture—Congresses. I. Owens, Lowell D. (Lowell Davis), 1931– . II. BARC Symposium VII Committee. III. Beltsville Agricultural Research Center. IV. Series.
S494.5.G44G46 1983 631.5′23 83-3252
ISBN 0-86598-112-4

83 84 85/ 10 9 8 7 6 5 4 3 2 1

Printed in the United States of America

Contributors and Their Affiliations

Frederick Ausubel
Massachusetts General Hospital
Department of Molecular Biology
Boston, Massachusetts, 02114

P. Stephen Baenziger
USDA–ARS
Plant Genetics and Germplasm Institute
Field Crops Laboratory
Beltsville, Maryland 20705

Lawrence Bogorad
Harvard University
The Biological Laboratories
16 Divinity Avenue
Cambridge, Massachusetts 02138

Donald D. Brown
Carnegie Institution of Washington
Department of Embryology
115 West University Parkway
Baltimore, Maryland 21210

G. Ram Chandra
USDA-ARS
Plant Genetics and Germplasm Institute
Seed Research Laboratory
Beltsville, Maryland 20705

Edward Cocking
The University of Nottingham
Department of Botany
University Park
Nottingham, NG7, 2RD
United Kingdom

Nina Fedoroff
Carnegie Institution of Washington
Department of Embryology
115 West University Parkway
Baltimore, Maryland 21210

Richard B. Flavell
Plant Breeding Institute
Maris Lane
Trumpington, Cambridge, CB2 2LQ
United Kingdom

J. Leslie Glick
GENEX Corporation
6110 Executive Boulevard
Rockville, Maryland 20852

Robert Goldberg
University of California, Los Angeles
Department of Biology
405 Hilgard Avenue
Los Angeles, California 90024

Richard A. Goldsby
Amherst College
Department of Biology
Amherst, Massachusetts 01002

John Kemp
Agrigenetics Corporation
Agrigenetics Research Park
5649 East Buckeye Road
Madison, Wisconsin 53716

Douglas Moore
USDA-ARS
Plum Island Animal Disease Center
Greenport, Long Island, New York 11944

S. Muthukrishnan
Kansas State University
Department of Biochemistry
Manhattan, Kansas 66506

Lowell D. Owens
USDA-ARS
Plant Physiology Institute
Tissue Culture and Molecular Genetics Laboratory
Beltsville, Maryland 20705

Robert A. Owens
USDA-ARS
Plant Protection Institute
Plant Virology Laboratory
Beltsville, Maryland 20705

Daryl R. Pring
USDA-ARS
Plant Pathology Department
University of Florida
Gainesville, Florida 32611

Gideon W. Schaeffer
USDA-ARS
Plant Physiology Institute
Tissue Culture and Molecular
Genetics Laboratory
Beltsville, Maryland 20705

Jozef Schell
Rijksuniversiteit—Gent
Laboratorium voor Genetics
K. L. Ledeganoksgtraat 35
B-9000 Gent, Belgium

Trevor A. Thorpe
The University of Calgary
Department of Biology
2500 University of Dr. N.W.
Calgary, Alberta, Canada T2N 1N4

Desh Pal S. Verma
McGill University
Department of Biology
Montreal, Quebec, Canada H3A 1B1

Richard H. Zimmerman
USDA-ARS
Horticultural Science Institute
Fruit Laboratory
Beltsville, Maryland 20705

Symposium Organization

Beltsville Agricultural Research Center (BARC)

P. A. PUTNAM, DIRECTOR

BARC SCIENCE SEMINAR COMMITTEE

Suzanne W.T. Batra, Chair
Murray R. Bakst, Vice Chair
Edward Allen
Gordon C. Carpenter
Albert B. DeMilo
Robert L. Jasper
John G. Moseman
Robert D. Romanowski
Richard M. Sayre
Christopher A. Tabor
William P. Wergin
Richard H. Zimmerman

BELTSVILLE SYMPOSIUM VII ORGANIZING COMMITTEE

Lowell D. Owens, Chair

PROGRAM

Lowell D. Owens, Chair
P. Stephen Baenziger
Dean E. Cress
Roger H. Lawson
Robert A. Owens
Vernon G. Pursel
Gideon W. Schaeffer
Lila O. Vodkin

ARRANGEMENTS

Freddi A. Hammerschlag, Chair
Judith A. Abbott
Austin Campbell
Gordon C. Carpenter
William S. Conway
Benjamin F. Matthews
Richard Sicher, Jr.
Charles Sloger

FINANCE

C. Jaycn Baker, Chair
Christopher A. Tabor
Norberta Schoene
Judy Calvert

PUBLICITY

Richard H. Zimmerman

PRESS RELATIONS

Ellen Mika

LOGO

Olivia C. Broome

Contents

[four] DNA Cloning

[five] Genetic Modification

[six] New Plants via Tissue Culture

Foreword

The annual Beltsville Symposium serves as a forum for examining recent developments in basic research that may aid in discovering solutions to agricultural problems. The seventh in this series focused on techniques developed in the past few years that promise to provide powerful new strategies for the genetic or phenotypic improvement of plants, microbes, and animals that are important to agriculture. These strategies include gene isolation and cloning; genome characterization; genetic modification through the transfer of DNA or organelles, protoplast fusion, or mutation and selection; hybridoma formation to produce monoclonal antibodies; and the design of gene vectors for microbial production of animal vaccines and hormones.

During May 16 to 19 of 1982, more than six hundred persons from nineteen countries assembled at the Beltsville Agricultural Research Center to hear twenty-two invited lectures and view forty-six poster presentations. While the symposium identified instances in which the new biotechnologies are being commercially exploited or are directly providing plant breeders with novel germplasm, it also made clear that the core genetic tool, recombinant DNA technology, is being used primarily in the discovery of new knowledge. Specifically, it is making possible investigations of how genes of higher organisms are structured and expressed and how their expression is controlled during the developmental stages of the organism. This knowledge, together with the identification of specific gene products, will form the basis for the truly remarkable advances of the future in the genetic improvement of life forms important to agriculture.

Paul A. Putnam, Director
Beltsville Agricultural Research Center

BELTSVILLE SYMPOSIA in AGRICULTURAL RESEARCH

A Series of Annual Symposia Sponsored by
THE BELTSVILLE AGRICULTURAL RESEARCH CENTER
Northeastern Region, Agricultural Research Service
United States Department of Agriculture

[7] Genetic Engineering: Applications to Agriculture

one
INTRODUCTION

1] How Modern Methods are Solving Biological Problems

by DONALD D. BROWN*

INTRODUCTION

Genetic biochemistry has become one of the most exciting fields of modern biology. The great transition in how we do genetics dates from the Watson-Crick discovery of the chemistry of DNA, but in the last ten to fifteen years a series of newly discovered methods has totally revolutionized the way genetics is used to solve biological problems. I shall summarize them very briefly. The basic tool today for working on genes from a chemical point of view is called molecular hybridization. It takes advantage of Watson-Crick base pairing complementarity and enables one to measure quantitatively the abundance of a gene, and the amount of the gene product RNA. It is used to isolate genes and to map genes on chromosomes. Molecular hybridization was developed in 1968 by Roy Britten and his colleagues at The Department of Terrestrial Magnetism of the Carnegie Institution of Washington. Next came the discovery of dozens of different enzymes that will cleave DNA or RNA in every imaginable way, phosphorylate DNA, remove one strand or the other, and splice DNA back together again. Improvements in protein fractionation including immunology have been essential.

Central to the new methods is recombinant DNA technology—the ability to clone any gene from any source in bacteria, and to grow it in massive amounts. For example, there is one part in 10^6 of the globin gene in each of us. If we did whole body homogenization of 500 persons, and even if we had techniques for isolating the globin gene from the rest of the DNA, we wouldn't obtain as much of the gene as we can from one liter of bacteria using recombinant DNA technology. The techniques are simple, and they provide homogeneous preparations of DNA fragments.

DNA sequencing, developed by Fred Sanger at Cambridge, now makes it easier to sequence DNA than any other macromolecule. One no longer

*Carnegie Institution of Washington, Department of Embryology, 115 West University Parkway, Baltimore, Maryland, 21210.

sequences proteins, but rather isolates the gene, sequences it, and from this predicts the amino acid sequence of the protein.

With increasing frequency researchers are reporting successes in gene transfer methods, i.e., introducing genes, often in a recombinant form, into living cells or organisms. Also, it is now possible to do a different kind of genetics, called genetics in vitro, or surrogate genetics. Purified genes in recombinant form can be deleted or mutated with enzymes and tested for their function in the test tube. With such techniques we can delimit regions in and around genes that are responsible for their function in the living cell. The next step is to identify molecules in cell extracts that influence the expression of these genes.

These methods have amalgamated the disparate fields of biology. During the past fifty years biochemists interested in developmental biology studied chickens, frogs, and sea urchins. Geneticists studied mainly *Drosophila* and, to a lesser extent, mouse. One might have preferred to have studied yeast, the closest eucaryote to *Escherichia coli* for a geneticist, but it doesn't develop very much. These new techniques now permit genetics with frogs and biochemistry with *Drosophila*. Previously, one scientist spent his entire career characterizing less than 20 mutants of frogs. Now in one day, with modern biochemical techniques, one can generate hundreds of mutants for any single gene and the next day characterize them by DNA sequencing. Likewise, until five or ten years ago there hadn't been a single protein purified from *Drosophila*—even with the hundreds of mutants that had been discovered. No single mutant had been found that correlated with a structural change in a protein.

It is clear that another major barrier is being overcome, that between animal biologists and plant biologists. The presidency of the Society of Developmental Biology alternated each year between an animal biologist and a plant biologist. Now everyone is coming together and speaking the same language. Today genetic experiments are done differently than ever before. Consider the old central dogma of genetic informational flow from DNA to RNA to protein, and the attending disciplines. Here is another amalgamation, because previously one worked as a DNA chemist or a protein chemist or an RNA chemist, but never all three. The new methods now permit the scientist to flow from one macromolecule to the other, starting at the protein and working back to the gene, or starting at the gene and working to the protein. Soon, organisms with simple genomes, such as *Drosophila,* yeast, or the nematode *Caenorhabditis elegans,* will have their entire genomes represented in libraries in recombinant DNA form. Genes will be located not by traditional genetic mapping but by restriction enzyme polymorphism. This technology is being applied today for yeast and particularly in *Drosophila,* where there were already some important genetic markers to serve as a start for this kind of library.

Two dimensional (2-D) gels can be used to fractionate complex protein mixtures, and computer programs can facilitate their analysis. It is feasible to consider describing the 2-D gel location of every single human protein. How can these methods be applied? In a disease such as cystic fibrosis, which is caused by a single genetic defect, and for which the product of the gene defect is unknown, one might ask whether there is a single altered or missing protein. This is brute force biochemistry. If one finds a protein that looks like a good candidate for the defective gene product, then it is easy to work back to the gene itself. The protein is purified from even the tiny amount present in a gel, and a partial amino acid sequence obtained at the picomole level. Using this partial sequence information a short polypeptide is synthesized against which antibody is made. The antibody is used to isolate the protein in large amounts for study. Alternatively, one synthesizes a DNA primer based on the amino acid sequence in that region. Using this primer and a population of total messenger RNA (mRNA), a complementary DNA (cDNA) molecule is synthesized, cloned in bacteria and used as a probe for the native gene. Thus, from a protein one can find its gene, and of course, from a gene we can find the gene product.

GENE EXPRESSION

Genes affecting development. There exists in *Drosophila* a fascinating group of mutants called homeotic mutants, that have extraordinary developmental significance. One single mutation changes an antenna to a leg. How can a single gene have such an extraordinary effect? David Hogness and his colleagues at Stanford, using the DNA library technology, have succeeded in isolating and mapping one such gene. What can one do with a milligram of a gene that causes an antenna to change to a leg? To begin with, one can find out by various modern methods whether the gene is transcribed into RNA and whether the RNA encodes a protein. The molecular explanation of how a tissue develops into a leg rather than an eye is not elucidated by the discovery of the gene product, but the latter enables the investigation of certain questions. Is this protein located on the cell surface? Is it involved with cell recognition? Is it a nuclear protein? Furthermore, when one has such a gene, and assuming gene transfer methods are available, one can mutate the gene, transfer it back into an organism, and then ask whether or not the phenotype has been altered in any significant way. Thus, rather than finding mutants by traditional means, the gene is mutated in vitro and reintroduced into the organism by gene transfer methods. This has been carried out successfully for *Drosophila,* as I will describe later.

Transcription initiation control. The study of the regulation of a gene's expression involves questions at two levels of control. First, what influences the correct starting of RNA synthesis and the correct stopping of RNA synthesis, i.e., the initiation and termination of transcription? Superimposed upon that level of control is the developmental control of genes. Why is the globin gene turned off in a skin cell when it works quite well in a red blood cell? Using in vitro systems we were able to show that initiation of transcription of the 5S ribosomal RNA gene depends on a region not on either side, but right in the middle of the gene. We now know from the work of R. G. Roeder and his colleagues in St. Louis that this internal control region in the middle of the gene is a binding site for a specific positive transcription factor; it does not recognize even the closest related eucaryotic genes, those coding for transfer RNAs. The protein binds to the center of the gene and somehow directs the RNA to start transcription accurately. When the polymerase reaches the end of the gene it recognizes the termination signal, terminating the transcription event. All of this information was accumulated by virtue of the fact that the gene could be cloned, mutated in vitro, and transcribed faithfully using cell extracts. Experiments on the developmental control of genes have coupled in vitro mutagenesis with gene transfer methods. Some success in reproducing developmental control in vitro and after gene transfer is beginning to yield information about the sequences in and around a gene that are responsible for controlling its expression.

A missing technology—the "functionator." From this discussion it is obvious that there is an enormous missing technology. The example of the mutant that caused the change of an antenna to a leg, or the example of cystic fibrosis, focused on the goal of identifying and isolating the protein that is altered or missing. Yet what is ultimately important is the function of that protein, i.e., its role in the molecular biochemistry of cystic fibrosis or in tissue determination. We would like to be able to take the protein out of the gel and reform its secondary structure and assay its funciton. If it is an enzyme, we would like to be able to determine by screening a variety of substrates that the protein is a kinase, or an ATPase, or that it possesses some other biochemical function of interest. We are some distance from that technology now, but I suggest that it will be one of the emerging technologies of the next decade, one that will help explain, in biochemical terms, genes whose phenotypes are clear now only in descriptive terms.

GENE ALTERATION

Most of what is now known about gene control in eucaryotes were total surprises at the time of their discovery. Two general kinds of control of gene expression are found in eucaryotic organisms—the alteration of genes on the onehand, and differential gene expression on the other. There are several different kinds of gene alteration: gene amplification, in which an organism makes more copies in a cell of one or a few genes to make more of a gene product; and genetic rearrangement, discovered first in maize by Barbara McClintock. Differential gene expression refers to the modulation of gene activity with no change in the genes themselves.

Amplification. Gene amplification was first discovered in our laboratory by Igor Dawid and myself and independently by Joe Gall at Yale for the ribosomal RNA genes of a frog. The purpose of the massive increase in the number of these genes that occurs in the growing oocyte is to synthesize more gene product. Since these discoveries in 1968, many more cases of gene amplification have been found. Now there are examples of genes which are amplified at specific stages in the development of an organism. Furthermore, it is now possible to carry out what I call "forced" gene amplification. It has long been known that if one administers to bacteria a drug that specifically inhibits some enzymatic system, the bacteria responds by making more genes for the enzyme and thereby becomes more resistant. R. T. Schimke and his colleagues demonstrated that animal cells in culture will amplify folate reductase genes in response to increasingly higher levels of the inhibitor methotrexate. Many examples of forced gene amplification are being reported. Methotrexate, for example, is a drug used to treat leukemia. With time, patients become resistant to methotrexate because, as would be predicted from these experiments, they have amplified their folate reductase genes. The isolation of a folate reductase gene that is resistant to this drug provides selective means for amplifying other genes. Using recombinant DNA technology, the desired gene is inserted next to the folate reductase gene. Following transformation with the cloned DNA, one selects for cells that are resistant to the drug. By maintaining this selection pressure one is able to transform specifically for just the desired gene and force its amplification to many copies. This will certainly be a very powerful method for the future.

Rearrangement. Studies on gene rearrangement began in the 1940s with the work of Barbara McClintock. Her work was ignored for a long time because its relevance was not appreciated. Now we know that organisms from bacteria to man are literally teeming with transposable elements. They occur in the middle of genes and around genes, and they are moving about. Almost certainly the majority of genetic mutations will be shown to be due not to single base changes, but to the insertion of transposable elements in or around genes. In my opinion, of all the current cancer theories the idea of transposable elements resulting in genetic rearrangements represents the best explanation for somatic mutations which could lead to cancer.

A well-documented story is that of antibody production. It is now known that at least part of the diversity of immunoglobulins is due to a genetic rearrangement programmed into the development of cells of the immune system. The cell takes two pieces of genetic material, neither of which functions by itself, and brings them together to form one single gene, which then encodes the production of a single polypeptide chain. In another example, transposition or rearrangement of genes confers upon parasitic trypanosomes the ability to avoid their host's defense mechanisms. Trypanosomes alter their surface antigens by rearranging the genes for these molecules at a high frequency, enabling resistant populations in an infected individual to escape the host's antibody response and to rebuild their population again.

Recently the transformation of *Drosophila* has been accomplished by Gerald Rubin and Allan Spradling in our department at the Carnegie Institution, using transposable elements. Since transposable elements have the unique ability of being able to move around in the genome, Rubin and Spradling predicted and confirmed that genes could be introduced into the chromosomes of *Drosophila* if they were associated with transposable elements. In fact, genes have been inserted into the flies' germ line, resulting in permanent stable transformation.

It may well be that transposable elements of one kind or another will be general vectors for transforming cells of almost any eucaryote in the future.

CHAOS IN THE EUCARYOTIC GENOME

Consider where basic research has brought us! Fifteen years ago I considered the eucaryotic genome to be organized with absolute precision, where every nucleotide mattered, where everything was in exactly the correct order and was quite stable. One gene followed another in perfect array. But the results from all of these recent experiments lead us to view that the eucaryotic genome is a colossal junkyard. It is difficult, sometimes, to find which gene is the normal functional one, because genes occur in multigene families where some function and others, called pseudogenes, are evolutionary relics that are nonfunctional. Given a set of eight or ten genes which are all slightly homologous to each other, it is not a trivial task to sort out which gene is the functional one.

"Excess" DNA. Satellite DNA, discovered in the early 1960s, is comprised of simple sequences of DNA, some as short as two nucleotides long, repeating over and over again, comprising in some organisms more than half of their entire genetic material. To this day no function has been found for satellite DNA. Generally, it is not transcribed into RNA, i.e., it doesn't code for a protein, although there are occasional exceptions to this rule.

Another mystery is the immense polyploidy, or high DNA content, of some organisms. Two plants may appear essentially identical, but one of them may have more than twice as much DNA in each nucleus as does the other. How did this happen? What is its purpose? Does it have a function? There is much speculation but not very much data. More than 20% of the human genome consists of dispersed repeated sequences - - - , highly repetitive copies of DNA sequences 300 to 1,000 nucleotides long that are dispersed all over the genome. Their function is also unknown.

Intervening sequences. The most remarkable and striking discovery made on eucaryotic genes in recent years is that of intervening sequences. Most genes for proteins are split in the genome. As many as 50 intervening sequences can exist in one gene. The collagen gene of a chicken is 35 kilobases long and codes for a final mRNA that is only a few percent of this length, due to the presence of more than 50 intervening sequences. The vast majority of the gene is intervening sequences. In order to make a functional mRNA, the entire 35 kilobases must be transcribed and the RNA transcripts with intervening sequences removed. What is the possible function of intervening sequences? A number of hypotheses have been proposed. Walter Gilbert has proposed that they have evolutionary value for the construction of new genes. When gene duplication occurs. sections of the duplicate gene could be used to construct a new hybrid gene. Evolution would be enhanced by the movement of the functional domains of genes. Another hypothesis has been that they are essential for the transport of RNA to the cytoplasm. Some evidence was found for this. If one removes one of the intervening sequences of the globin gene, the resulting RNA does not reach the cytoplasm. In another gene, however, removal of the intervening sequence has no effect on RNA transport. The presence of intervening sequences with concommitant splicing leads to many regulatory possibilities. For example, a gene with intervening sequences could lead to the synthesis of different proteins just by differential splicing of the mRNA. In one case it has been shown that differential splicing results in two slightly different proteins with important but distinct functions. But generally speaking this doesn't appear to happen. Perhaps intervening sequences have regions that affect the control of a gene's expression. Mutations of the globin gene at the splice junction result in the genetic disease thallasemia. The transcribed RNA cannot be processed, and no functional mRNA for globin is made. So, there can be dire consequences if the splicing process is destroyed. A fascinating example can be found in yeast mitochondria where the intervening sequence of a gene codes for a protein which then splices out the intervening sequence to make functional mRNA—rather a circular rationale for the presence of intervening sequences.

Implications. Never predicted, intervening sequences are still not understood five years after their discovery. I maintain that this situation has immense implications for how we do biological research and for how we should support biological research. There is no theoretical arm of biology. Biology is an experimental science. Success lies in new methods, hard work, and bright people. Consider the current arguments in the field of evolution—the "gradualists" versus the "punctuationalists." They can argue incessantly about how evolution has occurred—gradually, as was originally imagined by Darwin, or in quantum leaps. Maybe the long-sought-for missing links never existed. When one considers the junkyard nature of the eucaryotic genome, who could be surprised at whatever happens during evolution? Evolution could be gradual for a while, and then there might be an explosion of transposable elements, wiping out a whole species.

HOW TO ADVANCE BIOLOGICAL SCIENCE

How have we been so successful at this research enterprise? At the end of World War II the National Institutes of Health did something significant when they recognized that there were important societal problems to be solved. Everyone can list the important diseases to be cured or the important agricultural problems; that is not difficult. The question is how one solves them. The National Institutes of Health listed some societally important problems in biomedical research and gave them the names of institutes. But what they didn't do was to decide how scientists were going to attack these problems. Instead they decided to set up a competitive grants program and support individual investigators. And just as important, they recognized that young scientists had to be supported as soon as possible in their careers and given independence. They must be induced to follow their own scientific instincts. Implicit in this system is the assumption that bright people, if supported, will come up with good ideas. The old Manhattan project approach was discarded for the unqualified support of gifted scientists. This approach has made the United States preeminent in the area of biomedical research. It is the essence of the U.S. system of research support. People now come from all over the world to study in biomedical laboratories throughout this country. The U.S. system for biomedical research is unique in that young people become independent very early. But now, just as everyone is admitting that basic research is not irrelevant, the U.S. government is cutting back its support of the biological research enterprise that has been so successful. This has profound implications and leads to important questions for several of the constituencies comprising the audience of the Beltsville Symposium: members of the USDA, members of private industry, and scientists who may or may not have ties to industry. To members of the USDA, I ask why the competitive research grants system is so small and so reluctantly supported? Where are the postdoctoral fellowship training programs that will encourage beginning scientists to take up plant biology? There are young scientists trained as experts in animal viruses who would love to study plant viruses, but who will support them and who will fund them? Could not some fraction of the funds the USDA now appropriates to research in land grant programs be distributed by competition? I submit that the return on investment over the long run would be enormous. To members of private industry who are so excited about the potentials of biotechnology—the biotechnology which has resulted from years of basic studies at academic and nonprofit research institutions all over the country—you have a long-term self-interest, as does the USDA, in supporting and strengthening these institutions. There are dangers in the current perturbation of this delicate enterprise caused by emphasis on contract research, and by putting money into research superstars in areas important for short-term gains. Some fraction of your profits should be reinvested in the research system in a nontargeted way. The semiconductor industry is experimenting on how to do this right now. The chemical industry is considering it seriously.

The key to good research is competition and the support of young people. Some of us in the academic world have started a foundation, the Life Sciences Research Foundation, to try to encourage the funding of nontargeted research. Our first attempt is a postdoctoral fellowship program. We present ourselves essentially as an inexpensive peer review mechanism that will select the very best young scientists wherever they are in biology, studying whatever field interests them. We believe this is the time-tested way to improve biological research in this country. And I encourage you in industry to set aside some fraction of your profits for this purpose. Finally, I address those of you in academia, some with equity in companies, others who are consultants. Remember where the funds for your own scientific support and development came from, and demand that the companies you work for set aside a fraction of their profits for nontargeted research and training. It is essential that we consider now the long-term interests of this remarkable research enterprise. Because, after all, we need to not only make use of the present technology in the best possible way, but also to water the roots and lay the foundation for the biotechnology of ten years in the future.

two

MOLECULAR GENETICS

2] Chromosomal Variation at the Molecular Level in Crop Plants

by RICHARD B. FLAVELL*

Abstract

Many kinds of mutations accumulate in plant chromosomes. They frequently involve the amplification, deletion, rearrangement, and transposition of short segments of DNA. These molecular events, revealed by molecular biology and genetic engineering techniques, result in the "turnover" of many DNA sequences over evolutionary timescales. Knowledge of these sorts of mutations is bringing new understanding to some aspects and problems of plant breeding. The repeated sequences in chromosomes provide simple ways of detecting genetic variation in plant populations. Several methods are illustrated. Some general points concerning the molecular basis of variation in plant populations are discussed.

INTRODUCTION

Plant breeding involves the exploitation of genetic variation in plant chromosomes and its recombination into new permutations. To understand the nature and the origin of the variation the plant breeder uses, it is essential to be able to look at chromosomes and genes at the molecular level. The development of techniques in molecular biology and, in particular, the complete purification of large amounts of specific DNA sequences by cloning in bacteria, has made this possible. Even though the application of these techniques to plant DNA sequences is in its infancy, a considerable amount of general information has already accumulated, and we have a more detailed and, in many ways, different picture of the molecular basis of genetic variation than the one of only a few years ago.

Much of our current perspective on the structure of plant chromosomes and possible molecular origins of genetic variation has come from studying sequences that occur many times in the haploid genome (repeated sequences).

*Plant Breeding Institute Trumpington, Cambridge CB2 2LQ, UK.

GENETIC ENGINEERING: APPLICATIONS TO AGRICULTURE (Beltsville Symposium 7—Lowell D. Owens, ed.) Rowman & Allanheld, Totowa

This is because (a) they contribute most of the DNA (Flavell 1980), (b) they are technically easier to study, and (c) they are often not highly conserved and so tolerate all sorts of mutational events. Furthermore, on the assumption that the related sequences arose initially from a single sequence, characterization of the structural and organizational differences between individual members of the family provides valuable information on the kinds of molecular changes that occur and accumulate in chromosomal DNA. Extension of intraspecies studies of repeated sequences to groups of closely related species has taught us that much of the chromosomal DNA outside the coding sequences changes at a dramatic rate during evolution. Recurrent sequence amplification and deletion events can spread rapidly through populations and result in DNA "turnover" (Flavell 1980, 1982a; Thompson and Murray 1980; Dover et al. 1982). Old sequences are replaced by new ones. Sometimes the new sequences are different versions of the old sequences; sometimes they are unrelated to previously amplified sequences.

Chromosome sequence variation is also created by the movement ("transposition") of short pieces of DNA to new sites in the genome (Flavell, O'Dell, and Hutchinson 1981). This has been clearly established by finding closely related sequences in thousands of different positions on all chromosomes. Structurally similar sequences are often also at different positions in closely related species. The mechanisms producing these various kinds of "macromutations" are not yet well understood. Nevertheless, they have been major architects of chromosomal DNA organization throughout evolution.

To illustrate the kind of chromosomal variation coming from major changes in copy number of a sequence and its transposition, I include two examples. In the first, the positions of major arrays of a repeated sequence in the hexaploid wheat genome are shown (Fig. 2.1). The sequence is present in large blocks on *all* the chromosomes coming from the diploid ancestor of the B genome but only on chromosome 4 of the A genome. No large blocks of the sequence are present in the D genome chromosomes (Hutchinson and Lonsdale 1982). The presence of the same sequence on all chromosomes of the B genome diploid illustrates that sequences move between genomes. The much lower amounts of the sequence in the A and D genome diploids illustrate how all the chromosomes of a diploid can differ substantially from a close relative due to sequence amplification and dispersal. In fact, the blocks of the sequence in chromosome 4A are found only in hexaploid wheat, not in the present-day A genome diploids studied to date (Gerlach and Peacock 1980; Flavell 1982a; Hutchinson and Lonsdale 1982). These chromosomal changes, therefore, may have occurred after the origin of tetraploid or hexaploid wheat.

In the second example, a short sequence purified from *Triticum monococcum* has been hybridized to DNAs isolated from several *Aegilops* species, after restriction with the endonuclease Sau-3A and transfer to a nitrocellulose filter (Southern 1975). The different amounts of the sequence in the six species is very evident (see Fig. 2.2), and the different hybridization banding patterns result from amplification and/or deletion events occurring in one species and not another. This sequence is localized at a large number of sites in *Aegilops*

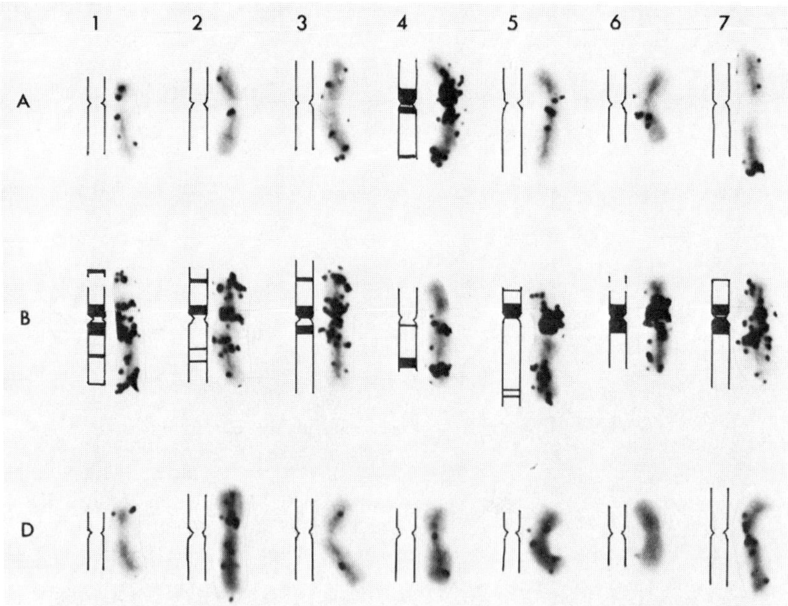

Figure 2.1. The distribution of a family of repeated sequences in the three diploid genomes of hexaploid wheat. A ^{3}H-labeled repeated sequence purified by molecular cloning from the wheat genome was hybridized to metaphase chromosomes and the sites of hybridization revealed by autoradiography. The chromosomes have been arranged into three (A, B & D) genome sets, each containing seven chromosomes. A schematic representation of each hybridization pattern is shown for each chromosome. (From Hutchinson and Lonsdale 1982.)

speltoides chromosomes but at very few in *Aegilops comosa* chromosomes. Therefore, from consideration of this sequence alone, the differences between the chromosomes of *Aegilops speltoides* and *Aegilops comosa* must involve a huge number of sites.

Only a small proportion of the DNA in plant chromosomes is made up of coding and other sequences essential for gene expression, replication, etc. (Flavell 1980). Most of the DNA appears to be "secondary" DNA (Hinegardner 1976). The "turnover" of DNA sequences during relatively short evolutionary timescales occurs mainly in the "secondary" DNA (Hinegardner 1976). This is presumably because natural selection prevents fixation of many of the variants that arise in the coding sequences. Thus, as species diverge the coding regions usually remain similar, but secondary DNA may diverge relatively rapidly.

For the subsequent sections of this chapter I have chosen a number of topics relevant to plant breeding where knowledge of these kinds of molecular changes is bringing new understanding and new opportunities.

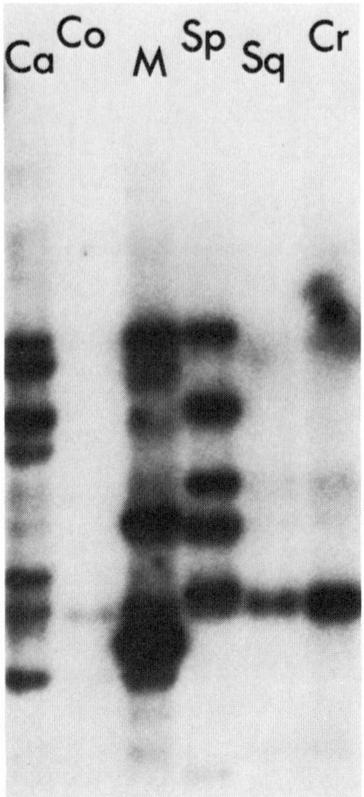

Figure 2.2 Hybridization patterns of a cloned DNA sequence from *Triticum monococcum* to a range of *Aegilops* DNAs. The *Aegilops* DNAs were restricted with the endonuclease Sau-3A, fractionated by electrophoresis on agarose, transferred to nitrocellulose (Southern 1975), and hybridized with a ^{32}P-labeled, cloned DNA sequence. The sites of hybridization were revealed by autoradiography. Ca = *Aegilops caudata,* Co = *Ae. comosa,* M = *Ae. mutica,* Sp = *Ae. speltoides,* Sq = *Ae. squarrosa,* Cr = *Ae. crassa.*

VARIATION IN GENOME AND CHROMOSOMAL SIZE

More than a 200-fold variation in genome size occurs among flowering plants (Bennett and Smith 1976). This variation is due mostly to the differential fixation of the products of DNA amplification events and changes in polyploidy. What determines the differential rates of fixation of repeated sequence DNA in different species is very complex (Dover et al. 1982; Flavell 1982a), but variation in the total mass of DNA does appear to have phenotypic consequences (Bennett 1972; Cavalier-Smith 1978). Increasing the total DNA mass appears to reduce the maximum rate of development through mitosis and meiosis, unless polyploidy is involved. It also results in larger cells and pollen (Bennett 1973).

The observation that crop plants with low DNA content are naturally adapted to equatorial latitudes, while species with higher DNA contents are naturally adapted to polar latitudes or high altitudes (Bennett 1976), suggests that geographical limits to the cultivation of specific species cannot be altered unless the breeder alters the DNA contents or overcomes the effects of DNA mass by other means, e.g., the creation of polyploids.

CHROMOSOME STRUCTURE AND THE EXPLOITATION OF INTERSPECIES HYBRIDS

Making crosses between different species has long been an activity of the plant breeder. Sometimes the aim has been to create stable interspecies hybrids as new crops, for example, triticale. On other occasions the aim has been to introduce one or a small number of characters from one species into another. All sorts of problems have arisen in the efficient exploitation of interspecies hybrids. Chromosomes of one parent are sometimes eliminated, somatic growth is sometimes poor due to genetic interactions, and sterility is common because of poor chromosome pairing at meiosis. The poor chromosome pairing and reduced recombination also make difficult the introduction of a few genes from one species into another.

Undoubtedly these interspecies incompatibilities have many molecular causes, but research over the past few years into the genomes of related cereal species suggests that the relatively rapid differential "turnover" of "secondary" DNA sequences during species divergence may be the molecular basis of many of the problems.

Where the chromosomes of two related species have diverged considerably, due to the amplification, deletion, and transposition of DNA sequences, it would be surprising if chromosome pairing and recombination were not affected in the F_1 hybrid between the two species. Studies of DNA sequence homology between species and chiasma frequency in hybrids of *Aegilops, Triticum, Secale,* and *Hordeum* have shown a crude correlation between a reduction in chiasma frequency and DNA sequence divergence (Flavell 1982a).

A developmental problem in the wheat x rye hybrid, triticale, can also be attributed, in part, to structural dissimilarities between the wheat and rye chromosomes. The developmental problem, high-lighted by the plant breeder as one of the special problems associated with breeding high-yielding triticale, is grain shriveling. The shriveling is due to abnormalities in nuclear division in the very early stages of endosperm development (Thomas et al. 1980). The abnormalities appear to result from the failure of some terminal regions of rye chromosomes to complete division at mitosis (Bennett 1977). These terminal regions are heterochromatic and contain very large numbers of a few repeated sequence families that are not present in the wheat chromosomes (Appels, Driscoll, and Peacock 1978); Bedbrook et al 1980; Appels et al. 1981). It therefore seems reasonable to conclude that the amplification and fixation of repeated sequences on the ends of all rye chromosomes is the molecular source of a developmental problem in the wheat x rye hybrid. This conclusion is supported by the studies on triticale lines that have deletions of some of these repeats. These lines have reduced grain shriveling and higher grain test weights and yields (Gustafson and Bennett 1982).

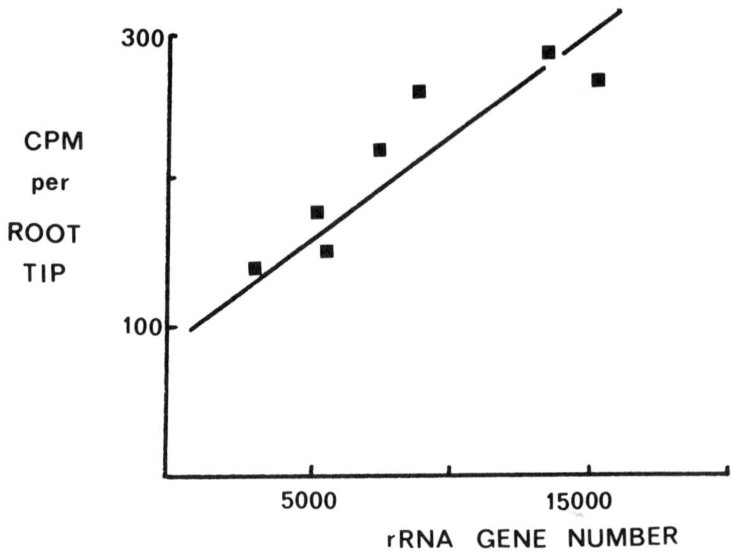

Figure 2.3. Detection of variation in rDNA using root-tip squashes. Root tips of wheat plants differing in the number of RNA genes (Flavell and O'Dell 1976) were squashed onto nitrocellulose, the DNA denatured in alkali, neutralized, and then baked at 80°C. The DNAs were then hybridized with ^{32}P-labeled plasmid DNA pTA71 containing rRNA genes from wheat (Gerlach and Bedbrook 1979). The amount of labeled rDNA bound to each squash was determined by scintillation counting. Each point is the mean of four replicates. (From J. Hutchinson, A. Abbot, and R. Flavell, unpublished.)

RAPID DETECTION OF VARIATION USING DNA PROBES

The variation within and between species due to the amplification and deletion of sequences is especially easy to assay using radioactive, repeated sequence, DNA probes. If this variation is closely linked to phenotypic traits the breeder wishes to select in his populations, then the assays provide a rapid means of identifying the desired plants without growing the plants beyond the small seedling stage. We (J. Hutchinson, A. Abbott and R. Flavell, unpublished) have recently developed a simple assay that enables thousands of plants to be assayed in a few days. Root tips or other pieces of tissue are squashed onto filter paper and, after denaturing the cellular DNA by dipping the filter paper in alkali and then neutralizing, the DNA is baked on to the paper. The paper is then incubated with a ^{32}P-labeled denatured DNA probe to allow specific hybridization of the probe to the denatured plant cell DNA. The amount of the probe bound to the filter paper, which measures the relative amount of the sequence in the cellular DNA on the filter paper, is determined by autoradiography or by scintillation counting. The value of the method to the plant

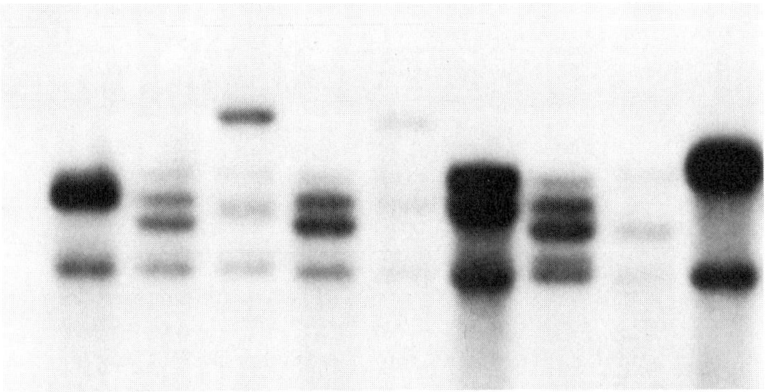

Figure 2.4. Variation at the rDNA loci in wheat, detected using restriction endonucleases. DNAs from different wheat varieties were cleaved with the restriction endonucleases Bam HI and EcoRI, fractionated on an agarose gel, transferred to nitrocellulose (Southern 1975), and hybridized with the plasmid pTA 71, which contains rRNA genes from wheat (Gerlach and Bedbrook 1979). The hybridization patterns were detected by autoradiography.

breeder will depend upon the number of suitable probes isolated by molecular cloning techniques, the variation in the amount of the sequence within the plant population, and the genetic linkage of the repeated DNA locus to important phenotypic characters. The method has already been used to distinguish maize seedlings with different cytoplasms conferring male sterility (Flavell 1982b) and wheat lines containing individual rye chromosomes (J. Hutchinson, A Abbott and R. Flavell, unpublished). To illustrate the use of the method, hybridization of ^{32}P-labeled rDNA to root tips of wheat lines differing in the number of rRNA genes is shown in Figure 2.3. Each point is the mean of four replicates. Even with this crude assay it appears that genotypes differing in rDNA by a factor of two could be distinguished readily. Because the assay uses only root tips it is nondestructive, and the selected young seedlings can be grown for further study.

It is unlikely that a large number of loci within a species can be monitored this way, because suitable quantitative variation in copy number of a sequence will rarely exist or be found. Sequence variation in individual genes or in sequences closely linked to a gene can be detected, however, using restriction endonucleases in combination with a labeled DNA probe. For these assays it is necessary to purify the DNA from each plant. The numbers of plants that can be conveniently assayed is therefore considerably fewer. Different hybridization patterns of purified ribosomal DNA genes to DNAs from wheat varieties is shown in Figure 2.4 (see also Flavell 1982b). The variation in the banding patterns is due principally to variation in the spacer DNA between the transcribed portions of the genes. There is clearly much variation in this region within wheat, and the region is therefore a useful one for following loci closely linked to it through breeding programs. The rDNA loci measured here are on the short arms of chromosomes 1B, 6B, and 1A (Flavell and O'Dell 1976).

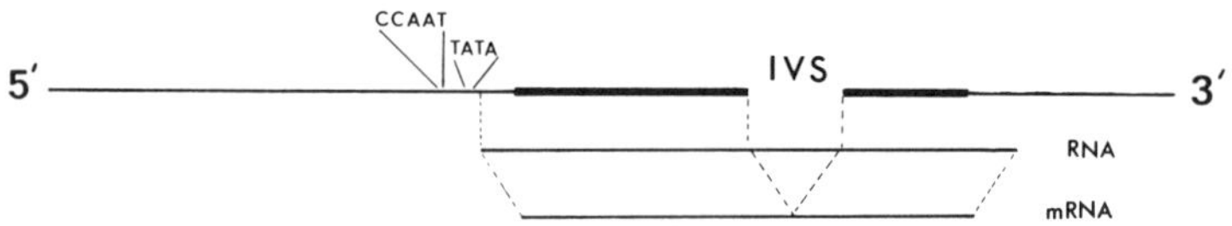

Figure 2.5. Schematic representation of a gene.

MOLECULAR BASIS OF VARIATION IN GENE EXPRESSION

It is vital to understand the relationships between gene structure and function if specific kinds of mutations are to be recognized by DNA assays or if genes with specific properties are to be constructed for reinsertion into plants. Some general features about sequences commonly found necessary for gene expression are emerging, especially from animal gene studies, but it will be necessary to study each gene of interest to gain specific detail. In Figure 2.5 the structure of a gene with one intervening sequence is schematically presented. Sequences related to TATA, about 30 base pairs (bp), and to CCAAT, about 80 bp from where RNA transcription is initiated, are important for the correct initiation of transcription (for example, see Grosveld et al. 1982). Other characteristic sequences determine the end points of transcription. Yet little is known about the role of other DNA sequences in and around genes that modulate the extent and timing of gene expression.

Mutations affecting gene expression can be expected to occur on either side of the gene or within the intron and the coding sequence, based on recent studies on some animal genes. In the 5S RNA genes of *Xenopus,* a region within the coding sequence is recognized by a specific protein, and this complex is essential for gene transcription. (for example, see Bogenhagen, Sakonju, and Brown 1980). The sequence in this region is also conserved in plant 5S RNA genes (Gerlach and Dyer 1980). Mutations in this region would be expected to modify the expression of the gene. It was therefore of interest to find a 15 bp duplication in this region in many 5S RNA genes of wheat (Gerlach and Dyer 1980). No transcript that was 15 bp longer could be detected in the plant. These genes may therefore be pseudogenes—nonfunctional copies of the gene. It is also of interest to note that the mutation is a duplication of a short DNA segment.

Because we know that amplification, deletion, and transposition of short segments of DNA are common, then it is reasonable to speculate that a high proportion of natural mutations will also be due to these events. Over the next few years the extent to which this is true will emerge from studies already underway on many loci in corn, soybean, etc., especially those in which the insertion of transposable controlling elements is known to be the basis of the variation in the timing and extent of gene expression (Fincham and Sastry 1974). If it is true, this will contribute a very important addition to our knowledge of the variation the plant breeder uses. It will also influence how we attempt to create more variation. Mutagens that create only a single-base change would clearly be inadequate for producing the *range* of variation similar to that found in nature.

Because DNA segments are duplicated over evolutionary timescales, it is to be expected that gene sequences will frequently be present in several or many copies, even though few of them may be active. Where multigene families occur, such as those encoding the ribosomal RNAs and storage proteins in cereals, variation in copy number is common and is probably due to unequal crossing over (Flavell and Smith 1974). Furthermore, the different copies evolve together (concerted evolution, Zimmer et al, 1980) due to unequal crossing over and/or gene conversion kinds of events (Smith 1976; Scherer and Davies 1980; Jeffreys 1982). This can result in the relatively rapid fixation of new variants throughout the members of a gene family, enhancing the divergence of homologous genes between different individuals and species (Dover et al. 1982).

Allelic variation at a locus, where the alleles can be distinguished by the extent to which they behave as dominant or recessive in different genetic backgrounds, is well known in plants. The molecular basis of dominance is not known, but while studying the structure of the ribosomal RNA genes in wheat we have found that genes behaving as dominant have methyl groups preferentially removed from certain cytosine residues, while the recessive genes (inactive ones) are methylated at these same sites. This state of methylation is dependent on the genetic background, because in the absence of the dominant alleles the recessive alleles are more active and become demethylated at the specific sites. This change in gene *structure,* dependent upon the number and type of alleles elsewhere in the genome, may be common and therefore may constitute another kind of molecular event that contributes to genetic variation in plant populations. How the methylation of specific sequences is genetically determined and how it relates to gene expression remains to be elucidated.

CONCLUDING REMARKS

The more we understand about the molecular basis of the variation that the plant breeder wishes to exploit, the greater the likelihood that new progress based upon molecular biology and genetic manipulation will emerge. This progress may in the longer term be the design of new genes but in the shorter term is likely to be the development of new techniques for detecting useful variation. The newer techniques of molecular biology and genetic engineering are clearly powerful new additions to the variety of methods already available for genetic analysis of our crop plants. It should be noted, however, that if the molecular biologist is to explore the fundamental basis of gene expression, then he will need genetic variation from the geneticists and breeders. A two-way exchange between breeders and molecular biologists is therefore clearly essential for the endeavors of both sides to realize their full potential.

LITERATURE CITED

Appels, R., E. S. Dennis, D. R. Smyth, and W. J. Peacock. 1981. *Two repeated DNA sequences from the heterochromatic regions of rye (Secale cereale) chromosomes*. Chromosoma (Berl.) 84:265–77.

Appels, R., C. Driscoll and W. J. Peacock. 1978. *Heterochromatin and highly repeated DNA sequences in rye (Secale cereale)*. Chromosoma 70:67–89.

Bedbrook, J. R., J. Jones, M. O'Dell, R. D. Thompson, and R. B. Flavell. 1980. *A molecular characterization of telomeric heterochromation in Secale species*. Cell 19:545–60.

Bennett, M. D. 1972. *Nuclear DNA content and minimum generation time in herbaceous plants*. Proc. R. Soc. (Lond.) *B* 181:109–35.

———. 1973. *Nuclear characters in plants*. Brookhaven Symposia in Biology, 25 pp. 344–66.

———. 1976. *DNA amount, latitude and crop plant distribution*. Environmental and Experimental Botany 16:93–108.

———. 1977. *Heterochromatin, aberrant endosperm nuclei and grain shriveling in wheat-rye genotypes*. Heredity 39:411–19.

Bennett, M. D., J. B. Smith. 1976. *Nuclear DNA amounts in angiosperms*. Phil. Trans. Roy. Soci. (Lond.) *B* 274:227–74.

Bogenhagen, D. F., S. Sakonju, and D. D. Brown. 1980. *A control region in the center of the 5S RNA gene directs specific initiation of transcription. II:The 3'* border of the region. Cell 19:27–35.

Cavalier-Smith, T. 1978. *Nuclear volume control by nucleoskeletal DNA, selection for cell volume and cell growth rate and the solution of the DNA C-value paradox*. J. Cell Sci. 34:247–78.

Dover, G. A., S. Brown, E. Coen, J. Dallas, T. Strachan, and M. Trick. 1982. *The dynamics of genome evolution and species differentiation*. Pages 343–72 in G. A. Dover and R. B. Flavell, eds., *Genome Evolution*. Academic Press, London.

Fincham, J. R. S., and G. R. K. Sastry. 1974. *Controlling elements in maize*. Ann. Rev. Genet. 8:12–50.

Flavell, R. B. 1980. *The molecular characterization and organization of plant chromosomal DNA sequences*. Ann. Rev. Plant Physiol. 31:569–96.

———. 1982a. *Amplification, deletion and rearrangement: Major sources of variation during species divergence*. Pages 301–24 in G. A. Dover and R. B. Flavell, eds., *Genome Evolution*. Academic Press, London.

———. 1982b. *Recognition and modification of crop plant genotypes using techniques of molecular biology*. In I. Vasil, K. Frey, and W. Scowcroft, eds., *Frontiers of Plant Breeding*. Academic Press.

Flavell, R. B., and M. O'Dell. 1976. *Ribosomal RNA genes on homoeologous chromosomes of groups 5 and 6 in hexaploid wheat*. Heredity 37:377–85.

Flavell, R. B., M. O'Dell, and J. Hutchinson. 1981. *Nucleotide sequence organization in plant chromosomes and evidence for sequence translocation during evolution*. Cold Spring Harbor Symp. Quant. Biol. 45:501–508.

Flavell, R. B. and D. B. Smith. 1974. *Variation in nucleolus organizer rRNA gene multiplicity in wheat and rye*. Chromosoma (Berl.) 47:327–34.

Gerlach, W. L., and J. R. Bedbrook. 1979. *Cloning and characterization of ribosomal RNA genes from wheat and barley*. Nucl. Acids Res. 7:1869–86.

Gerlach, W. L., and T. A. Dyer. 1980. *Sequence organization of the repeating units in the nucleus of wheat which contain 5S rRNA genes*. Nucl. Acids Res. 8:4851–65.

Gerlach, W. L., and W. J. Peacock. 1980. *Chromosomal locations of highly repeated DNA sequences in wheat.* Heredity 44:269–76.

Grosveld, G. C., E. de Boer, C. K. Shewmaker, and R. A. Flavell. 1982. *DNA sequences necessary for transcription of the rabbit* β globin gene in vivo. Nature (Lond.) 295:120.

Gustafson, J. P., and M. D. Bennett. 1982. *The effect of telomeric heterochromatic from Secale cereale L. on triticale (X Tritico secale Wittmarck).* Can. J. Genet. Cytol. 24:83–92.

Hinegardner, R. 1976. *Evolution of genome size.* Pages 179–99 in F. J. Ayala, ed., *Molecular Evolution.* Sinauer Assoc., Sunderland, Mass.

Hutchinson, J., and D. Lonsdale. 1982. *The chromosomal distribution of cloned highly repetitive sequences from hexaploid wheat.* Heredity 48:371–76.

Jeffreys, A. J. 1982. *Evolution of globin genes.* Pages 157–76 in G. A. Dover and R. B. Flavell, eds., *Genome Evolution.* Academic Press, London.

Scherer, S., and R. W. Davis. 1980. *Recombination of dispersed repeated DNA sequences in yeast.* Science 209:1380–84.

Smith, G. P. 1976. *Evolution of repeated DNA sequences by unequal crossover.* Science 191:528–35.

Southern, E. M. 1975. *Detection of specific sequences among DNA fragments separated by gel electrophoresis.* J. Mol. Biol. 98:503–17.

Thomas, J. B., P. J. Kaltsikes, J. P. Gustafson, and D. G. Roupakias. 1980. *Development of kernel shriveling in triticale.* Z. Pflanzenphysiol. 85:1–27.

Thompson, W. F., and M. G. Murray. 1980. *Sequence organization in pea and mung bean DNA and a model for genome evolution.* Pages 31–45 in D. R. Davies and D. A. Hopwood, eds., *4th John Innes Symposium.* John Innes Institute, Norwich, UK.

Zimmer, E. A., S. L. Martin, S. M. Beverley, Y. W. Kan, and A. C. WIlson. 1980. *Rapid duplication and loss of genes coding for the α-chains of hemoglobin.* Proc. Nat. Acad. Sci. USA 77:2158–62.

3] The Structure and Expression of the *Shrunken* Locus in Maize Strains with Mutations Caused by the Controlling Element *Ds*

by N. FEDOROFF and D. CHALEFF*

ABSTRACT

Molecular studies have been carried out on mutations caused by the transposable element *Ds* at the *Shrunken (Sh)* locus in maize. Three mutant alleles have been examined, and all have been found to have structural alterations in the immediate vicinity of the transcription unit. The structural alterations map near the 5′ end of the transcription unit in two of the mutants. The *Sh*-encoded sucrose synthetase is not detectable in immature endosperm tissue of either mutant strain, but kernels contain low levels of what appears to be a normal transcript of the locus. The third mutant contains a rearranged transcription unit in which the 5′ and 3′ ends of the transcription unit are separated by an insertion or rearrangement. Aberrant transcripts of the locus are present in immature kernels of this mutant strain. The transcripts are slightly shorter than the *Sh*-encoded sucrose synthetase mRNA, are missing the coding sequence beyond the rearrangement breakpoint, and encode two polypeptides that are antigenically related to sucrose synthetase, but have a slightly higher mobility on denaturing gels. A derivative of this mutant has also been examined and found to have a similar genetic lesion.

INTRODUCTION

Controlling elements in maize transpose, cause insertion mutations, and are associated with the origin of chromosome breaks and rearrangements. At least six distinct families of controlling elements have been identified, of which three have been subjected to extensive genetic analysis (see McClintock 1965;

*Department of Embryology, Carnegie Institution of Washington, 115 W. University Pky., Baltimore, Maryland 21210.

GENETIC ENGINEERING: APPLICATIONS TO AGRICULTURE (Beltsville Symposium 7—Lowell D. Owens, ed.) Rowman & Allanheld, Totowa

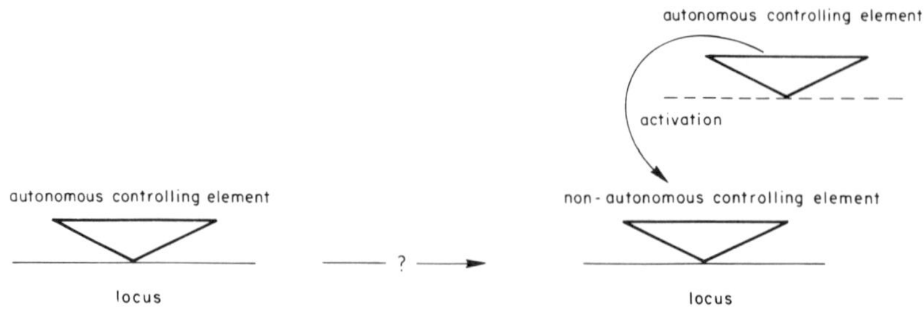

Figure 3.1. A diagrammatic representation of the relationship between an autonomous and a nonautonomous controlling element in maize. An autonomous element activates transposition and other genetic changes involving a nonautonomous element. A nonautonomous element can insert at a locus in such a way that expression of the locus is inhibited; transposition of the nonautonomous element away from the locus can be accompanied by reexpression of the locus. The possibility that a nonautonomous element can arise directly by mutation from an autonomous element is represented by an arrow interrupted by a question mark to indicate the absence of direct evidence for the implied sequence relationship.

Fincham and Sastry 1974; and Fedoroff 1983 for reviews). These include the *Activator-Dissociation* family of elements that are responsible for the mutations at the *Shrunken (Sh)* locus described here (McClintock 1951). Within a given family are elements that can transpose autonomously and elements that cannot. As diagrammed in Figure 3.1, an autonomous element can activate a nonautonomous element to transpose. Genetic evidence has shown that nonautonomous elements can be derived from or directly replace autonomous elements (McClintock 1965), and they may therefore represent transposition-defective mutant elements. *Dissociation (Ds)* designates a nonautonomous element that was first identified by virtue of its ability to provide a specific site of chromosome breakage and acentric-dicentric formation (McClintock 1946, 1947). Chromosome breakage and acentric-dicentric formation at *Ds* require the simultaneous presence of an autonomous element, designated *Activator (Ac)* for its ability to activate chromosome breakage at *Ds* (McClintock 1947). *Ds* can also transpose in the presence of *Ac* and insert at a variety of loci to cause mutations (McClintock 1951, 1956b). *Ds* insertion mutations are stable in the absence of *Ac*, but in its presence revert or further mutate both somatically and germinally. Somatic mutation is manifested by the appearance of genotypically and phenotypically altered sectors of cells within the organism (for illustration see McClintock, 1965, and Fedoroff, 1983). Both the somatic and germinal reversions of *Ds* insertion mutations are commonly associated with transposition of the element away from the locus (McClintock 1951, 1965).

Several unstable mutant alleles of the *Shrunken (Sh)* locus have been described whose origin is associated with the *Ds* element (McClintock 1952, 1953, 1954, 1955, 1956a). The alleles are designated *sh-m6233*, *sh-m5933*, and *sh-m6258*. The unstable alleles of the *Sh* locus arose in a strain containing a non-transposing *Ds* element just distal to the *Sh* locus on the short arm of chromosome 9. The *Ds* element in the progenitor strain did not transpose to genetically distinguishable sites, but did behave as a local mutagen, causing mutations at loci immediately adjacent to its site of insertion (McClintock 1952, 1953). Although most such mutations proved stable, several unstable alleles of the *Sh* locus and the adjacent *Bronze* locus were isolated from strains with a non-transposing *Ds* distal to the *Sh* locus (McClintock 1953, 1954, 1956a). The unstable mutations revert in the presence of *Ac* without concomitant transposition of the element away from the locus (McClintock 1953). Revertants remain unstable, mutating to recessive alleles in the presence of the autonomous element. The behavior of these mutations is therefore quite different than the behavior of many other *Ds* insertion mutations and suggests that they arise and revert by local chromosomal rearrangements, such as inversions, although the involvement of extremely short-range transposition events cannot be excluded.

STRUCTURE OF THE *Sh* LOCUS IN THE *sh-m* STRAINS

We have examined the structure and expression of the *Sh* locus in the *sh-m5933, sh-m6233,* and *sh-m6258* strains (Chaleff et al. 1981; Fedoroff, Mauvais, and Chaleff 1983). An additional strain, designated *sh-m6795,* was also examined. The *sh-m6795* strain was derived as a spontaneous recessive mutant from a revertant of the *sh-m6258* strain (McClintock, personal communication). Structural studies on these mutants have also been carried out by Burr and Burr (1980, 1981) and Döring, Geiser, and Starlinger (1981). The *Sh* locus in maize encodes sucrose synthetase, which catalyzes the reversible conversion of sucrose to fructose and UDP-glucose, the glucose donor in starch biosynthesis (Chourey and Nelson 1976). The molecular weight of the sucrose synthetase monomer is 92 kilodaltons (kD), and it is encoded by a 3-kilobase (kb) mRNA (Fedoroff, McCormick, and Mauvais 1980; Burr and Burr 1980, 1981; Wöstemeyer et al. 1981). The structural studies summarized here were carried out using a cloned complementary DNA (cDNA) copy of a portion of the sucrose synthetase mRNA to probe genomic DNA (Fedoroff, McCormick and Mauvais 1980; Chaleff et al. 1981; Fedoroff, Mauvais, and Chaleff 1983). Each mutant strain was found to have a different structural alteration at the *Sh* locus. As diagrammed in Figure 3.2b and c, changes in the location of restriction endonuclease cleavage sites at the left end of the *Sh* locus (Fig. 3.2a) near the 5′ end of the transcription unit were detected in strains *sh-m6233* and *sh-m5933*. The altered region is represented by an open box in Figure 3.2b and c, and uncertainty with respect to the location of the juncture between the *Sh* and foreign sequence is represented by the discontinuous portion of the box. The two strains differ in the location of the juncture between the *Sh* locus and foreign sequences, as indicated. It is not known whether the sequence alterations lie within or outside of the 5′ end of the transcription unit in either case. Strains *sh-m6258* and *sh-m6795* have similar, but not identical, structural alterations that affect the structure of the transcription unit itself. In both strains, the transcription unit is interrupted within an intervening sequence near the 3′ end of the sequence encoding the mRNA. The

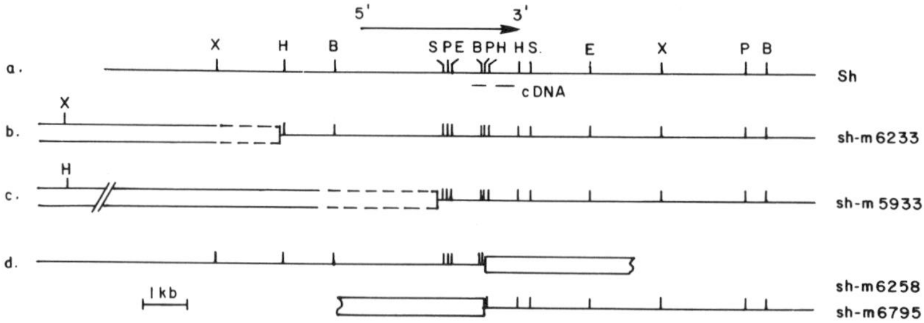

Figure 3.2. Partial restriction endonuclease cleavage site maps of the *Shrunken* locus in several strains with unstable mutations and in the progenitor strain. The restriction maps were constructed from the results of blot hybridization studies on genomic DNA of the indicated strains probed with a cloned cDNA copy of part of the mRNA encoded by the *Sh* locus (Chaleff et al. 1981; Fedoroff, Mauvais, and Chaleff 1983). The locations are shown for sites cleaved by Xba I (X), Hind III (H), Bgl II (B), Sst I (S), Pst I (P) and BstE II (E). The arrow shows the direction of transcription. The location of the sequences represented in the cDNA probe is indicated below the diagram in a. The mRNA sequence represented in the cDNA spans at least one intervening sequence near the 3′ end of the gene; its location and approximate length are indicated. The portion of the locus that shows alterations in the location of restriction endonuclease cleavage sites is represented by open boxes for the mutant strains designated *sh-m6233* (b), *sh-m5933* (c), *sh-m6258* and *sh-m6795* (d). The broken portion of the boxes in b and c represents the uncertainty with respect to the endpoint of the sequence alteration detected at the *Sh* locus in *sh-m6233* and *sh-m5933*. Although the differences are not indicated in the figure, the *Sh* loci in strains *sh-m6258* and *sh-m6795* are distinguishable on the basis of the location of restriction endonuclease cleavage sites within the foreign sequence adjacent to the 5′ end of the rearranged locus.

interruption maps to the same location in both strains, although the strains are distinguishable on the basis of the location of restriction enzyme cleavage sites within the foreign sequence interrupting the gene. It is unclear whether the gene is altered due to an insertion or a rearrangement, such as inversion, since a single restriction endonculease fragment bearing gene sequences both 5′ and 3′ to the foreign DNA has not yet been detected.

Expression of the altered *Sh* locus in immature kernels of the *sh-m* strains was investigated. None of the strains contained detectable amounts of the native sucrose synthetase encoded by the *Sh* locus in immature endosperm tissue (Table 3.1), although all contained small amounts of the sucrose synthetase encoded by the minor, distantly related second sucrose synthetase gene (Chourey 1981; McCormick, Mauvais, and Fedoroff 1982). None of the strains showed high levels of inactive sucrose synthetase-like polypeptides detectable by immunoprecipitation with antiserum against purified sucrose synthetase (Table 3.1). Immature endosperm tissue of both strains *sh-m6258* and *sh-m6795* did contain very small amounts of two polypeptides that are immunologically related to sucrose synthetase, but have molecular weights of 82 and 85 kD. The immunologically related 82- and 85-kD proteins were detected among in vitro translation products of immature kernel poly A^+ mRNAs of both of these

Table 3.1 Expression of the *Sh* Locus in Immature Kernels of Mutant Strains Whose Origin Is Associated with the Controlling Element *Ds*

	Native *Sh*-encoded sucrose synthetase	*Sh*-encoded mutant polypeptides	Sucrose synthetase mRNA
sh-m5933	–	–	0.5%–1%
sh-m6233	–	–	0.5%–1%
sh-m6258 and *sh-m6795*	–	82 and 85 kD	10%–25%, structure aberrant

Note: A hyphen indicates that such proteins were not detectable within the limits of sensitivity of the assays employed.

strains as well (Mauvais, Chaleff, and Fedoroff, 1983). The abundance of the aberrant sucrose synthetase-like polypeptides was much greater among in vitro translation products than it was in immature endosperm, suggesting that the aberrant polypeptides are unstable in vivo (Mauvais, Chaleff, and Fedoroff 1983). Strains *sh-m6233* and *sh-m5933* contain small amounts (0.5%–1%) of an mRNA that comigrates with the 3-kb sucrose synthetase mRNA (Table 3.1; Fedoroff, Mauvais, and Chaleff 1983). By contrast, strains *sh-m6258* and *sh-m6795* contain 10% to 25% of the normal amount of an mRNA that is approximately 2.8 kb in length and which lacks homology with the sucrose synthetase locus beyond the insertion or rearrangement breakpoint within the coding sequence (Fig. 3.1). It has been shown by hybrid-selected translation that the aberrant mRNAs present in these strains encode the immunoprecipitable 82- and 85-kD proteins (Fedoroff, Mauvais, and Chaleff 1983).

DISCUSSION

The results of the studies on strains *sh-m6258* and *sh-m6795* reveal the relationship between the mutant phenotype and the underlying genetic lesion. It has been found that both mutant alleles have a rearrangement that maps within the transcription unit and results in a physical separation of its 5′ and 3′ ends. Transcription of the 5′ end of the locus is relatively unaffected by the rearrangement, while the sequence on the 3′ side of the rearrangement breakpoint is not transcribed, suggesting that transcription terminates within the foreign DNA that has been introduced at the locus. The aberrant transcripts of the rearranged *Sh* locus present in immature kernels of strains *sh-m6258* and *sh-m6795* encode two polypeptides that are immunologically related to the sucrose synthetase monomer, but appear to be smaller by 7 and 10 kD. Very small amounts of the aberrant polypeptides can be detected in immature endosperm tissue of these strains, suggesting that the polypeptides are not stable. Thus the mutant phenotype of both strains is consequent on the inability of the rearranged locus to yield a structurally normal sucrose synthetase. Whether the aberrant polypeptides that are encoded by the transcript of the locus present in these strains are enzymatically inactive or whether the mutant polypeptides are enzymatically active, but highly unstable, is not known. The presence of two aberrant polypeptides in immature endosperm of both strains is also unexplained.

The relationship between the structural alterations detected at the *Sh* locus in strains *sh-m5933* and *sh-m6233* and expression of the locus cannot be determined unequivocally from the available data. The structural changes are confined to the left or 5′ end of the locus, but it is not yet known whether the transcription unit is altered in either mutant strain. The observation that immature kernels of both strains contain small amounts of what appears to be a normal mRNA is consistent with the possibility that the structural alterations affect initiation of transcription, but it does not exclude other explanations.

Both genetic and molecular evidence show that the structural alterations observed in these strains are associated with the *Ds* element. The mutations are stable in the absence of the *Ac* element and revert in its presence (McClintock 1953). Reversion is not, however, accompanied by transposition of the *Ds* element to a genetically distinguishable site; revertants give rise to further recessive mutations at the same locus, as in the case of the origin of *sh-m6795* from an *Sh* revertant of strain *sh-m6258* (McClintock, personal communication; N. Fedoroff, unpublished observations). Molecular studies on these related strains reveal extremely similar genetic lesions and indicate that the *sh-m6795* arose as the result of a structural change at exactly the same site as in the original *sh-m6258* strain. Molecular studies have been initiated on several *Sh* strains derived from the *sh-m6233* and *sh-m5933* strains in the presence of an *Ac* element. The preliminary results obtained with several of the revertants show that the restoration of the *Sh* phenotype is associated with restoration of the restriction enzyme cleavage site map characteristic of the *Sh* allele from which the mutant alleles were derived, as well as with a duplication of part of the locus (N. Fedoroff and J. Mauvais, unpublished observation). Thus, the observed structural alterations are relevant to the mutant phenotype and show the *Ac*-activated reversion characteristic of *Ds* mutations. It is not yet clear whether the *Ds* mutations in the *sh-m* strains are attributable to insertion of *Ds* element or are the result of a *Ds*-mediated chromosomal rearrangement. However, the genetic properties of the *sh-m* strains and the molecular data that have been obtained so far suggest that the mutations are attributable to chromosomal rearrangements.

LITERATURE CITED

Burr, B., and F. Burr. 1980. *Detection of changes in maize DNA at the Shrunken locus due to the intervention of Ds elements*. Cold Spring Harbor Symp. Quant. Biol. 45:463–65.

—, 1981. *Controlling-element events at the Shrunken locus in maize*. Genetics 98:143–56.

Chaleff, D., J. Mauvais, S. McCormick, M. Shure, S. Wessler, and N. Fedoroff. 1981. *Controlling elements in maize*. Carnegie Inst. Wash. Year Book 80:158–74.

Chourey, P. S. 1981. *Genetic control of sucrose synthetase in maize endosperm*. Mol. Gen. Genet. 184: 373–76.
Chourey, P.S., and O. E. Nelson. 1976. *The enzymatic deficiency conditioned by the shrunken-1 mutations in maize*. Biochem. Gen. 14:1041–55.
Döring, H. P., M. Geiser, and P. Starlinger. 1981. *Transposable element Ds at the shrunken locus in Zea mays*. Mol. Gen. Genet. 184:377–80.
Fedoroff, N. 1983. *Controlling elements in maize*. In J. Shapiro, ed., *Mobile Genetic Elements*. Academic Press, New York (in press).
Fedoroff, N., S. McCormick, and J. Mauvais. 1980. *Molecular studies on the controlling elements of maize*. Carnegie Inst. Wash. Year Book 79:51–62.
Fedoroff, N., J. Mauvais, and D. Chaleff. 1983. *Molecular studies on mutations at the Shrunken locus in maize caused by the controlling element Ds*. J. Mol. Appl. Genet. (in press).
Fincham, J. R. S., and G. R. K. Sastry. 1974. *Controlling elements in maize*. Ann. Rev. Genet. 8:15–50.
McClintock, B. 1946. *Maize genetics*. Carnegie Inst. Wash. Year Book 45:176–86.
—, 1947. *Cytogenetic studies of maize and Neurospora*. Carnegie Inst. Wash. Year Book 46:146–52.
—, 1951. *Chromosome organization and genic expression*. Cold Spring Harbor Symp. Quant. Biol. 16:13–47.
—, 1952. *Mutable loci in maize*. Carnegie Inst. Wash. Year Book 51: 212–19.
—, 1953. *Mutation in maize*. Carnegie Inst. Wash. Year Book 52:227–37.
—, 1954. *Mutations in maize and chromosomal aberrations in Neurospora*. Carnegie Inst. Wash. Year Book 53:254–60.
—, 1955. *Controlled mutation in maize*. Carnegie Inst. Wash. Year Book 54:245–55.
—, 1956a. *Mutation in maize*. Carnegie Inst. Wash Year Book 55:323–32.
—, 1956b. *Controlling elements and the gene*. Cold Spring Harbor Symp. Quant. Biol. 21:197–216
—, 1965. *The control of gene action in maize*. Brookhaven Symp. Biol. 18:162–84.
McCormick, S., J. Mauvais, and N. Fedoroff. 1982. *Evidence that the two sucrose synthetase genes in maize are related*. Mol. Gen. Genet. 187:494–500.
Wöstemeyer, J ., U. Behrens, A. Merckelbach, M. Müller, and P. Starlinger. 1981. *Translation of Zea mays endosperm sucrose synthase mRNA in vitro*. Eur. J. Biochem. 114:39–44.

4] The Organization and Expression of Maize Plastid Genes

by LAWRENCE BOGORAD,* EARL J. GUBBINS,† ENNO T. KREBBERS,* IGNACIO M. LARRINUA,* KAREN M. T. MUSKAVITCH,* STEVEN R. RODERMEL,* and ANDRÉ STEINMETZ**

ABSTRACT

The maize chloroplast genome is a circle of about 139,000 base pairs (bp). It contains two 22,000 bp-long inverted repeats; each repeat contains, among other genes, a set of genes for ribosomal RNAs. Among the maize plastid genes that have been located and sequenced in our laboratory are those for a number of tRNAs, the large subunit of the carbon dioxide-fixing enzyme ribulosebisphosphate carboxylase, and the genes for the beta and epsilon subunits of the maize chloroplast coupling factor for photophosphorylation (CF_1). A number of unidentified open reading frames have also been located, although their functions have not been determined. (The genes for the 23 and 16S rRNAs and for the two tRNAs found in the spacer between these two genes have been sequenced by Kossel and his coworkers.)

Maize plastid genes identified thus far exhibit several features reminiscent of eubacterial genes, such as the types of RNA-polymerase recognition sequences (promoters), sequences comparable to Shine-Dalgarno 16S rRNA recognition sequences, the use of the universal code (in contrast to codon usage in mitochondria of yeast and animals). On the other hand, several tRNA genes of maize have now been shown to carry large introns and, unlike eubacterial tRNA genes, the 3' CCA is not included in the coding region.

One example of each of two expression classes of maize plastid genes has been identified and studied, and the levels of mRNA for these genes appear to be regulated. Photogene 32 is a thylakoid membrane component expressed strongly during light-induced development but little, if at

*Department of Cellular and Developmental Biology, Harvard University, 16 Divinity Avenue, Cambridge, Massachusetts 02138; †Amoco Research Center, P.O. Box 400, Mail Station B1, Naperville, Illinois 60566; and **IBMC, 15, rue René Descartes, F-67084, Strasbourg, CEDEX, France.

GENETIC ENGINEERING: APPLICATIONS TO AGRICULTURE (Beltsville Symposium 7—Lowell D. Owens, ed.) Rowman & Allanheld, Totowa

all, in dark-grown plants. Transcripts of the gene for the large subunit of ribulosebisphosphate carboxylase are abundant in bundle sheath cells of maize leaves but are absent or virtually missing from mesophyll cells of the same leaf.

Maize chloroplast DNA dependent-RNA polymerase plus the S factor preferentially transcribes plastid genes over bacterial plasmid genes in vitro. Furthermore, this enzyme system exhibits preferential transcription of some chloroplast genes over others.

INTRODUCTION

The average dream of a genetic engineer is to find a gene that "needs attention," to get it out of its normal environment, to improve it, and to put it back into the genome. Other variations of the dream, are to replace a poor or only satisfactory gene with a superior one, or to add to a genome a gene for a new function from another species or a different kingdom—i.e., to introduce an alien gene. The alien gene needs to be dressed in clothes that will make it indistinguishable from a native in the foreign environment. The gene will need to be taught local customs, and it will be necessary to slip it into the foreign environment without the normal defense mechanisms coming into play. Finally, the genetic engineer hopes to sit back and observe that the alien gene hasn't been discovered and rooted out. Watching it flourish and express it itself, thus changing the organism into which it has been placed, provides the final satisfaction.

To realize the latter kind of dream, it is necessary to know exactly what the inhabitants of the foreign environment look like and which features may be so characteristic and unmistakable as to preclude the use of aliens of certain sizes or shapes; of aliens that cannot be reshaped; or of aliens who can't be retrained in the new language. It is also necessary to understand the intimate habits of the natives in detail and to learn the structure of the society well enough to alter the alien to be able to assume a specific role in the society—a regulator in the capitol, a member of the proteinaceous reserve.

Cells of green plants, like those of other eucaryotes, contain distinctive separated sets of nuclear and mitochondrial genes but, in addition, they have plastids carrying still another genome. Mitochondria and plastids are the products of both nuclear and organelle genes. An especially striking feature of this interdependence is the observation that multimeric enzymes and structures of organelles are comprised of both nuclear-coded and organelle-coded elements (Bogorad 1975; Bogorad et al. 1981).

We have concentrated on the analysis of the maize chloroplast genome and the expression of its genes. We have been trying to find out what genes are in the chloroplast, how these genes are dressed, what turns them on, and what keeps them silent. From such experiments we hope to learn about genes that may be usefully altered or moved around in ways that will teach us about biology of eucaryotic cells and will likely make modifications of organisms possible.

THE MAIZE PLASTID CHROMOSOME

The maize plastid chromosome is a circle of approximately 139,000 base pairs (bp) containing two large inverted repeated sequences. Each of the latter is about 22,000 bp (Bedbrook and Bogorad 1976; Bedbrook, Kolodner, and Bogorad 1977).

Each chloroplast may contain about fifty times 139,000 bp of DNA, but the fact that it has been possible to find a single solution to the circle problem indicates that all, or virtually all, of the DNA molecules in a single chloroplast have the same sequence. That is, the genome in maize is 139,000 bp, but there may be about 50 molecules of that size in each plastid.

GENES FOR RIBOSOMAL RNAs

Figure 4.1 shows the locations on the maize chloroplast chromosome of recognition sites for the restriction endonucleases Sal I and Bam HI. The locations of the genes for the 16S, 23S, 4.5S, and 5S rRNAs of the maize chloroplast chromosome are also shown. One copy of this set of genes is in each of the large, inverted, repeated sequences. The order of this group of genes (in the direction of their transcription) is 16S, spacer, 23S, 4.5S, and 5S. Koch, Edwards, and Kossel (1981) have shown that the spacer contains genes for isoleucine and alanine tRNAs. Overall, the orientation of the genes for rRNAs and the positions of the genes for the two tRNAs resemble that seen in *Escherichia coli*. The DNA sequences for the maize 16S and 23S rRNA genes show considerable homology to the corresponding *E. coli* genes (Schwarz and Kossel 1980; Edwards and Kossel 1981).

The presence of two large, inverted, repeated sequences, each containing a set of genes for plastid rRNAs, is common to all plants studied, with a few notable exceptions. For example, plastid chromosomes of *Vicia faba* (Koller and Delius 1980) and of *Pisum sativum* (Palmer and Thompson 1981) lack the inverted repeated sequence, and the rRNA genes are present only once per chromosome. The *Euglena* plastid chromosome, on the other hand, contains three sets of genes for rRNAs plus an extra copy of the 16S rRNA gene arranged in tandemly repeated segments (Gray and Hallick 1978; Rawson et al., 1978; Jenni and Stutz 1979). But most other plants studied—from eucaryotic algae to flowering plants—exhibit the pattern seen in *Zea mays* plastid DNA: two large inverted repeated sequences each carrying a set of genes for rRNAs.

GENES FOR PLASTID TRANSFER RNAs

Maize chloroplast DNA contains a full set of transfer RNAs (R. Selden, A. Steinmetz, L. McIntosh, L. Bogorad, G. Burkard, M. Mubumbila, E. Crouse,

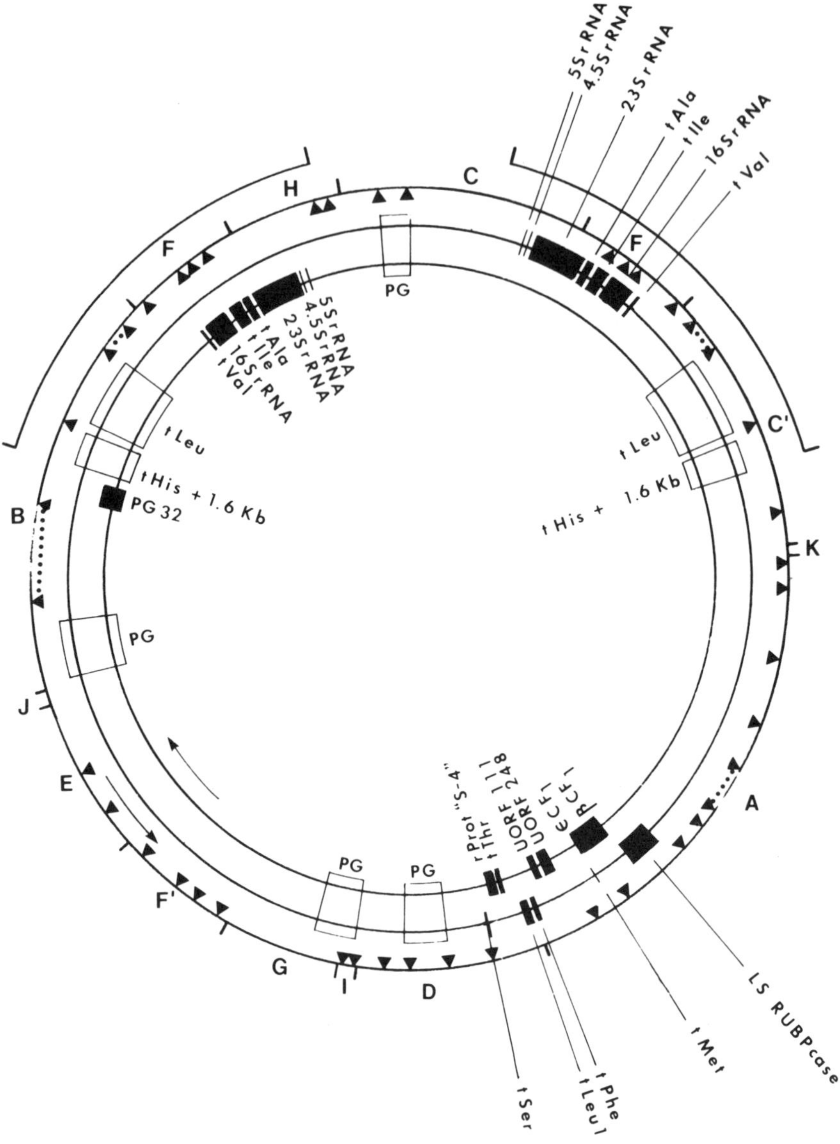

Figure 4.1. A map of the circular maize plastid chromosome showing, on the outer circle, recognition sites for restriction enzymes Sal I (fragments generated by this enzyme are designated by capital letters) and by Bam HI (black triangles). It also shows the locations of the two large inverted repeated segments (at the top of this figure) and the positions of genes for rRNAs. The directions of transcription of the two inner circles, representing the two DNA strands, are shown by arrows. Among the other genes shown on this diagram are those for transfer RNAs that have been sequenced, with the exception of the tRNA valine gene that contains a large intron and lies between epsilon CF_1 and UORF (unidentified open reading frame) 248. The location of photogene 32 is shown on the inner circle, and the approximate locations of other light-inducible genes that have not yet been assigned to a DNA strand are indicated by boxes labeled PG. rProt "S-4" is a gene with a very strong sequence homology to *E. coli* ribosomal gene S4. (A. Steinmetz, A. Subramanian, and L. Bogorad, unpublished.)

and J. Weil, unpublished). A number of these have been sequenced (Koch, Edwards and Kossel 1981; Steinmetz et al. 1983; A. S. Steinmetz, E. Krebbers, and L. Bogorad, unpublished). In general the tRNAs show quite high overall homology to comparable tRNAs of *E. coli*. Unlike the bacterial tRNAs, however, the 3' terminal CCA is not encoded in the gene—it must be added after transcription. An especially interesting feature of maize plastid tRNA genes is the presence of introns hundreds of nucleotides in length at various places in the anticodon loop of transfer RNAs for alanine and isoleucine (Koch, Edwards, and Kossel 1981), for leucine (Steinmetz, Gubbins, and Bogorad 1982), and for valine (E. Krebbers, A. Steinmetz, and L. Bogorad, unpublished).

GENES FOR PLASTID PROTEINS

Ribulose bisphosphate carboxylase (Rubpcase) is the enzyme that catalyzes the condensation of carbon dioxide with ribulose bisphosphate to produce two molecules of phosphoglyceric acid—the substrate for photosynthetic carbon reduction. This large enzyme is comprised of eight each of two different-sized polypeptide chains. The gene for the large polypeptide—the LS (large subunit) of Rubpcase—is transmitted uniparentally from the maternal parent, while the gene for the small subunit of RuBPCase is distributed in a Mendelian pattern in tobacco (Kawashima and Wildman 1972). The uniparental maternal inheritance of the LS could be taken to mean that the gene is located in an organelle that is transmitted only maternally.

The relatively small size of the maize chloroplast chromosome permitted us to employ a direct method for physically identifying and localizing genes for certain plastid components. The first gene we sought (Coen et al. 1977) was that for the LS. To locate the gene on the maize chloroplast chromosome, fragments of the plastid DNA were cloned in *E. coli* and used as templates to direct the synthesis of ^{35}S-methionine–labeled polypeptides in a linked transcription-translation system in vitro. The objective was to identify the fragment of DNA carrying the LS gene by its capacity to direct the synthesis of an identifiable polypeptide product. In this case the product was to be identified initially by its size and by its ability to be precipitated by an antibody against Rubpcase.

In preliminary experiments (Coen 1978) maize plastid DNA was used as a template with several different combinations of components in the linked transcription-translation system: *E. coli* RNA polymerase plus an *E. coli* translation system; *E. coli* RNA polymerase plus a wheat germ translation system; *E. coli* RNA polymerase with a rabbit reticulocyte translation system; wheat germ RNA polymerase plus an *E. coli* translation system, etc. The preliminary assay was for production of polypeptides of about 50 kilodaltons (kD), i.e., the approximate size of the LS. The radioactive polypeptide products were asssayed by polyacrylamide gel electrophoresis in sodium dodecylsulfate. From these tests of various combinations, the *E. coli* RNA polymerase plus rabbit reticulocyte lysate transcription-translation system was selected.

To identify which particular portion of the chloroplast chromosome contained the LS gene, it was necessary to have substantial amounts of DNA to use as templates. Ample amounts of chloroplast DNA sequences could be obtained most conveniently by cloning fragments of the chloroplast chromosome in *E. coli* and using that DNA to direct the linked transcription-translation. Thus, the next step in the work was to determine which restriction endonucleases could be used for cloning the DNA without disrupting the genes of the 50-kD size class, including LS. At the time, we were using three cloning vehicles: pmB 9 for cloning fragments generated by Eco RI; RSF 1030 for cloning fragments generated by Bam HI; and pML 21 for cloning fragments generated by Sal I. To determine whether Eco RI, Bam HI, or Sal I cut the DNA within the gene of interest, thus preventing it from directing the synthesis of the full-sized polypeptide, maize plastid DNA was digested with one of these three restriction endonucleases and then used as a template in the linked transcription-translation system. Digestion with Bam HI did not interfere with the capacity of the plastid DNA to direct the synthesis of a 50-kD polypeptide (or polypeptides) by the linked transcription-translation system (Coen et. al. 1977). On the other hand, digestion with Eco RI rendered the DNA ineffective as a template for production of polypeptides of this size. Using this information, various Bam HI-generated fragments of maize plastid DNA cloned in RSF 1030 were used as templates. The ninth-largest fragment (Bam 9) was found to direct the synthesis of a polypeptide whose size, as estimated on polyacrylamide SDS gels, matched that of the LS made in vivo by maize seedlings supplied ^{35}S-methionine. That this in vitro product corresponded to LS was supported by showing that the in vitro product is immunoprecipitated by antibody against Rubpcase and that radioactivity of the immunoprecipitate is decreased by addition of nonradioactive maize leaf Rupbcase to the in vitro reaction mixture before introduction of the antibody. Further evidence for the identity was provided by comparisons of the products of limited proteolysis of the polypeptide synthesized in vitro and in vivo (Coen et al. 1977), and the identification was confirmed by analyses of the tryptic digestion products of the two polypeptides (Bedbrook et al. 1979).

The 4.3-kilobase pair (kbp) Bam fragment 9 bearing the LS gene was analyzed further in two steps. First, after establishing a restriction endonuclease recognition site map of Bam 9 for the enzymes Bgl II, Eco RI, and Pst I, it was found that digestion of the DNA fragment with Eco RI or Pst I would abolish its ability to direct the synthesis of the 50-kD LS polypeptide. From this observation it was concluded that cuts made by these restriction endonucleases were within the gene. (The result of digestion with Eco RI confirmed the work in which total chloroplast DNA digested with Eco RI was not useful as a template for synthesis of the polypeptide.) On the other hand, digestion of the cloned DNA with the restriction enzyme Bgl II had no effect on the ability of the DNA to direct the synthesis of LS. This permitted further delineation of the coding region and subsequent work, using S_1 nuclease mapping (Link and Bogorad 1980), further defined the coding area and also revealed the presence on this fragment of part of a gene coding for a 2.2-kb RNA approximately 300 nucleotides away and transcribed in the opposite direction (i.e., from the other

```
                                                                       LS RUBPcase
                                                                           -60 from ATG
                                                                              *
-60       -50       -40       -30       -20       -10             →→→→→→→
 .         .         .         .         .         .              .
ATGAGTTGATATTAATTAAATATCATTTTTTTTAGATTTTGCAAAGGTTTCTTTCGCCT AATCCT
CTCTGTTGACAGCAATCTATGCTTCACAGTAGTATAT ATTT
       .        .        .            .
      -30      -20      -10           →→→→→→→→→βε CF1
                                      *
                                      -300 from ATG
```

Figure 4.2. Untranscribed DNA sequences lying upstream of the transcription initiation start sites for the LS Rubpcase (upper) and beta-epsilon of CF_1 (lower) genes. Numbering is from the site of initiation of transcription. The two sequences are aligned at their presumptive "−35" sequences. These sequences have very high homology to one another, but the spacing between these and the much less conserved "10 sequences," as well as the distances from the "−10 regions" to the starts of transcription, all vary. These two genes are under different expression controls in vivo and are transcribed to different levels in vitro.

strand of DNA) from the LS gene. (The gene for the 2.2-kb RNA has been shown to code for the beta and epsilon subunits of CF_1 [Krebbers et al. 1982] as described below.)

All the information outlined above was utilized subsequently for determining the DNA sequence of the LS gene of maize (McIntosh, Poulsen, and Bogorad, 1980). This was the first gene for a plant protein to be sequenced, and the process provided a good deal of information about this particular maize plastid gene, several features of which have been found subsequently to be common to plastid genes: (a) The universal nucleotide code for amino acids is used in plastids in contrast to the different codon systems used by yeast, Neurospora and mammalian mitochondria (Barrell et al. 1980; Bonitz et al. 1980; Heckman et al. 1980). (b) Five nucleotides upstream from the translation initiation codon for methionine on the expected mRNA transcript is the nucleotide sequence GGAGG that is complementary to a sequence near the 3' terminus of maize chloroplast 16S rRNA (Schwarz and Kossel [1980] had earlier sequenced the maize 16S rRNA gene). *E. coli* mRNAs also contain, at about the same position, a sequence complementary to one near the 3' terminus of 16S rRNA. This "Shine-Dalgarno" sequence is believed to play a role in the binding of mRNA to a ribosome (Shine and Dalgarno 1974; Steitz and Jakes 1975). (c) Transcription of the LS gene starts 63–64 nucleotides upstream from the ATG initiation codon. Examination of the maize LS gene in the region around 10 and around 35 nucleotides upstream of the transcription start site (i.e., in the "−10" and "−35" regions) failed to reveal sequences comparable to those seen on procaryotic genes in the same "promoter" region (i.e., RNA polymerase recognition sequences) or to those observed regularly on nuclear genes in the −30 and −80 regions. Nevertheless, some sequences resembling procaryotic −10 sequences are found at nucleotide regions around −25 and −35 from the transcription initiation site (i.e., 25 and 35 nucleotides upstream of the initiation site) and a 6-nucleotide–long sequence strongly resembling those seen in the −35 regions of procaryotic genes is present at −51 to −56 (Fig. 4.2). (d) Sequences present in the gene on both sides of the site corresponding to the 3' terminus of the mRNA could form stem and loop

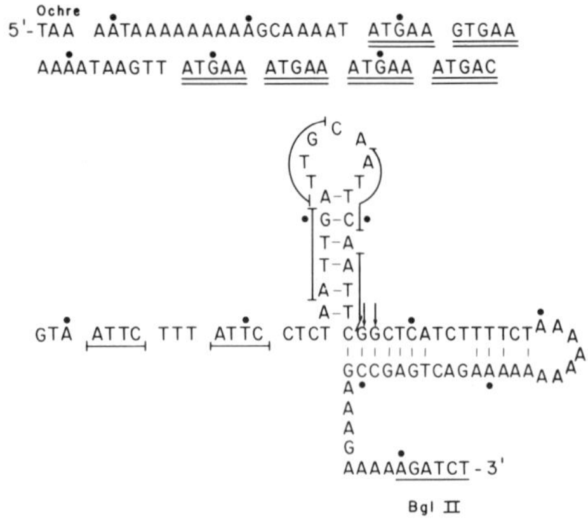

Figure 4.3. The DNA sequence at the 3′ terminus of maize chloroplast gene for the large subunit of Rubpcase. Short sequences (ATGAA, GTGAA, ATGAC) that have double underlines bear strong nucleotide homology to one another; those marked by single underlines form another set of homologous sequences (ATTC, ATTG). Two possible stem and loop structures are shown; the upright one would probably not be very stable. The transcript formed in vivo terminates at about the position marked with arrows (C. Poulson, L. McIntosh, and L. Bogorad, unpublished). It is not known whether this is the site of termination of transcription or of subsequent RNA processing.

structures which might play roles in either transcription termination or mRNA processing (Fig. 4.3).

Sequencing of other chloroplast genes for tRNAs and proteins has permitted the accumulation of data on variations in the Shine-Dalgarno sequences of the protein genes and conserved sequences that may be promoters associated with these and other genes.

The availability of a cloned maize plastid DNA fragment carrying part of the 2.2-kb RNA gene and all of the adjacent LS gene permitted an experiment to be carried out to determine whether transcripts of these two genes are present in bundle sheath and mesophyll cells of maize. Maize is a C-4 plant and thus contains Rubpcase in plastids of its bundle sheath but not mesophyll cells.

In these experiments, mesophyll cells were separated from bundle sheath strands enzymatically, and RNA was prepared from the two preparations. Cross-contamination was estimated at about 3% (Link, Coen, and Bogorad 1978). The RNA samples were made radioactive by labeling with ^{32}P in vitro and then hybridized against fragments of a cloned maize chloroplast DNA segment containing the LS and 2.2-kb genes. Fragments containing sequences coding for each of these two genes were separated electrophoretically for the hybridizations. Transcripts for both of these genes were abundant in bundle sheath RNA, but only the 2.2-kb RNA was present in mesophyll cells. Thus, these two adjacent genes are transcribed differently in mesophyll and in bundle sheath cells. The mechanism for the discrimination remains to be determined. The DNA sequences 5' to the two transcribed regions are shown in Figure 4.2, aligned at conserved sequences resembling those at the −35 regions of eubacterial genes. The sequences in these regions may be involved in RNA polymerase interaction. The regions of the LS and 2.2-kb genes shown in Figure 4.2 are different from one another: in sequence (particularly in conserved sequences closest to the sites of transcription initiation); in the distances from conserved sequences (underlined) to transcription start sites; in the distances between the "−35" sequences of the two genes (the distance for the LS gene is an unusually long 21 nucleotides for a number of plastid genes the distance is more commonly 17–18 nucleotides); and in the presence at −35 to −40 in the LS gene of a sequence resembling a "−10" sequence. The significance, if any, of the differences remains to be determined. Another comparison of interest is between the sequences upstream of the transcription and translation start sites of LS genes of the C-4 plant maize (McIntosh, Poulsen and Bogorad 1980) and the C-3 plant spinach (Zurawski et al. 1981), in which LS is not restricted to certain photosynthetic cell types of the leaf (Fig. 4.4). The mRNA for spinach LS starts at about 179 nucleotides upstream of the translation start site (Zurawski et al., 1981) and thus extends 115 nucleotides beyond the 5' end of the maize mRNA. It is interesting that the sequence in the "−10" region of the maize gene (at −87 to −92 in relation to the translation initiation site) is considerably different from that of the spinach gene (Fig. 4.4), while the "−35" type sequence of the maize gene (positions −115 to −120) is completely conserved in the spinach gene. But, again, we do not yet know which, if any of, the differences or similarities are critical.

Photogenes. Leaves of dark-grown angiosperms are usually pale yellow because they contain no chlorophyll, although they have a small amount of another green pigment, the chlorophyll precursor protochlorophyllide. Their plastids, designated etioplasts, lack the characteristic thylakoids of mature functional chloroplasts, but contain instead one or more paracrystalline prolamellar bodies (Bogorad 1981). Chloroplast maturation begins upon illumination: chlorophyll is produced; additional chloroplast membranes are produced; thylakoids and grana are formed. Saturation hybridization of maize plastid DNA with RNA from etioplasts reveals that approximately 70% to 80% of the plastid chromosome (assuming that the coding is equivalent to one strand of DNA) is transcribed in etioplasts, and approximately 90% to 95% is transcribed in mature, photosynthetically competent chloroplasts, if one takes into account the presence of the large inverted repeated sequences (L. Haff and L. Bogorad, unpublished).

To identify specific sequences that are transcribed during light-induced development but not in darkness, RNA was isolated from etioplasts and from plastids at various stages of light-induced development. The RNA was labeled in vitro with ^{32}P and hybridized against electrophoretically separated fragments of maize plastid DNA produced by digestion with restriction endonu-

```
-220        "-35"          -200          "- 10"        ↓ ↓        -170      -160
CGGTTGGGTTGCGCC-ATAT-ATATGAAAGAGTATACAATAATG-ATGTATTTGGCGAATCAAATA
CTATAGTAAAATTCCTATAGGATA-GAAC--GTACACAGGG-TGTATACATAAAT-GAATGAAA--
-220                      -200                                -170

      -150                                                        -110
CATGGTCTATTAAGG-AACCAT------------TTTGATTAGTTGATAAT-ATTAATTGAGAATT
CATATTAC-TTAACTTAAGCATACTCCTTTTTTATTTAATGAGTTGATATTAATTAAATATCATTT
                                                "-35"
 -160                     -140                          -110

    -100                                                           -40
TGATGAAAGATTGCTATAAAAGGTTTCATTAAGGCCTAATTTATGTCGAGTAGACCTTGTTGCTTT
TTTTT--AGATTTTTGG-AAAGGTTTCTTTC--GCCTAAT-CCTATGGAGTTGTCCCTGT-GCT-T
                                          ↑↑
    "-10"                         -70          -60                -40

        -30                       -10         +1
GTTGTAAAAA-TTAAAATTTGAAGTTGTAGGGAGGGACTTATG                 SPINACH
GT-GTGAATTCTTAATTCATGA-GTTGTAGGGAGGGACTTATG***************  MAIZE
        -30                       -10         +1

 -30                  -10
TTGCGCcatatatatgaaagagTATACAataatga                          SPINACH
                                 ↓↓
TTGATAttaattaaaTATCATttttttTAGATTtttggaaaggtttctttcgcctaa    MAIZE
                                                        ↑↑
    -50        -40        -30
```

Figure 4.4. Upper: DNA sequences upstream of translation initiation (position +1) of the spinach (Zurawski et al. 1981) and maize (McIntosh et al. 1980) genes for LS Rubpcase. (The portions of the two genes that code for amino acids are symbolized by a row of asterisks.) Nucleotides are numbered with reference to translation initiation, designated position +1. Transcription is initiated at approximately −63 to −64 in the maize gene (marked by two upward facing arrows) and at about −178 to −179 in the spinach gene. Regions of homology between the two sequences are marked by lines between the two sequences. Sequences reminiscent of conserved −10 and −35 promoter regions of bacterial genes are indicated by double overlines for spinach and double underlines for maize. *Lower:* Nucleotide sequence upstream of the transcription initiation sites for the two genes. In this figure numbering is with respect to transcription, rather than translation, initiation.

cleases. Other aliquots of the RNA samples were used to direct polypeptide synthesis in vitro, using the rabbit reticulocyte translation system (Bedbrook et al. 1978). The most conspicuous change in the pattern of hybridization of the RNA samples to maize plastid DNA fragments was to Bam fragment 8—little hybridization to this chloroplast DNA sequence by RNA from etioplasts was observed but strong hybridization by RNA from mature plastids. Increasing degrees of hybridization were exhibited by RNA from plastids at various stages of light-induced development. A parallel pattern of change was seen in the RNA translation experiments: a 34.5-kD polypeptide that was a conspicuous product of synthesis using chloroplast RNA as a template was hardly detectable or not seen in etioplast RNA-directed protein synthesis. Increasing amounts of this protein were produced when the templates were RNAs from plastids at various stages of light-induced development.

Concurrently, a study of thylakoid membrane synthesis in vivo during light-induced development was being carried out by Grebanier, Steinback, and Bogorad (1979). A number of polypeptide components of thylakoid membranes are present in etioplasts, but an additional set is produced during light-induced development. Among the latter, a polypeptide of 32 kD is conspicuous. In still another set of experiments we were conducting at the same time (Grebanier et al. 1978), we were identifying polypeptides produced by isolated maize plastids in vitro. The major membrane-associated product of protein synthesis by isolated maize chloroplasts was found to be a 34.5-KD protein. Comparison of the products of partial proteolysis (Cleveland et al. 1977) of this in vitro–synthesized product with a 32-kD thylakoid membrane protein that is heavily labeled in vivo, and that appears during light-induced plastid development, led to the conclusion that the 34.5-kD species is probably a precursor of the 32-kD one. Thus, isolated plastids appear to synthesize and insert into their membranes a 34.5-kD polypeptide that is processed to a 32-kD form in vivo but is not processed in any great amount by the isolated maize chloroplasts. From analyses of patterns of digestion of the 34.5- and 32-kD proteins in thylakoids exposed to pronase or trypsin, it could be seen that the newly synthesized 34.5-kD species is "completely susceptible" to digestion by these proteases, while one (or more) fragments(s) of the 32-kD form is protected from digestion under these conditions. This led to the conclusion that the isolated maize chloroplast does not appear to be able to fully integrate the newly made protein into the membrane—full integration could require the processing step that is not carried out much in isolated maize plastids.

A segment of maize plastid DNA fragment Bam 8 containing the gene for the 32-kD polypeptide (designated photogene 32) has been cloned and shown to direct the synthesis in vitro of a polypeptide that corresponds to the 34.5-kD protein made by isolated plastids (Bogorad et al. 1980).

The photogene 32 product has subsequently been identified as the azido-atrazene binding protein (Steinbeck et al. 1981) that is the determinant of the electron transport function of a bound plastoquinone molecule in the photosystem II complex (Arntzen et al. 1983).

Additional photogenes have been mapped to a number of sites on the maize chloroplast chromosome by refinements of the procedures used for locating photogene 32. The approximate locations of some of these additional photogenes are shown in Figure 4.1. They are not all grouped in one region—there is no single photogene operon, although we do not know yet whether some photogenes are grouped. Current research is focused on DNA sequence analyses of photogenes 32 and other photogenes to determine whether they have common features that may play roles in the photoregulation of their transcription.

Subunits of CF_1. Another problem of interest in understanding plastid development relates to the integrated production of polypeptides, i.e., the apparently coordinated synthesis of polypeptide subunits of multimeric enzymes or structures. The chloroplast coupling factor for photosynthetically driven ATP production is an example. Two complexes of proteins associated with thylakoid membranes are directly involved: the membrane embedded complex CF_0 and the 9-nm–diameter spheres attached to CF_0 units, designated CF_1, made up of five polypeptides designated alpha, beta, gamma, delta, and epsilon. Although the stoichiometry of the subunits is not positively established, in part

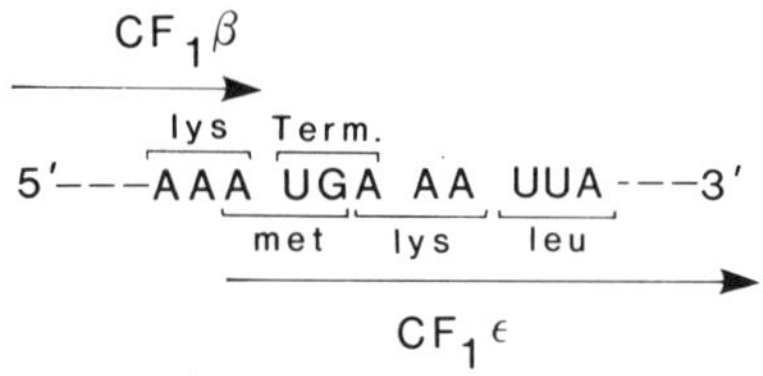

Figure 4.5. The region of fusion between the genes for the beta and epsilon subunits of maize CF_1 as they would appear in the RNA transcript. The methionine initiating codon for the epsilon gene is formed from the terminal A of the lysine, together with the U and G of the termination codon for the beta gene (Krebbers et al. 1982).

because of instability of the complex, McCarty (1979) favors the stiochiometry 2:2:1:1:2. We have found recently (Krebbers et al. 1982) that the 2.2-kb RNA gene that is adjacent to but divergently transcribed from the LS gene is the gene for the beta subunit of CF_1 fused with the gene for the epsilon subunit, as shown in Figure 4.5.

These two genes are separated from one another by about 20 nucleotides in the *E. coli* chromosome. The fusion of the genes for the beta and epsilon polypeptides in the *Zea mays* chromosome may ensure at least roughly equal production of the two subunits. It should be noted that equal production of the two subunits need not imply a 1:1 stoichiometry in the final complex, since the half life of the two subunits may differ. In any event, the beta and epsilon subunit genes are structurally integrated, and the 2.2-kb transcript is the fused message for both proteins. The alpha subunit, on the other hand, is located some distance away, and some other control signals are needed to assure roughly stoichiometric production of the alpha with beta and epsilon polypeptides. The nature of the DNA sequences around the alpha subunit polypeptides coding region will be of considerable interest in the quest for an understanding of possible mechanisms by which gene expression is coordinated.

CONSERVED DNA SEQUENCES IN AND AROUND TRANSCRIBED REGIONS: CONTROL SEQUENCES?

DNA sequences have now been accumulated from enough plastid genes to permit meaningful analyses for conserved sequences at positions that may be involved in the regulation of transcription or translation. A consensus conserved sequence 5′ upstream of demonstrated or estimated transcription start sites for 15 genes, including genes for tRNAs and proteins, is shown in Figure 4.6 The deviations from this consensus sequence are great. For example, promoter-like conserved sequences in the region upstream of the transcription start site for maize LS (Fig. 4.2) are at positions that do not correspond to

Figure 4.6. A consensus sequence for conserved nucleotides in the −10 and −35 regions for 15 plastid genes, including several maize plastid tRNA genes and maize protein genes.

those in procaryotic or eucaryotic nuclear genes. Also, within the conserved regions, in comparison to T_7A_2 and *E. coli* −10 and −35 sequences, DNA sequences range from a complete match between the A_2 promoter of bacteriophage T_7 and the T_7A_2 promoter for phenylalanine tRNA, to great divergence in histidine tRNA, which has only two positions in common with T_7A_2 in the −35 region and four in the −10 region. The significance of these differences remains to be elucidated.

As described in the discussion on the LS gene, a short nucleotide sequence just upstream of the translation start site is complementary to a sequence close to the 3′ end of the 16S rRNA of maize plastids. A Shine-Dalgarno-like sequence just like this is not found universally in maize plastid genes for proteins. For example, in the case of the beta-epsilon CF_1 gene there is sequence of nucleotides complementary to the 16S rRNA but to a different portion of the rRNA, as shown in Figure 4.7. In bacteria the nature of the complementary sequence and its distance from the translation start site are believed to play roles in regulation of translation. Experimental analyses using a homologous plastid-based translation system will be necessary before this question can be resolved for plastid translation.

MAIZE PLASTID DNA-DEPENDENT RNA POLYMERASE

We have been studying the maize plastid DNA-dependent RNA polymerase as a component of the transcriptional control system in plastid development. This enzyme was solubilized in 1971, and its subunit structure was determined later (Bottomley, Smith, and Bogorad 1971; Smith and Bogorad 1974; Kidd and Bogorad, 1979). The enzyme is active in synthesizing RNA when the subunit complex is 180, 140, 110, 90, and 40 kD (Smith and Bogorad 1974); but until such subunits can be disassembled, reassembled, and tested for polymerase subfunctions, one cannot say for this, or other complex multimeric enzymes, which of the component polypeptides are essential or what their functions may be. At earlier stages of purification many more polypeptides are, of course, associated with the enzyme, and some of these may be important in transcription of genes in vivo.

The nuclear RNA polymerase II of maize was purified at about the same time as the chloroplast enzyme in order to compare the two (Strain, Mullinix, and Bogorad 1971; Mullinix, Strain, and Bogorad 1973; Kidd and Bogorad 1980); it was found to include polypeptides of 180, 160, 43, 28, and 22 kD, plus a few other probable subunits. Comparisons of tryptic digestion patterns of

```
MAIZE CHLOROPLAST GENES.
16S RNA RECOGNITION SEQUENCES(?).
                ....AAATTATGTGATAATTATG..... 3' β CF1
                            .....
16S rRNA 3' End  UUUCCUCCACUAGGUC.............
                     .....
            ....TTGTAGGGAGGGACTTATG........ 3' LS RuBPcase
```

Figure 4.7. Sequences upstream of translation initiation start sites *(ATG)* for beta CF_1 and LS Rubpcase genes of maize in relation to sequences at the 3′ end of the 16S rRNA. These two genes are complementary to different nucleotides near the 3′ end of the 16S rRNA. It remains to be determined whether these differences play any roles in the relative frequencies or rates of translation of these messages.

roughly corresponding-sized subunits of maize plastid and maize nuclear type II polymerases revealed that all are distinctive (Kidd and Bogorad 1979). This analysis was carried out for the plastid and nuclear polymerase subunit pairs 180–180 kD; 140–160 kD; 43–42 kD; and 28–27 kD.

For purposes of purification, an assay of simply the rate of incorporation of nucleotides into RNA is adequate, but it is inadequate for analysis of functions that may be related to transcriptional regulation. We set out to look for characteristics and activities that might be more meaningful in terms of selective transcription of chloroplast genes. Maize plastid RNA polymerase was solubilized from sucrose gradient-purified maize plastid membranes, and the preparation was fractionated by passage through a column of DE 52, using a gradient of KC1 as an eluant. The bulk of the enzyme activity appeared at the expected position, eluting with about 0.2 M KC1. This enzyme was highly active with calf thymus DNA (or other linear DNAs) in accordance with our earliest observations (Bottomley, Smith, and Bogorad 1971). But since we were interested in looking at activities in vitro with specific chloroplast genes as templates—templates that had not been available up to that time—we also tested the activity of the enzyme with chimeric plasmids containing maize chloroplast genes that had been cloned in *E. coli.* The activity of the 0.2 M KC1 enzyme fraction from DE 52 with this template was much lower than when calf thymus DNA (or linearized chimeric plasmid DNA) was used (Jolly and Bogorad 1980). Yet a fraction eluting with 0.5 M KC1 proved to strongly stimulate the transcription of the chimeric plasmid DNA—about 5- to 15-fold depending upon the preparation—but had no effect on the rate of transcription of linear DNA. The active component of this fraction was purified and designated the S-factor; it is a polypeptide of about 27 kD (Jolly and Bogorad 1980).

We next attempted to determine whether the S-factor stimulated transcription of all portions of the chimeric plasmid equally or whether some sequences were transcribed more actively than others. In the first experiments, the analysis was carried out by simply digesting the chimeric plasmid with various restriction enzymes that would separate chloroplast DNA sequences from fragments containing vehicle DNA and then testing, by Southern hybridization

(Southern 1975), with the radioactive product of the polymerase reaction to assay relative transcription. We found that in the absence of the S-factor there was strong transcription of both vehicle and chloroplast DNA sequences. In the presence of the S-factor, on the other hand, very strong preferential transcription of some of the chloroplast DNA sequences and an apparent reduction in the transcription of plasmid vehicle DNA sequences was noted.

A more quantitative approach to the problem was then developed. To one set of nitrocellulose filters we attached pZmc 150 DNA. This is a chimeric plasmid comprised of maize plasmid Eco RI fragment *1* inserted into the bacterial plasmid pMB 9. To another set of filters pMB 9 DNA was fixed. pZmc 150 DNA was used as a template to produce radioactive RNA by the maize plastid RNA polymerase. Aliquots of the in vitro products were hybridized to sets of filters containing either pZmc 150 or pMB 9. The relative transcription of pMB 9 DNA and maize plastid DNA sequence DNA could be calculated from the quantitative hybridization data. In the absence of the S-factor the level of transcription from pMB9 and Eco *1* sequences was about the same. On the other hand, in the presence of the S-factor, transcription of the chloroplast DNA sequence was about eight times greater than that of the vehicle DNA.

In the earliest experiments, using Southern hybridization to estimate the relative transcription of portions of the template, we found that the transcriptional preference of the enzyme plus S-factor was affected by the conformation of the template. If we used template DNA that was supercoiled, the preference for transcription of plastid DNA sequences was strong; if relaxed circular DNA was the template, preferential transcription was nearly lost. This was confirmed in quantitative experiments by treating the cloned pZmc 150 DNA with an unwinding enzyme from *E. coli* (omega) prior to its use as a template. Under these circumstances the relative transcription of the maize sequence to the vehicle sequence was only about 2.5:1, rather than about 8:1. Thus, the preferential transcription of maize plastid DNA sequences by the homologous DNA-dependent RNA polymerase depends both upon the presence of the S-factor and on the supercoiled conformation of the DNA itself.

The DNA-dependent RNA polymerse of *E. Coli* uses pZmc 150 as a template in the same way as maize plastid RNA polymerase without the S-factor—i.e., vehicle and plastid DNA sequences in pZmc 150 are transcribed about equally. Furthermore, the S-factor has no effect on the specificity of the enzyme, nor can it replace the sigma factor of the *E. coli* polymerase.

In addition to having the capacity to discriminate between chloroplast and bacterial plasmid DNA sequences, the maize plastid RNA polymerase plus S-factor system is capable of distinguishing between different chloroplast genes in vitro. For example, as already noted, the LS gene is about 300 nucleotides away and divergently transcribed from the beta-epsilon gene of CF_1. All of the LS gene and part of the beta-epsilon CF_1 gene are contained on the Bam HI–generated fragment 9 of the maize plastid chromosome. When supercoiled DNA containing these sequences is introduced into the maize RNA polymerase–S-factor system, the LS gene is transcribed about three times more than the beta-epsilon CF_1 gene. In the absence of the S-factor, both genes are transcribed about equally (Jolly et al. 1981). It is of special interest to note that in vivo the two genes are transcribed in a ratio of about 3 to 1—i.e., close to the in vitro system's transcription ratio.

CONCLUSIONS

The maize plastid chromosome is tightly packed with genes. In addition to features described here, overlaps between divergently transcribed genes have been observed (Schwarz, et al. 1981) and, in general, genes found to date are generally quite close to one another. The genes themselves bear marked resemblances to those of eubacteria, but tRNA genes have eucaryotic features as well. The 3′ terminal CCA is not in tRNA genes, and large introns have been found in four tRNA genes.

One example of a gene limited to expression during light-induced development (photogene 32) and a gene expressed in bundle sheath but not mesophyll cells (the LS of Rubpcase) have been located and studied. Additional examples of such genes remain to be identified and analyzed to determine whether there are any familial characteristics in the sequences that flank the transcribed regions—sequences that may have roles in the control of transcription.

The in vitro maize plastid transcription system using maize chloroplast DNA-dependent RNA polymerase is likely to be useful in reconstructing the events that occur in vivo and in learning about mechanisms for control of gene expression.

ACKNOWLEDGMENTS

We are grateful to the National Science Foundation, the Competitive Research Grants Office of the U.S. Department of Agriculture, and the National Institute of General Medical Sciences for research grant funds which, in part, permitted this work to be done. The research was also supported in part by the Maria Moors Cabot Foundation for Botanical Research of Harvard University. Karen M.T. Muskavitch and I.M. Larrinua were recipients of NRSA Postdoctoral Fellows awarded by the National Institute of General Medical Sciences. A. Steinmetz is Charge de Recherche at C.N.R.S., France, and was the recipient of a NATO Research Grant.

LITERATURE CITED

Arntzen, C.J., S.C. Darr, J.E. Mullet, K.E. Steinback, and K. Pfister. 1983 (n.t.) In B. Trumpower, ed. *Function of Quinones in Energy Conserving Systems*. Academic Press, N.Y. (forthcoming).

Barrell, B.G., S. Anderson, A.T. Bankier, M.H.L. DeBruijn, E. Chen, A.R. Coulson, J. Drouin, I.C. Eperon, D.P. Nierlich, B.A. Roe, F. Sanger, P.H. Schreier, A.J.H.

Smith, R. Staden, and I.G. Young. 1980. *Different pattern of codon recognition by mammalian mitochondrial tRNAs*. Proc. Nat. Acad. Sci. USA 77: 3164–66.

Bedbrook, J.R., and L. Bogorad. 1976. *Endonuclease recognition sites mapped on Zea mays chloroplast DNA*. Proc. Nat. Acad. Sci. USA 73: 4309-13.

Bedbrook, J.R., D.M. Coen, A. Beaton, L. Bogorad, and A. Rich. 1979. *Location of the single gene for the large subunit of ribulosebisphosphate carboxylase on the maize chloroplast chromosome*. J. Biol. Chem. 254: 905–10.

Bedbrook, J.R., R. Kolodner, and L. Bogorad. 1977. *Zea mays chloroplast ribosomal RNA genes are part of a 22,000 base pair inverted repeat*. Cell 11: 739–49.

Bedbrook, J.R., G. Link, D.M. Coen, L. Bogorad, and A. Rich. 1978. *Maize plastid gene expressed during photoregulated development*. Proc. Nat. Acad. Sci. USA 75: 3060–64.

Bogorad, L. 1975, *Evolution of organelles and eukaryotic genomes*. Science 188: 891–98.

Bogorad, L. 1981. *Chloroplasts*. J. Cell Biol. 91: 256s–270s.

Bogorad, L., S.O. Jolly, G. Link, L. McIntosh, Z. Schwarz, and A. Steinmetz. 1980. *Studies of the maize chloroplast chromosome*. Pages 87–96 in T. Bucher, W. Sebald, and H. Weiss, eds., *Biological Chemistry of Organelle Formation*. Springer Verlag, Berlin.

Bonitz, S.G., R. Berlani, G. Coruzzi, M. Li, G. Macino, F.G. Nobrega, M.P. Nobrega, B.E. Thalenfeld, and A. Tzagoloff. 1980. *Codon recognition rules in yeast mitochondria*. Proc. Nat. Acad. Sci. USA 77: 3167–70.

Bottomley, W., H.J. Smith, and L. Bogorad. 1971. *RNA polymerases of maize: Partial purification and properties of the chloroplast enzyme*. Proc. Nat. Acad. Sci. USA 26: 2412-16.

Cleveland, D.W., S.G. Fischer, M.W. Kirschner, and U.K. Laemmli. 1977. *Peptide mapping by limited proteolysis in sodium dodecyl sulfate and analysis by gel electrophoresis*. J. Biol. Chem. 252: 1102–1106.

Coen, D.M. 1978. *Identification and mapping of protein coding sequences in maize chloroplast DNA*. Ph.D. dissertation, Mass. Institute of Technology.

Coen, D.M., J.R. Bedbrook, L. Bogorad, and A. Rich. 1977. *Maize chloroplast DNA fragment encoding the large subunit of ribulosebisphosphate carboxylase*. Proc. Nat. Acad. Sci. USA 74: 5487–91.

Edwards, K., and H. Kossel. 1981. *The rRNA operon from Zea mays chloroplast: Nucleotide sequence of 23S rDNA and its homology with E. coli 23S rDNA*. Nucl. Acids. Res. 9: 2853-68.

Gray, P.W., and R.B. Hallick. 1978. *Physical mapping of the Euglena gracilis chloroplast DNA and ribosomal RNA gene region*. Biochemistry 18: 284–90.

Grebanier, A.E., D.M. Coen, A. Rich, and L. Bogorad. 1978. *Membrane proteins synthesized but not processed by isolated maize chloroplasts*. J. Cell Biol. 78: 734–46.

Grebanier, A.E., K.E. Steinback, and L. Bogorad. 1979. *Comparison of molecular weights of proteins synthesized by isolated chloroplasts with those which appear during greening*. Plant Physiol. 63: 436-39.

Heckman, J.E., J. Sarnoff, B. Alzner-DeWeerd, S. Yin, and U.L. RajBhandary. 1980. *Novel features in the genetic code and codon reading patterns in Neurospora crassa mitochondria based on sequences of six mitochondrial tRNAs*. Proc. Nat. Acad. Sci. USA 77: 3159–63.

Jenni, B., and E. Stutz. 1979. *Mapping of a DNA sequence complementary to 16S rRNA outside of the three rRNA sets*. FEBS Lett. 102: 95–99.

Jolly, S.O., and L. Bogorad. 1980. *Preferential transcription of cloned maize chloroplast DNA sequences by maize chloroplast RNA polyerase*. Proc. Nat. Acad. Sci. USA 77: 822–26.

Jolly, S.O., L. McIntosh, G. Link, and L. Bogorad. 1981. *Differential transcription in vivo and in vitro of two adjacent maize chloroplast genes: The large subunit of ribulosebisphosphate carbyoxylase and the 2.2–kilobase gene.* Proc. Nat. Acad. Sci. USA 78: 6821–25.

Kawashima, N., and S.G. Wildman. 1972. *Studies on fraction I protein. IV Mode of inheritance of primary structure in relation to whether chloroplast or nuclear DNA contains the code for a chloroplast protein.* Biochim. Biophys. Acta 262: 42-49.

Kidd, G.H., and L. Bogorad. 1979. *Peptide maps comparing subunits of maize chloroplast and type II nuclear DNA-dependent RNA polymerase.* Proc. Nat. Acad. Sci. USA 76: 4890–92.

———. 1980. A facile procedure for purifying maize chloroplast RNA polymerase from whole cell homogenates. Biochim. Biophys. Acta 609: 14-30.

Koch, W., K. Edwards, and H. Kossel. 1981. *Sequencing in the 16S-23S spacer in a ribosomal RNA operon of Zea mays chloroplast DNA reveals two split tRNA genes. Cell 25: 203–14.*

Koller, B., and H. Delius. 1980. *Vicia faba chloroplast DNA has only one set of ribosomal RNA genes as shown by partial denaturation mapping and R-loop analysis.* Mol. Gen. Genet. 178: 261–69.

Krebbers, E.T., I.M. Larrinua, L. McIntosh, and L. Bogorad. 1982. *The maize genes for the beta and epsilon subunits of the photosynthetic coupling factor CF_1 are fused.* Nucl. Acids Res. 10: 4985–5002.

Link, G., and L. Bogorad, 1980. *Sizes, locations and directions of transcription of two genes on a cloned maize chloroplast DNA sequence.* Proc. Nat. Acad. Sci. USA 77: 1832–36.

Link, G., D.M. Coen, and L. Bogorad. 1978. *Differential expression of the gene for the large subunit of ribulose bisphosphate carboxylase in maize leaf cell types.* Cell 15: 725–31.

McCarty, R.E. 1979. *Roles of a coupling factor for photophosphorylation in chloroplasts.* Ann. Rev. Plant Physiol. 30: 79-104.

McIntosh, L., C. Poulsen, and L. Bogorad. 1980. *Chloroplast gene sequence for the large subunit of ribulosebisphosphate carboxylase of maize.* Nature 288: 556–60.

Mullinix, K.P., G.C. Strain, and L. Bogorad. 1973. *RNA polymerases of maize. Purification and molecular structure of DNA-dependent RNA polymerase II.* Proc. Nat. Acad. Sci. USA 70: 2386–90.

Palmer, J.D., and W.F. Thompson. 1981. *Rearrangements in the chloroplast genomes of mung bean and pea.* Proc. Nat. Acad. Sci. USA 78: 5533-37.

Rawson, J.R.Y., S.D. Kushner, D. Vapnek, V.N.K. Alton, and C.L. Boerma. 1978. *Chloroplast ribosomal RNA genes in Euglena gracilis exist as three clustered tandem repeats.* Gene 3: 191–209.

Schwartz, Z., and H. Kossel. 1980. *The primary structure of 16S rDNA from Zea mays chloroplast is homologous to E. coli 16S rRNA.* Nature (Lond.) 283: 739–42.

Schwartz, Z., S. Jolly, A. Steinmetz, and L. Bogorad. 1981. *Overlapping divergent genes in the maize chloroplast chromosome and in vitro transcription of the gene for $tRNA^{His}$.* Proc. Nat. Acad. Sci. USA 78: 3423–27.

Shine, J., and L. Dalgarno. 1974. *The 3'-terminal sequence of Escherichia coli 16S Ribosomal RNA: Complementarity to nonsense triplets and ribosome binding sites.* Proc. Nat. Acad. Sci. USA 71: 1342–46.

Smith, H.J., and L. Bogorad. 1974. *The polypeptide subunit structure of the DNA-dependent RNA polymerase of Zea mays chloroplasts.* Proc. Nat. Acad. Sci. USA 71: 4839–42.

Southern, E.M. 1975. *Detection of specific sequences among DNA fragments separated by gel electrophoresis.* J. Mol. Biol. 98: 503-17.

Steinback, K.E., L. McIntosh, L. Bogorad, and C.J. Arntzen. 1981. *Identification of the triazene receptor protein as a chloroplast gene product.* Proc. Nat. Acad. Sci. USA 78: 7463–67.

Steinmetz, A., E.J. Gubbins, and L. Bogorad. 1982. *The anticodon of the maize chloroplast gene for tRNA^{Leu}UAA is split by a large intron.* Nucl. Acids Res. 10: 3027–37.

Steinmetz, A., E.T. Krebbers, Z. Schwarz, E.J. Gubbins, and L. Bogorad. 1983. *Nucleotide sequences of five maize chloroplast transfer RNA genes and their flanking regions.* J. Biol. Chem. (in press).

Steitz, J., and K. Jakes. 1975. *How ribosomes select initiator regions in mRNA: Base pair formation between the 3′ terminus synthesis in Escherichia coli.* Proc. Nat. Acad. Sci. USA 72: 4734–38.

Strain, G.C., K.P. Mullinix, and L. Bogorad. 1971. *RNA polymerases of maize: Nuclear RNA polymerases.* Proc. Nat. Acad. Sci. USA 68: 2647–51.

Zurawski, G., B. Perrot, W. Bottomley, and P.R. Whitfeld. 1981. *The structure of the gene for the large subunit of ribulose 1,5–bisphosphate carboxylase from spinach chloroplasts.* Nucl. Acids Res. 9: 3251–80.

5] Mitochondrial DNA Plasmids and Cytoplasmic Male Sterility

by D. R. PRING*

ABSTRACT

Plasmid-like linear or circular mitochondrial DNAs have been found in fertile and cytoplasmic male-sterile maize and sorghum, ranging in size from ca. 1 to 8 kilobase pairs (kbp). Linear molecules of 6.4 and 5.4 kbp are associated with one source of male sterility in maize, and linear molecules of 5.7 and 5.3 kbp are found in one source in sorghum. Hybridization experiments indicate limited homology between the maize and sorghum plasmid-like DNAs. Smaller molecules are also associated with sources of male sterility in maize, and are found in male-sterile sorghum. Hybridization data indicate the maize DNAs may have their origins in normal, fertile cytoplasm mitochondrial DNA (mt DNA). Analyses of spontaneous male-fertile plants from male-sterile lines indicate a concomitant integration of sequences of the plasmid-like DNAs into the principal mtDNA genome, demonstrating a transposon-like capability. Investigations with other sources of male sterility in maize show a gradation of homology to the plasmid-like DNAs, which may suggest a correlation with the biology and genetic behavior of male-sterile maize cytoplasms.

INTRODUCTION

Cytoplasmic male sterility (cms) of crop plants is widely used in the production of hybrid seed and is essentially mandatory for production in species with perfect flowers. Cytoplasmic male sterility is used in the commercial production of maize, sorghum, pearl millet, sugarbeet, and sunflower hybrid seed, for instance. Of these species only maize, with imperfect flowers spatially separated, is amenable to hybrid production by manual emasculation.

An agronomic rationale for investigations of cms is the documented susceptibility of maize lines in the T or Texas source of cms to two fungal pathogens

*U.S. Department of Agriculture, Agricultural Research Service, Department of Plant Pathology, University of Florida, Gainesville, Florida 32611.

GENETIC ENGINEERING: APPLICATIONS TO AGRICULTURE (Beltsville Symposium 7—Lowell D. Owens, ed.) Rowman & Allanheld, Totowa

and their toxins. The T source is one of three major groups of male-sterile cytoplasms in maize, and the remaining two, the C and S groups, do not share this disease susceptibility. For further details the reader is referred to several reviews of cms and its relationship to disease susceptibility (Duvick 1965; Ullstrup 1972; Levings and Pring 1979; Kemble and Pring 1982). It is interesting to note that the T cytoplasm additionally confers sensitivity to methomyl, a systemic insecticide (Humaydan and Scott 1977).

CYTOPLASMIC MALE STERILITY IN MAIZE

Examination of organelle DNAs. Isolation and characterization of mitochondrial (mt) and chloroplast (ct) DNAs from normal (N) and male-sterile cytoplasms have generally implicated the mitochondrion as a carrier of determinants conditioning cms. Restriction endonuclease fragment analyses of mtDNA indicated strict maternal inheritance (Levings and Pring 1976; Pring and Levings 1978) and demonstrated that each source of cms, the T, C, and S sources, is characterized by a unique and distinct mtDNA.

Evidence implicating the mitochondrion in both the disease susceptibility and cms of the T source was obtained through examination of tissue-culture–derived plants, either nonselected or selected by the addition of toxins from the pathogens to the culture media (Gengenbach, Green, and Donovan 1977; Brettell, Thomas, and Ingram 1981). Disease-resistant, male-fertile plants were obtained in both investigations. A plausible explanation for these occurrences is the selection and amplification of a normal-type mtDNA during tissue culture or the regeneration process, resulting in an essentially normal cytoplasm phenotype. Studies of the progeny in each case, however, indicated that the mtDNA of the regenerated plants was unlike that of known N cytoplasm mtDNA and was very similar to that of parental T cytoplasm (Gengenbach et al. 1981; Brettell, Conde, and Pring 1982). Variation in the mtDNA among progeny of regenerated plants was extensive, suggesting a tendency toward a high frequency of rearrangement as a result of tissue culture or regeneration. It is relevant to emphasize that amplification of the altered mtDNA is inherent in the detection of such events. Limited data to date indicate that restriction fragments that distinguish the parental T cytoplasm mtDNA are not detectable in the altered progeny (Kemble and Pring 1982).

Maize plasmid-like DNAs. A major impetus for studies on plasmid-like DNAs in higher plants was provided by the investigations of Laughnan and associates (Laughnan and Gabay 1978), who described instability of the S cytoplasm of maize, expressed as spontaneous, heritable reversions of this cytoplasm to male fertility. Associated with this phenomenon were effects on nuclear loci that suggested a possible episomal model as an explanation for the events. The reader is referred to a recent review for an extensive treatment of genetic and biochemical data related to the phenomena (Laughnan, Gabay-Laughnan, and Carlson 1981). Plasmid-like linear DNAs were subsequently isolated from the S cytoplasm of maize (Pring et al. 1977), and these appeared to be intimately involved in the instability of the S cytoplasm. Examination of spontaneous male-fertile revertants revealed that the plasmid-like DNAs S-1 and S-2 were absent as free molecules, but that at least part of the S-2 DNA sequences were detectable as an integrated part of the principle genome (Levings et al. 1980).

It is now apparent that repeated DNA sequences are associated with these plasmid-like molecules in the free state and also with the molecule from whence they possibly originated—the N cytoplasm mtDNA of maize. S-1 and S-2 DNAs share terminal inverted repeat sequences (Levings and Pring 1979) and are characterized by a ca. 1400 basepair (bp) common sequence (Kim et al. 1982). N, male-fertile mtDNA shares extensive homology with S-1 and S-2 DNAs (Spruill, Levings, and Sederoff 1980; Thompson, Kemble, and Flavell 1980; Koncz et al. 1981), while mtDNA from T, C, and S sources of cms share less homology (Thompson, Kemble, and Flavell 1980; Koncz et al. 1981).

Recently it was established that N cytoplasm mtDNA is characterized by extensive homology to the plasmid-like DNAs in two separated regions of the genome, and that adjacent to each site is a 26-kb repeated sequence (Lonsdale, Thompson, and Hodge 1981). These data were obtained by cloning 32 to 47-kbp segments of the mt genome with cosmid cloning vectors. When these cosmid clones were used to compare genome structure of the T, C, and S cytoplasm mtDNA in the S-1 and S-2 regions, it was observed that each mtDNA was indistinguishable outside the S-1 and S-2 sequence sites; i.e., the genomes were very similar (D. R. Pring and D. M. Lonsdale, unpublished). Within areas adjacent to S-1 and S-2 sequences, the S cytoplasm mtDNA shared more similarity with N mtDNA than did the T and C cytoplasms. Although there is currently no basis for relating these differences to the genetic and biological behavior or these cytoplasms, it is clear that the cytoplasms differ in areas of mtDNA known to exhibit transpositional activity.

Our studies of the structure of the S-1 and S-2 regions of the maize mitochondrial genome have included examinations of entries within the N and C groups. Variations occur within cytoplasms designated N. From examination of 14 N cytoplasm mtDNAs by use of cosmid probes, three groups emerged, varying exclusively with a probe that spans the S-1 region (J. M. McNay, D. R. Pring, and D. M. Lonsdale, unpublished). Lines A188, W182BN, and F6 each have a smaller *Bam*H1 fragment bearing major homology to S-1. The line Black Mexican is also characterized by altered S-1 homology fragments. We have not observed variation using a probe that spans the S-2 region. In addition to changes in the S-1 homology region, N cytoplasm mtDNA is characterized by other rearrangements (Levings and Pring 1977) that apparently do not carry homology to either of the cosmid probes.

Another manifestation of mtDNA variation is reflected in the size of small DNAs associated with the mitochondrion (Kemble and Bedbrook 1980; Kemble, Gunn, and Flavell 1980). When N, C, T, and S cytoplasm mtDNAs are compared, most N cytoplasms exhibit a 2.30-kbp fragment, as do C and S cytoplasms, while T cytoplasm mtDNA is characterized by a slightly smaller, ca. 2.1-kbp fragment (Fig. 5.1A). Surprisingly, A188 (N) also exhibits a ca. 2.1-kbp rather than a 2.30-kbp DNA (Fig. 5.1A, 3), a feature also characteristic of W182BN (N) and F6 (N) (J. M. McNay, D. R. Pring, and D. M. Lonsdale, unpublished). The 2.30-kbp DNA is known to share homology to S-2; when S-2 DNA is nick-translated and hybridized to these DNAs (Fig. 5.1B), the 2.1-kbp

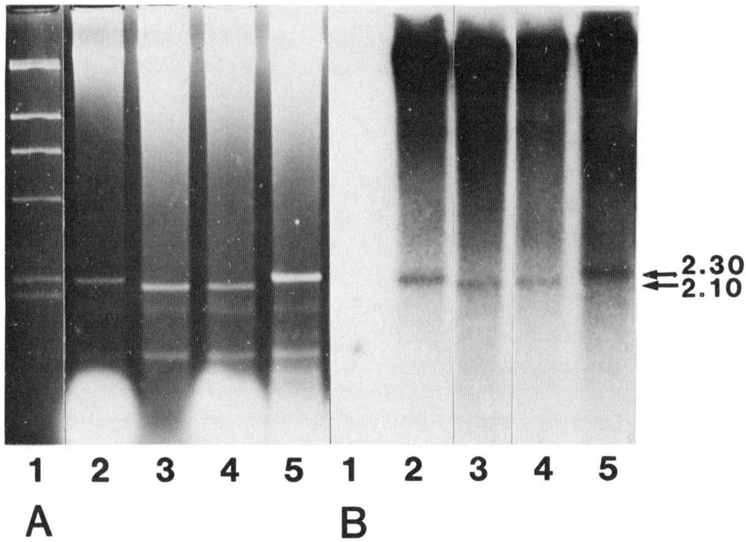

Figure 5.1. Agarose gel electrophoresis (A) and hybridization of nick-translated S-2 DNA (B) to (1) *Hind*III-digested bacteriophage lambda DNA, and undigested mtDNAs from (2) Wf9 (N), (3) A188 (N), (4) Wf9 (T), and (5) Wf9 (C). Molecular sizes of lambda fragments are 27.5, 23.1, 9.42, 6.56, 4.36, 2.32, and 2.02 kbp.

fragments in A188 (N) and WF9 (T) appear to share equal homology to S-2. T cytoplasm mtDNA contains little homology to S-1 or S-2 DNA (Thompson, Kemble, and Flavell 1980; Koncz et al. 1981); part of this homology is found in the 2.1-kbp DNA (Fig. 5.1B, 4).

The C group of male-sterile cytoplasms can be differentiated into three subgroups by restriction analysis (Pring, Conde, and Levings 1980). Restriction fragments of a number of these subgroups were blotted onto nitrocellulose and hybridized to the cosmid clones; a fourth subgroup, designated PR, was clearly identified on the basis of unique hybridization to these probes. Certain of the subgroups showed variation in homology to cosmids spanning the S-1 and S-2 regions (D. R. Pring, A. Lusby, V. E. Gracen, and D. M. Lonsdale, unpublished).

In these studies it became apparent that differences between N, C, T, and S cytoplasms include not only variation in the S-1 and S-2 regions, but also important variation in areas removed from these sites. Until these genomes can be effectively mapped and assigned functional or transcriptional activity, conclusions concerning a role of the S-1 and S-2 sequences in cms or other phenomena must be treated with caution. Although the apparent transposition of S-2 sequences into the principle mtDNA genome of S cytoplasm maize, concomitant with spontaneous reversion to male fertility, provides strong evidence for involvement of these molecules in the phenomenom, definitive and rigorous evidence will be required to unambiguously establish a causal relationship. In the mutants there are not only new restriction fragments that share homology with S-2 sequences, but also several new fragments that do not share homology and several fragments missing altogether (Levings et al. 1980).

Recent observations on related plasmid-like DNAs also confound a simple interpretation of the relationship of these DNAs to cms in maize. Weissinger et al. (1982) described the R-1 and R-2 plasmid-like DNAs from RU cytoplasms of Latin America races of maize and showed that the R-2 DNA shared extensive homology with S-2 DNA, but that the larger (ca. 7.4 kbp) R-1 DNA consisted of perhaps an entire S-1 molecule plus an additional ca. 1 kbp of unrelated DNA sequences. And most important, these cytoplasms were male-fertile. Any obvious explanation for an obligate role of the S-1 and S-2 DNAs in cms must consider the nonsterile nature of the RU cytoplasms. Two plasmid-like DNAs were also reported from five of six *Zea diploperennis* accessions, which appear to be larger than the analogous S-1, R-1 or S-2, R-2 DNAs (Timothy et al. 1982). These results indicate that plasmid-like DNAs occur frequently in *Zea;* interpretation of their possible roles in the molecular bases of cms in maize is not possible at this time.

CYTOPLASMIC MALE STERILITY IN SORGHUM

Examination of organelle DNAs. The utilization of cms in the production of hybrid sorghum relies extensively on the milo source of cms (Harvey 1977). A number of potentially useful alternative cytoplasms have been identified. Restriction endonuclease fragment analyses of many of these alternative sources revealed extensive mtDNA heterogeneity among the entries. Among 23 entries examined, eight different mtDNA patterns were apparent (Pring, Conde, and Shertz 1982; Conde et al. 1982), while only three ctDNA groups emerged. In an examination of six KS series cytoplasms, Conde et al. (1982) observed two distinct patterns; KS35, KS36, and KS37 exhibited mtDNAs indistinguishable from each other but clearly different from KS34, KS38, and KS39. The latter three were indistinguishable from the milo cytoplasm mtDNA. These distinct patterns were observed in fertility restoration trials involving nine pollen parents, at two locations during two years. The comparison of mtDNA restriction patterns and fertility restoration patterns suggests that mtDNA analyses had predictive value in identifying cytoplasms that exhibited unique genetic behavior.

Sorghum plasmid-like DNAs. One of the 23 sorghum cytoplasms examined, IS1112C, yielded evidence of mitochondrial plasmid-like DNAs (Pring, Schertz, and Conde 1981). Since the IS1112C entry derived from an Indian variety called Nilwa, we have identified these DNAs as N-1 and N-2 plasmid-like DNAs. When compared to the S-1 and S-2 DNAs of maize, the N-2 and S-2 molecules exhibited very similar electrophoretic migration characteristics, while the N-1 DNA migrated faster than the maize S-1 DNA (Fig. 5.2). The molecular size of N-1 is 5.7 kbp, while N-2 DNA is about 5.3 kbp, slightly lower than a value of 5.4 kbp for S-2 (Pring et al. 1982). Examination of a

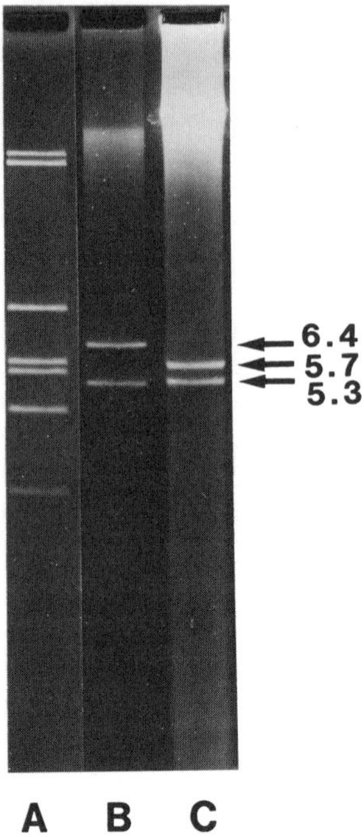

Figure 5.2. Agarose gel electrophoresis of (A) *Eco*RI-digested bacteriophage lambda DNA, (B) undigested mtDNA from the S cytoplasm of maize, and (C) undigested mtDNA from IS1112C sorghum. Arrows mark molecular sizes of S-1 (6.4 kbp), N-1 (5.7 kbp), and N-2 (5.3 kbp). Molecular sizes of lambda fragments are 24.7, 21.2, 7.42, 5.81, 5.65, 4.86, and 3.52 kbp.

number of lines in this cytoplasm indicates that the quantity of N-1 and N-2, relative to the mainband mtDNA, is lower than that observed for the maize plasmid-like DNAs.

Since the molecular sizes are close to those of the maize DNAs, an obvious question is whether the sorghum DNAs share homology with the maize DNAs. Extensive homology would indicate that a similar function could perhaps be assigned to these molecules. DNA-DNA hybridization experiments with nick-translated N-1 and N-2 DNAs (Pring et al. 1982) revealed trace homology between N-1 and S-1 or S-2, and detectable homology between N-2 and S-2. We conclude that the molecules may share homology, but this homology is certainly much less than that shared by the S-1 and S-2 DNAs and the R-1 and R-2 plasmid-like DNAs of maize recently described by Weissinger et al. (1982).

Hybridization studies of the N-1 and N-2 DNAs with sorghum mtDNAs suggest that the kafir normal cytoplasm mtDNA does not share extensive homology to the plasmid-like DNAs (C. D. Chase and D. R. Pring, unpublished). This is in contrast to the maize hybridization results, which indicate homology among all normal cytoplasm mtDNAs examined to date. Kafir sorghum is of African origin, while IS1112C was collected in India; certain of the Indian normal cytoplasm mtDNAs appear to carry sequences homologous to the plasmid-like DNAs.

Sorghum mtDNAs also include a number of low molecular weight DNAs, similar to those described in maize. A 2.25-kbp molecule was observed with mobility similar to an analogous DNA in maize. At least four additional molecules, with molecular sizes of 1 to 1.4 kbp, were also detected in the IS1112C cytoplasm. The specificity and distribution of these molecules among male-fertile and male-sterile cytoplasm mtDNAs is currently under study.

OBSERVATIONS ON OTHER PLASMID-LIKE DNAS

Zea and *Sorghum* species are not the sole examples of mtDNAs that include small, discrete components. Circular mtDNAs of 1.3 to 1.5 kbp were recently reported in sugarbeet (Powling 1981). In all fertile cytoplasms, DNAs of 1.3 and 1.4 kbp were observed, and additionally, either a 1.45- or 1.5-kbp DNA. MtDNA from male-sterile cytoplasms contained only the 1.5-kbp DNA. In this case, therefore, a missing small DNA was characteristic of male-sterile cytoplasms, instead of additional DNAs as shown in certain of the sorghum and maize sources of cms. It seems possible that these molecules have their origins in excision events in the large mtDNA and that amplification allows their detection. It is interesting that mtDNAs from male-fertile or -sterile sugarbeet also exhibited possible multimers of a small repeating unit, forming a "ladder" of species (Powling 1981). Similar observations have been made by Dale (1981) in *Phaseolus* and N cytoplasm maize tissue cultures. These events are somewhat analogous to the behavior of mtDNA in senescent *Podospora* (Cummings, Belcour, and Grandchamp 1979) and other fungi, where apparent deletion is followed by amplification of the remaining mtDNA sequences. Whatever their origin, the plant mtDNAs may be capable of extensive replication because of origins of replication or other related processes. It is clear that the nucleus plays a major role in transpositional events in the case of the two maize molecules, as well as in determining their relative proportions (Laughnan, Gaby-Laughnan, and Carlson 1981). These data strongly indicate that complex nuclear-mitochondrial interactions are operative in the replication or degradation of these DNAs, and additionally, in determining their behavior in cells.

CONCLUSIONS

A variety of small, discrete DNAs are associated with mitochondria of *Zea, Sorghum, Beta,* and *Phaseolus*. Variation in size or composition of some of these molecules is associated with cytoplasmic male sterility. Transpositional activity of sequences of the maize plasmid-like DNAs, concomitant with conversion to a male-fertile condition, suggests that other male-sterile systems may be similarly associated with mobile genetic elements. Structural alterations of the principal mtDNA of sources of cms in maize in regions carrying homology to the maize plasmid-like DNAs suggest a correlation between these alterations and the genetic behavior of the cytoplasms. Studies to determine whether a causal relationship exists, or whether other small mtDNAs are involved in the expression of the sterility trait, would be facilitated by the establishment of genome maps of these mtDNAs. The determination of transcriptional and translational activities of the DNAs and their correlation with the physiological and biochemical processes of pollen abortion will be required to elucidate the molecular bases of cms in higher plants.

LITERATURE CITED

Brettell, R. I. S., M. F. Conde, and D. R. Pring. 1982. *Analysis of mitochondrial DNA from four fertile maize lines obtained from a tissue culture carrying Texas cytoplasm.* Maize Genet. Coop. Newsl. 56:13–14.

Brettell, R. I. S., E. Thomas, and D. S. Ingram. 1981. *Reversion of Texas male-sterile cytoplasm maize in cluture to give fertile, T-toxin resistant plants.* Theor. Appl. Genet. 58:55–58.

Conde, M. F., D. R. Pring, K. F. Schertz, and W. M. Ross. 1982. *Correlation of mitochondrial DNA restriction endonuclease patterns with sterility expression in six male-sterile sorghum cytoplasms.* Crop Sci. 22:536–39.

Cummings, D. J., L. Belcour, and C. Grandchamp. 1979. *Mitochondrial DNA from Podospora anserina. II. Properties of mutant DNA and multimeric circular DNA from senescent cultures.* Mol. Gen. Genet. 171:239–50.

Dale, R. M. K. 1981. *Sequence homology among different size classes of plant mtDNAs.* Proc. Nat. Acad. Sci. USA 78:4453–57.

Duvick, D. N. 1965. *Cytoplasmic pollen sterility in corn.* Adv. in Gen. 13:1–56.

Gengenbach, B. G., J. A. Connelly, D. R. Pring, and M. F. Conde. 1981. *Mitochondrial DNA variation in maize plants regenerated during tissue culture selection.* Theor. Appl. Genet. 59:161–67.

Gengenbach, B. G., C. E. Green, and C. M. Donovan. 1977. *Inheritance of selected pathotoxin resistance in maize plants regenerated from cell cultures.* Proc. Nat. Acad. Sci. USA 74:5113–17.

Harvey, P. H. 1977. *Sorghum germplasm base in the U.S.* Corn and Sorghum Res. Conf., 1977, pp. 186–98.

Humaydan, H. S., and E. W. Scott. 1977. *Methomyl insecticide selective phytotoxicity on sweet corn hybrids and inbreds having Texas male-sterile cytoplasm.* HortSci. 12:312–13.

Kemble, R. J., and J. R. Bedbrook. 1980. *Low molecular weight circular and linear DNA in mitochondria from normal and male-sterile Zea mays cytoplasm.* Nature (Lond.) 284:565–66.

Kemble, R. J., R. E. Gunn, and R. B. Flavell. 1980. *Classification of normal and male-sterile cytoplasms in maize. II. Electrophoretic analysis of DNA species in mitochondria.* Genetics 95:451–58.

Kemble, R. J., and D. R. Pring. 1982. *Mitochondrial DNA associated with cytoplasmic male sterility and disease susceptibility in maize carrying Texas cytoplasm.* In Y. Asada, W. R. Bushnell, S. Ouchi, and C. P. Vance, eds., *Physiological and Biochemical Bases of Plant Infection.* Japan Scientific Society Press, Tokyo, and Springer-Verlag, New York (forthcoming).

Kim, B. C., R. J. Mans, M. F. Conde, D. R. Pring, and C. S. Levings, III. 1982. *Physical mapping of homologous segments of mitochondrial episomes from S male-sterile maize.* Plasmid 7:1–14.

Koncz, C., J. Sümegi, A. Udvardy, M. Racsmány, and D. Dudits. 1981. *Cloning of mtDNA fragments homologous to mitochondrial S2 plasmid-like DNA in maize.* Mol. Gen. Genet. 183:449–58.

Laughnan, J. R., and S. J. Gabay. 1978. *Nuclear and cytoplasmic mutations to fertility in*

S male-sterile maize, Pages 427–46 in D. B. Walden, ed., *Maize Breeding and Genetics*. John Wiley & Sons, New York.

Laughnan, J. R., S. J. Gabay-Laughnan, and J. E. Carlson. 1981. *Characteristics of cms-S reversion to male fertility in maize*. Stadler Symp. 13:93–114.

Levings, C. S., III, B. D. Kim, D. R. Pring, M. F. Conde, R. J. Mans, J. R. Laughnan, and S. J. Gabay-Laughnan. 1980. *Cytoplasmic reversion of cms-S in m*

Levings, C. S., III, and D. R. Pring. 1976. *Restriction endonuclease analysis of mitochondrial DNA from normal and Texas cytoplasmic male-sterile maize*. Science 193:158–60.

———. 1977. *Diversity of mitochondrial genomes among normal cytoplasms of maize*. J. Hered. 68:350–54.

———. 1979. *Molecular bases of cytoplasmic male sterility in maize*. Pages 171–93 in J. G. Scandalios, ed., *Physiological Genet*. Academic Press, New York.

Lonsdale, D. M., R. D. Thompson, and T. P. Hodge. 1981. *The integrated forms of the S1 and S2 DNA elements of maize male-sterile mitochondrial DNA are flanked by a large repeated sequence*. Nucl. Acids Res. 9:3657–69.

Powling, A. 1981. *Species of small DNA molecules found in mitochondria from sugar-beet with normal and male sterile cytoplasms*. Mol. Gen. Genet. 183:82–84.

Pring, D. R., M. F. Conde, and C. S. Levings, III. 1980. *DNA heterogeneity within the C group of maize male-sterile cytoplasms*. Crop Sci. 20:150–62.

Pring, D. R., M. F. Conde, and K. F. Schertz. 1982. *Organelle genome diversity in sorghum: Male-sterile cytoplasms*. Crop Sci. 22:414–21.

Pring, D. R., M. F. Conde, K. F. Schertz, and C. S. Levings, III. 1982. *Plasmid-like DNAs associated with mitochondria of cytoplasmic male-sterile sorghum*. Mol. Gen. Genet. 186:180–84.

Pring, D. R., and C. S. Levings, III. 1978. *Heterogeneity of maize cytoplasmic genomes among male-sterile cytoplasms. Genetics 89:121–36.*

Pring, D. R., C. S. Levings, III, W. W. L. Hu, and D. H. Timothy, 1977. *Unique DNA associated with mitochondria in the "S"-type cytoplasm of male-sterile maize*. Proc. Nat. Acad. Sci. USA 74:2904–2908.

Pring, D. R., Schertz, K. F., and M. F. Conde. 1981. *Plasmid-like and other cytoplasmic DNA characteristics of male-sterile sorghum*. Sorghum Newsl. 24:132.

Spruill, W. M., Jr., C. S. Levings, III, and R. R. Sederoff, 1980. *Recombinant DNA analysis indicates that the multiple chromosomes of maize mitochondria contain different sequences*. Devel. Gen. 1:363–78.

Thompson, R. D., R. J. Kemble, and R. B. Flavell. 1980. *Variations in mitochondrial DNA organization between normal and male-sterile cytoplasms of maize*. Nucl. Acids Res. 8:199–2008.

Timothy, D. H., C. S. Levings, III, W. W. L. Hu, and M. M. Goodman. *Zea diploperennis may have plasmid-like mitochondrial DNAs*. 1982. Maize Genet. Coop. Newsl. 56:133–34.

Ullstrup, A. J. 1972. *The impacts of the southern corn leaf blight epidemics of 1970–1971*. Ann. Rev. Phytopathol. 10:37–50.

Weissinger, A. K., D. H. Timothy, C. S. Levings, III, W. W. L. Hu, and M. M. Goodman. 1982 *Unique plasmid-like mitochondrial DNAs from indigenous maize races of Latin America*. Proc. Nat. Acad. Sci. USA 79:1–5.

three

COMMERCIAL APPLICATIONS

6] Utilization of Genetically Engineered Microorganisms for the Manufacture of Agricultural Products

by J. LESLIE GLICK, M. VIRGINIA PEIRCE, DAVID M. ANDERSON, CHARLES A. VASLET, and HUMG-YU HSIAO*

ABSTRACT

Genetically engineered microorganisms will soon play a significant role in the manufacture of products used in agriculture and in products obtained from agriculture. Many products currently used in crop and livestock production, such as pesticides, herbicides, vitamins, and antibiotics, can be produced in the future more economically, and in higher yields, by appropriately developed microbial strains. Furthermore, agricultural products not widely used at present due to prohibitive manufacturing costs will become more available, profoundly affecting agricultural practices. For example, microbially produced animal growth hormones and vaccines to prevent diseases affecting livestock are already being tested. In addition, it is anticipated that microbial production of expensive amino acids, such as tryptophan and threonine, could reduce their respective cost per pound ten- to twentyfold, thereby stimulating demand for them as feed supplements. Products obtained from agriculture, such as high-cost plant chemicals used in the manufacture of drugs, may also be produced more economically by genetically engineered microorganisms. The feasibility of the development of these new processes depends on technical and economic criteria and will require interdisciplinary cooperation.

INTRODUCTION

Genetic engineering technology may be applied several different ways to the manufacture of agriculture-related products. Microorganisms that naturally synthesize such products may be genetically altered in order to allow these products to be made more economically by fermentation. In addition, microor-

*Genex Corporation, 6110 Executive Blvd., Rockville, Maryland 20852.

ganisms may be altered so that they make products normally made only by higher organisms. A third application is to include one or more microbial enzymatic steps, optimized by genetic engineering, with a chemical process to assure specificity of key reactions.

Those products currently used in agriculture that are amenable to production by genetically engineered microorganisms include the following: antibiotics, growth stimulants, vitamins, amino acids, feed preservatives, single-cell protein, vaccines, herbicides, insecticides, and plant growth regulators.

OVERVIEW

Antibiotics are widely used in animal husbandry because they promote feed utilization efficiency, increase the rate of weight gain, and prevent disease. In 1979 about $300 million was spent on antibiotics for this purpose. Although the pathways for microbial biosynthesis of some antibiotics are complex and involved in overall microbial metabolism, genetic engineering techniques could be applied. For example, in the case of the tetracyclines these techniques could be used to increase intracellular pools of rate-limiting intermediates. Yet the future use of antibiotics in animal feeds is uncertain because of increasing concern about selecting for antibiotic-resistant microorganisms.

Because feed costs represent about 25% of total production costs of beef feedlot operations in the United States, growth stimulants are widely used to increase feed utilization efficiency. Four growth stimulant products currently employed contain anabolic steroids. For example, estrogen derivatives stimulate increased growth hormone secretion in cattle. In one study, calves treated with a combination of an androgen plus estrogen converted 18% more feed protein into body protein than untreated controls (Reynolds 1980). Although it may be possible to produce these steroids by genetically engineered microorganisms, growing concern about human consumption of hormone-treated meat makes the future market uncertain. This concern has already resulted in the 1979 banning of diethylstilbestrol for use as a growth stimulant.

Another growth stimulant, monensin, is the major feed product in use today. Monensin is an antibiotic, but it also increases feed utilization efficiency by affecting amino acid balance in the rumen. This compound is synthesized naturally by *Streptomyces* and is commercially produced by fermentation. Genetic engineering could probably be used to improve yields or product ratios in this fermentation.

The world market for vitamins for use in animal feeds was $185 million in 1981. As many as ten different vitamins are included in animal rations, several of which could be produced by genetically engineered microorganisms. Two examples, riboflavin and vitamin B_{12}, are produced commercially by microbial fermentation processes (Florent and Ninet 1979; Perlman 1979). Feed-grade riboflavin costs around $24 per lb, while vitamin B_{12} can cost more than $4,000 per lb. The cost of manufacture of both of these vitamins would be lower if produced by genetically engineered strains. Also added to animal feeds are the amino acids lysine and methionine. The world market for these two amino acids alone amounts to $400 million annually.

Feed preservatives, such as sorbic acid, are currently derived from petrochemicals. Sorbic acid also occurs naturally in plants and may be amenable to microbial production.

For more than fifteen years, the technology has been developing for using microorganisms as an inexpensive source of protein for supplementing animal as well as human diets. A variety of feedstocks have been evaluated, ranging from petrochemicals to waste effluents from various industrial processes. Imperial Chemical Industries (ICI) has applied genetic engineering techniques to improve the conversion by *Methylophilus* of methanol to cellular carbon (Windass et al. 1980). ICI is now the major producer of commercially available single-cell protein.

Vaccines represent another agricultural product category already beginning to be affected by genetic engineering. Vaccines produced by genetically engineered microorganisms to prevent foot-and-mouth disease (FMD) are currently being developed separately by Genentech, in collaboration with the USDA–Agricultural Research Service (Kleid et al. 1981), and by Biogen (Fildes, presentation at the Symposium on Microbiological Technology, Japan Management Assoc., Nov. 10–13, 1981). These FMD vaccines should be on the market within a couple of years. Intervet International, a division of the Dutch company Akzo, already has begun to market a vaccine using antigens produced by genetically engineered *Escherichia coli* to prevent scours, a diarrhea in young pigs and calves (Schuuring 1982). A similar product developed by Cetus is also about to be marketed (Cape, presentation at the Symposium on Microbiological Technology, Japan Management Assoc, Nov. 10–13, 1981). Improved vaccines against parasites, such as *Eimeria* that cause coccidiosis, may also result from genetic engineering technology.

Another animal disease that may be prevented by a product of genetic engineering is shipping fever, which has been estimated to cause an annual economic loss of $100 million (Irwin, Melendy, and Hutcheson 1980). Shipping fever appears to have a complex etiology, possibly beginning with a viral infection that leaves the animal susceptible to fatal bacterial infection. Shipping fever could be prevented by prophylactic treatment with a vaccine against the common bacterial agents, such as *Pasteurella,* or with an antiviral agent, such as bovine interferon. Genentech recently announced the cloning of bovine interferon genes.

Despite current levels of pest control, about one-quarter of the world's crops are destroyed or made less marketable by insects and other pests. Pesticide research today is focusing on the development of highly specific agents that are relatively harmless to man and animals. As much as 90% of today's pesticides are synthetic chemicals. Genetic engineering may be utilized to develop specific enzymatic process steps that should improve the specificity of chemical syntheses for certain existing products. Furthermore, the manufacture of natural biological pesticides, such as *Bacillus thuringiensis,* may be improved by genetic engineering technology.

Plant growth regulators represent the final category of products currently used in agriculture that may be manufactured by means of genetically engineering microorganisms. The functions and modes of action for many of these compounds are still under investigation, but a few cytokinins have already been found to be useful in breaking bud and seed dormancy. Some of the cytokinin genes have been found to be associated with bacterial plasmids (Moore 1979), making them more accessible for improved production by means of genetic engineering.

Most important, the application of genetic engineering technology is expected to change agricultural practices by making available products not widely used because of prohibitive manufacturing costs. Two such product lines are under development at Genex Corporation: animal growth hormones, and relatively expensive amino acids, such as tryptophan and threonine.

ANIMAL GROWTH HORMONES

The use of growth hormones instead of the growth stimulants mentioned earlier provides two advantages: (a) the growth hormones are more species-specific, and (b) they are proteins and therefore are destroyed in the human digestive tract. These factors will give animal growth hormones a competitive edge over steroid growth stimulants when large-scale production becomes economically feasible.

Identified and named on the basis of its observed biological activities, growth hormone (also called somatotropin or somatotropic hormone) is one of an array of peptide and protein hormones produced by the vertebrate pituitary gland located at the base of the brain (Turner 1960). Growth hormone is required for normal, balanced growth of preadult animals. Its growth-promoting effects are due primarily to its ability to increase the retention of nitrogen (as protein) in tissue. It does so by increasing membrane transport of amino acids into cells and by stimulating translation at the ribosomal level, two events essential for protein synthesis. Growth hormone also causes an increase in mitotic activity and cell division. The increased body weight observed in otherwise normal animals following growth hormone treatment is due to an actual increase in tissue protein and salts—not to an increased deposition of fat (Machlin 1972). These effects have immediate relevance to the science of animal husbandry, where growth hormone can be used as a natural means of increasing milk and meat production in domestic stocks (Machlin 1973; Peel et al. 1981). There is also preliminary evidence that it can increase wool production in sheep. Genex is one of a number of genetic engineering companies which are developing animal growth hormone (AGH) products.

Small amounts of pure AGH have been isolated from several different animal species and biochemically characterized. In general, AGHs are single-chain, unmodified globular proteins with molecular weights of 22,000–24,000 daltons. The mature hormone is 190–200 amino acids long. Complete amino acid sequence information is available for human, bovine, ovine, equine, and rat growth hormones (Dayhoff 1976; Seeburg et al. 1977; Martial et al. 1979; and Miller, Martial, and Baxter 1980). A partial sequence of porcine hormone is also known (Wilhelmi and Mills 1972). These data show AGHs to be highly conserved proteins with long homologous stretches in their primary amino acid sequences. Differences in chemical and physical properties among these

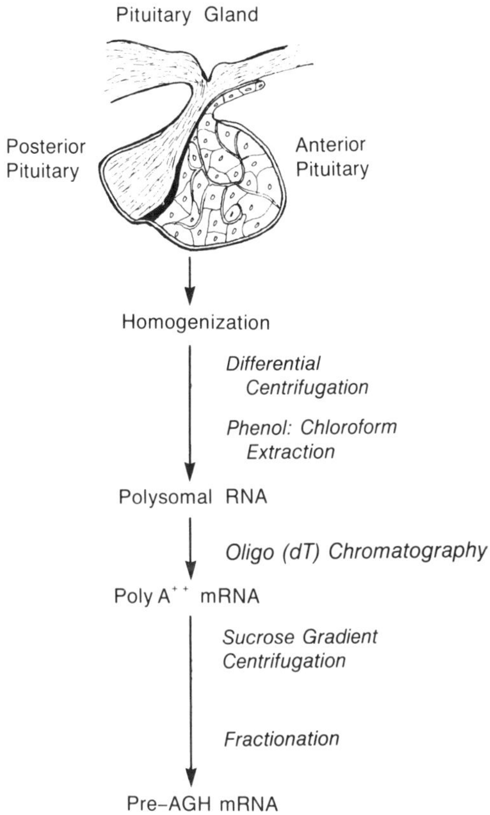

Figure 6.1. Isolation of pre-AGH mRNA.

AGHs are minor. As is characteristic of most secreted proteins, AGHs have a 26-amino acid "signal" or "pre" sequence at their N-terminus, which is presumably responsible for the passage of the newly synthesized hormone through the membrane of the cell. This "pre" peptide is eventually cleaved to produce the mature hormone (Davis and Tai 1980).

While current isolation methods yield sufficient quantities of AGH to conduct basic research, large-scale commercial preparation of these hormones requires methods that are less time-consuming and costly. Again, genetic engineering technology now makes it possible to obtain bacterial strains capable of producing high yields of active AGHs. Because AGHs are highly conserved, single polypeptide chains of moderate size, requiring no post-translational modifications of any amino acid residues (e.g., phosphorylation or glycosylation), they are well suited to production in genetically engineered bacterial strains. The major steps necessary for constructing an AGH-producing microorganism are:

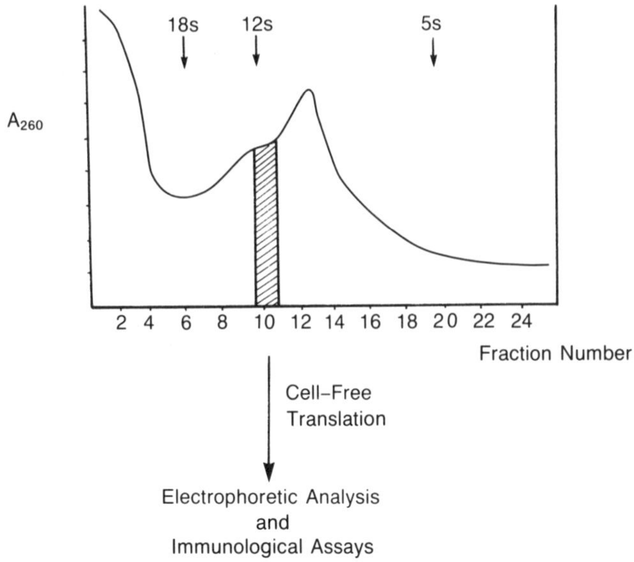

Figure 6.2. Sucrose gradient fractionation of pre-AGH mRNA.

1. isolation and recovery of pre-AGH mRNA from pituitary glands;
2. in vitro synthesis of double-stranded complementary DNA (ds-cDNA), using pre-AGH mRNA as a template;
3. insertion of the ds-cDNA into a suitable cloning vector and amplification of that vector in a bacterial host;
4. identification and recovery of the cloned AGH gene; and
5. insertion of the cloned gene into an appropriate expression vector and transformation of a suitable production microorganism.

Cloning animal growth hormone genes. At Genex, we have cloned the genes coding for bovine, porcine, and ovine growth hormones. The following procedures were applied to all three of these hormones:

The initial step in the construction of our AGH-producing microorganisms was the isolation of pre-AGH mRNA from fresh pituitary glands (Fig. 6.1). These were obtained at a local abattoir within one hour after the animals were sacrificed. The glands were removed and frozen immediately in liquid nitrogen. The frozen tissue was homogenized, and polysomal RNA was obtained by differential centrifugation and conventional phenol:chloroform extraction. Polyadenylated mRNA, which represents approximately 2%–5% of the polysomal RNA preparation, was isolated by standard oligo(dT) column chromatography (Aviv and Leder 1972). Pre-AGH mRNA was further purified from the entire mRNA population by sucrose density gradient centrifugation and fractionation (Fig. 6.2). Those fractions enriched for pre-AGH mRNA were

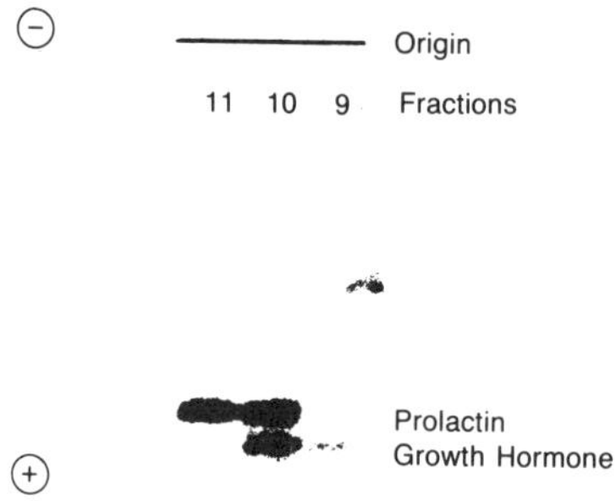

Figure 6.3. SDS-slab gel electrophoresis of bovine mRNA fractions. Fraction 11 contains pre-prolactin mRNA. Fraction 10 contains pre-prolactin mRNA and pre-bovine growth hormone mRNA. Fraction 9 contains pre-bovine growth hormone mRNA.

localized by cell-free translation in a rabbit reticulocyte system (Pelham and Jackson 1976) and identification of translation products as pre-AGH. Identification methods included SDS gel electrophoresis to establish molecular weight, as well as immunological assays with the appropriate anti-AGH antisera. Figure 6.3 is an SDS slab gel of the reticulocyte translation products of some of the bovine mRNA fractions. Fraction 11 contains a 26,000-dalton protein which should be pre-prolactin. Fraction 10 contains pre-prolactin plus a protein of pre-bovine growth hormone molecular weight. Fraction 9 contains only pre-bovine growth hormone.

The enriched mRNA from Fraction 10 was used for the synthesis of ds-cDNA by RNA-dependent DNA polymerase (reverse transcriptase) and the DNA polymerase I Klenow fragment (Klenow and Henningsen 1970; see Fig. 6.4). S1 nuclease digestion (Vogt 1973) was used to cleave the hairpin loop formed during ds-cDNA synthesis and to trim away the poly(dT) primer. Terminal transferase (Chang and Bollum 1971) catalyzed the addition of poly(dC) tails to ds-cDNA molecules of appropriate size (600 base pairs). Similarly, complementary poly(dG) tails were added to *Pst*I-cut DNA of the tet^r-amp^r cloning vector pBR322, and the two "tailed" DNA species were annealed. The recombinant plasmids were introduced by transformation into an *E. coli* host strain. Because *Pst*I cuts pBR322 in the ampicillin-resistance gene, hybrid plasmids containing inserted DNA should be amp^s and tet^r. Therefore, bacteria containing such recombinants were selected by their acquired resistance to tetracycline and continued ampicillin sensitivity. A significant fraction of the transformants obtained from this procedure were expected to have full-length AGH gene inserts because of the enriched AGH mRNA population used to synthesize the cDNA.

A modification of the Grunstein-Hogness in situ hybridizaiton method (Grunstein and Wallis 1979) was employed to screen amp^s-tet^r bacterial trans-

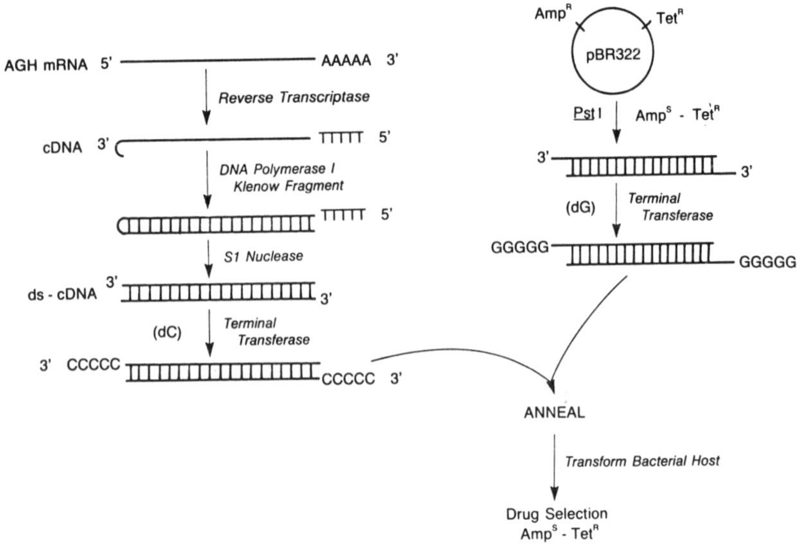

Figure 6.4. Animal growth hormone cDNA clone construction.

formants for pre-AGH–containing sequences (Fig. 6.5). Three different probes, listed below in order of preference, were used to select AGH-gene–containing recombinants from a cDNA clone bank.

1. ^{32}P-labeled cDNA prepared from AGH mRNA from the same species.
2. ^{32}P-labeled AGH mRNA from the same species.
3. ^{32}P-labeled synthetic DNA probe, which shares sequence homology with all AGH genes of interest.

When an AGH cDNA clone from any one species was identified, it too was used to screen for AGH DNA clones in the clone banks of other species, because of the extensive sequence homology among AGH genes from different species. For each Grunstein-Hogness screening, DNA from each colony (or group of colonies) was fixed to discrete zones on a nitrocellulose filter. The DNA was then denatured to form single strands, and the filter was incubated under defined conditions in hybridization medium containing a radioactively labeled probe. Following hybridization, unbound probe was washed from the filter, and colonies containing DNA to which the probe hybridized were identified by autoradiography (Fig. 6.6).

To demonstrate unambiguously that the inserted DNA of the selected clones coded for AGH, the nucleotide sequences of the DNA inserts were determined. Because the amino acid sequences of the AGHs of several species are known, a comparison with the nucleotide sequence of the cloned AGH DNA allows one to determine not only if the DNA codes for AGH, but if the entire coding sequence is present on the cloned fragment. A number of rapid DNA

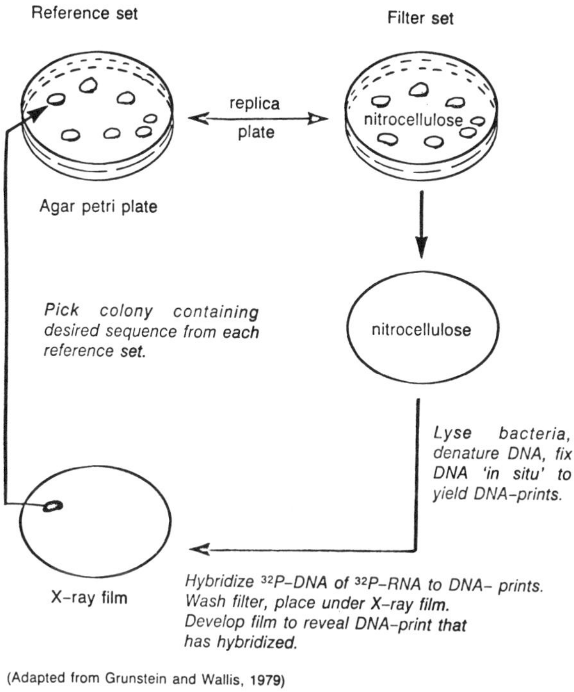

Figure 6.5. Colony hybridization procedure.

sequencing methods are currently available (Maxam and Gilbert 1977; Sanger, Nicklen, and Coulson 1977).

Expression. Cloned, sequenced AGH genes must be transferred to an expression vector and manipulated such that they are expressed at a high level under the control of a bacterial promoter. The production of mature AGHs can be monitored by polyacrylamide gel electrophoresis of whole-cell protein extracts. In addition, species-specific radioimmunoassays can be employed to quantitate the percentage of total bacterial protein being synthesized as AGH.

The techniques employed for the expression of eucaryotic genes introduced into bacteria were developed primarily in the gram-negative organism *E. coli*. *E. coli,* however, does not secrete many proteins into the culture medium and, therefore, the cells must generally be lysed before the expressed product of a cloned gene may be isolated. This is not a serious problem for the economical use of genetically engineered strains to produce high-cost proteins on an industrial scale. Nevertheless, the low yield and high costs of recovering intracellular proteins makes the production of low-cost proteins in *E. coli* economically prohibitive.

Furthermore, the *E. coli* cell wall contains endotoxin, which is released when the cell is disrupted. For this reason, purification of AGH must include

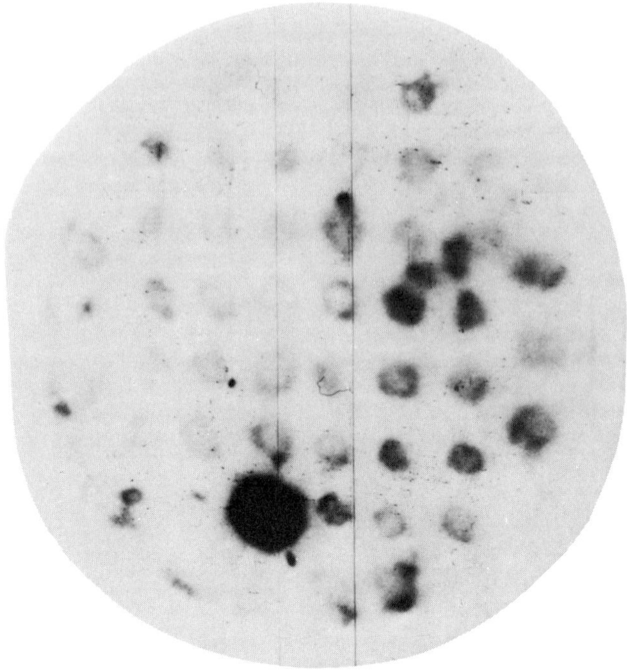

Figure 6.6. Colony of cells containing AGH gene identified by autoradiography.

thorough elimination of this toxic contaminant. Also, the AGH produced in *E. coli* will be likely to have one additional amino acid, methionine, at the N-terminus, because the genetic signal to start transcription in bacteria encodes methionine.

In contrast, a gram-positive organism, like *Bacillus subtilis,* is capable of secreting many proteins into the culture medium. If the AGH coding sequence carried on a *Bacillus* plasmid is linked to either the natural AGH signal sequence or a heterologous signal sequence, it is likely that *B. subtilis* will properly remove the signal peptide and secrete AGH into the medium. Techniques for using gram-positive production microorganisms that express eucaryotic genes are currently being developed in many laboratories.

Figure 6.7 compares the economics of producing AGH as an intracellular protein versus that of producing it as an extracellular protein. While in both cases the unit cost of AGH decreases as the level of expression increases, it is clear that in the generally achievable range of high expression (1%–10% of protein for intracellular production and 1–10 g/1 for extracellular production), the unit cost is significantly lower for extracellular production than for intracellular production at both low (1,000 kg/yr) and high (50,000 kg/yr) levels of production of AGH.

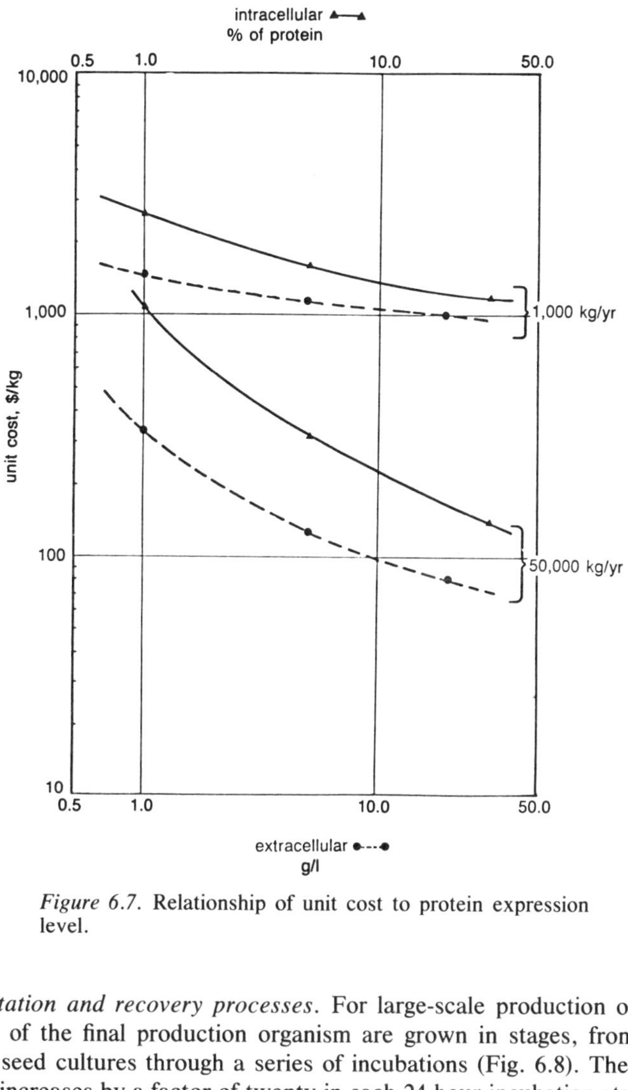

Figure 6.7. Relationship of unit cost to protein expression level.

Fermentation and recovery processes. For large-scale production of AGH, cultures of the final production organism are grown in stages, from small-volume seed cultures through a series of incubations (Fig. 6.8). The culture volume increases by a factor of twenty in each 24-hour incubation stage. The fermentation medium consists of corn syrup, autolyzed yeast extract, ammonia, and salts. Most of the ingredients are premixed and pumped into a fermentor. After addition of any supplemental ingredients and adjusting the pH, the final-stage seed inoculum is also pumped into the fermentor.

Each fermentation batch has a duration of about 2.5 days, with six hours allowed for turnaround. One fermentor could thus serve for 128 batches per year, even allowing for down time due to maintenance and repairs. During the first quarter of the fermentation, rapid culture growth conditions are maintained. When a high cell titer is achieved, culture conditions are adjusted to promote a high level of production of AGH. A fermentor capable of producing

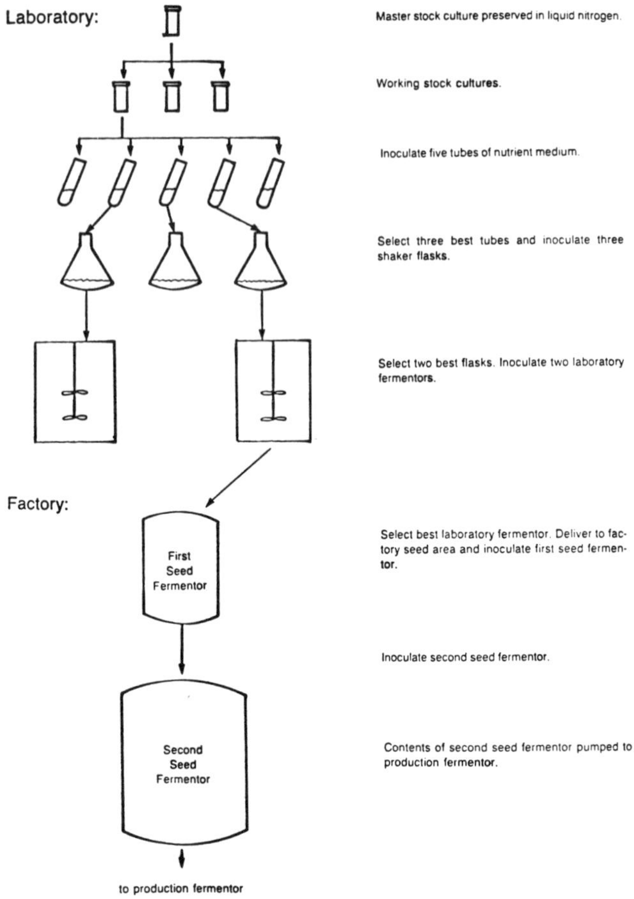

Figure 6.8. Seed development.

50,000 liters per batch could yield over 50,000 kg per year of AGH (assuming 80% recovery), if the microorganisms that were employed secreted AGH at a final concentration of 10 g/1.

The recovery process for fermentations using microorganisms that secrete AGH into the culture medium involves selective precipitations to remove impurities and separate the AGH, followed by purification of the AGH by anion exchange. The product is recovered as a buffered solution because AGH is susceptible to denaturation if lyophilized. This AGH solution is suitable for use in a delivery device.

Three of the steroid growth stimulants mentioned earlier are delivered to animals through a device implanted in the animal's ear. Figure 6.9 depicts one example of such a device that has recently become available (Anon. 1982). The steroid is incorporated into the matrix of the implant device so that it diffuses

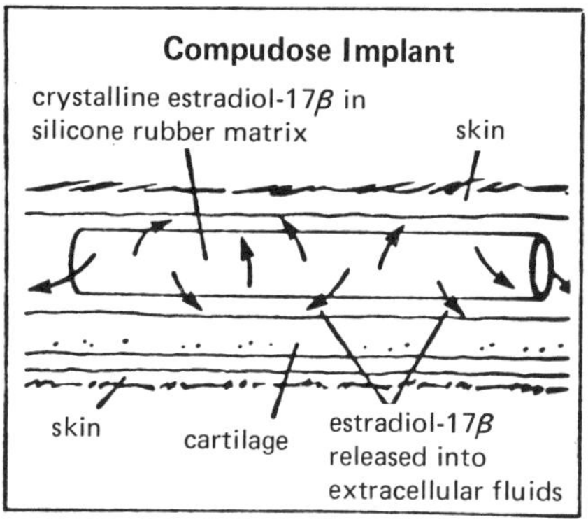

(used with permission from *Feedstuffs*)

Figure 6.9. Example of a delivery device for the sustained release of steroid hormones.

out at a fairly constant level for a defined period of time when the animals are in the feedlot. Preliminary work has been done to devise a delivery device for the sustained release of high molecular weight compounds such as AGH.

TRYPTOPHAN

Essential amino acids are required by man and other monogastric animals and must be provided in the diet, either as free amino acids or as constituents of dietary protein. The biological value of dietary proteins reflects the amounts and relative proportion of essential amino acids.

Currently, poultry and swine in the United States receive corn/soybean meal diets supplemented with the two most limiting essential amino acids, lysine and methionine, as free amino acids (Itoh et al. 1974). The third most limiting amino acid in these diets, tryptophan, is not currently added as a free amino acid because of high production costs. It is anticipated that the use of genetically engineered microorganisms will make tryptophan readily available for feed supplementation. Its actual use will depend on price and availability of corn and soy meal.

Tryptophan synthesis. Tryptophan can be synthesized by three routes: chemical synthesis, fermentation, and biocatalysis. The first process for chemical synthesis was demonstrated in 1907, and many different chemical processes have been published since then (Shirai et al. 1974). One drawback of chemical synthesis is that it is not stereospecific. Thus, a mixture of DL-tryptophan is produced, of which only one-half is the biologically active L-isomer.

The second route for tryptophan synthesis is fermentation, using any of a variety of microorganisms growing on molasses or other sugar-containing feedstocks. Tryptophan biosynthesis in microorganisms requires a number of enzymes, some of which are coordinately expressed in operons (Platt 1979). The tryptophan operon and the common aromatic amino acid biosynthetic pathway in *E. coli* are highly regulated by several different control mechanisms. This tight regulation keeps the cell from using energy for tryptophan synthesis unless it is required for cellular metabolism. That the tryptophan operon is tightly regulated reflects that it is one of the most energy-expensive amino acid syntheses for the cell. Microorganisms of other genera, such as *Corynebacterium,* have been found to produce higher levels of L-tryptophan by fermentation (Nakayama and Hagino 1971). Furthermore, mutant microorganisms of several genera have been isolated, over the past twenty years, that have altered regulatory controls and produce even higher levels of tryptophan by fermentation (Nakayama and Hagino 1974). Nevertheless, this route has not yet yielded commercially satisfactory titers.

The third possible route for tryptophan synthesis is to employ enzymes as biocatalysts. Two enzymes, tryptophanase and tryptophan synthase, exist in many microorganisms and can catalyze the synthesis of tryptophan from supplied precursors. The physiological function of tryptophanase is to catalyze the hydrolysis of L-tryptophan to indole, pyruvate, and ammonia. This reaction is reversible, however, and will yield tryptophan if indole, pyruvate, and ammonia are provided. Tryptophanase is susceptible to inhibition by tryptophan as well as indole, so high concentrations of pyruvate and ammonia must be provided. The instability of pyruvate, the problems of optimizing the kinetics of a reaction involving three components, and the reversibility of the reaction impair the commercial application of this process.

The second enzyme, available in microorganisms, that can catalyze the synthesis of L-tryptophan is tryptophan synthase. This enzyme catalyzes the terminal step in the biosynthesis of L-tryptophan, i.e., the conversion of indole-3-glycerol phosphate (InGP) and L-serine to L-tryptophan and glyceraldehyde-3-phosphate. Although this is the physiological reaction, the enzyme also efficiently converts indole and serine to L-tryptophan and water. The tryptophan synthase reaction is irreversible. Like tryptophanase, it is inhibited by indole, but this inhibition can be overcome by slow feeding of indole during the reaction. A major advantage of the tryptophan synthase reaction is that one may attain very high yields of tryptophan (in excess of 90% based on indole). Indeed, this reaction is the one of choice at Genex for the basis of a commercial route of tryptophan synthesis. The major disadvantage of this route is that it requires the amino acid L-serine, which is expensive to produce by current methods. For this reason, we are also designing a biocatalytic process for production of serine from inexpensive feedstocks (see below). The production of serine will then be coupled to the production of tryptophan.

Native tryptophan synthase is a tetrameric protein of two dissimilar subunits encoded by the two tryptophan operon genes *trpA* and *trpB* (Fig. 6.10) (Crawford 1980). The product of the *trpA* gene, termed α subunit, and the product of the *trpB* gene, termed β subunit, have been purified to homogeneity and extensively characterized. Two α subunits combine with a β_2 dimer to form a very stable $\alpha_2\beta_2$ complex, stabilized mainly by hydrophobic interactions (Dicamelli, Balbinder, and Lebowitz 1973).

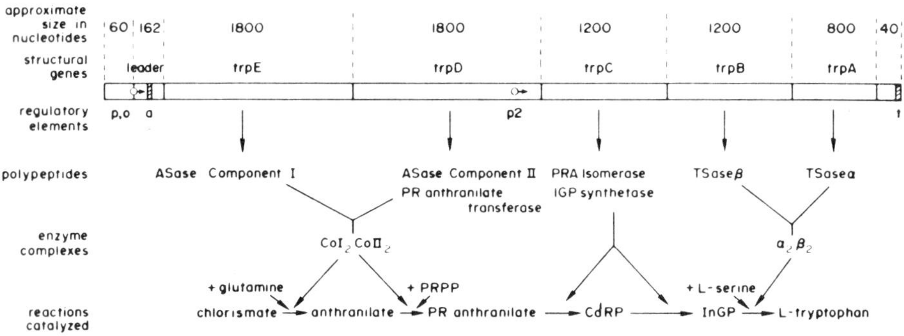

ASase = Anthranilate synthetase
PRA = Phosphoribosyl anthranilate
TSase = Tryptophan synthase
PRPP = Phosphoribosylpyrophosphate
CdRP = 1-(Carboxyphenylamino)-1-deoxyribulose-5-phosphate
InGP = Indole-3-glycerol phosphate

(Adapted from Platt, T., 1978)

Figure 6.10. Tryptophan operon.

The $\alpha_2\beta_2$ complex catalyzes the conversion of InGP and serine to tryptophan and glyceraldehyde-3-phosphate in two steps. First, the α subunit converts InGP to indole, which remains enzyme-bound. Second, the β_2 dimer combines serine and indole to make tryptophan. Therefore, indole can be substituted for InGP as a precursor for tryptophan synthesis. Although the β_2 dimer alone will catalyze the conversion of indole to tryptophan, the presence of the α_2 dimer in the $\alpha_2\beta_2$ complex stimulates β_2 activity more than 70-fold. For this reason, our process involves the $\alpha_2\beta_2$ complex.

Cloning and expressing tryptophan synthase genes. At Genex, we have cloned the tryptophan synthase genes of *E. coli* into a proprietary multicopy expression vector, pGX145, which contains bacterial transcription and translation control sequences. Our aim has been to introduce into a nonpathogenic, production strain of *Salmonella typhimurium* a vector capable of directing the controlled production of increased levels of tryptophan synthase. The *S. typhimurium* strain was chosen because it lacks tryptophanase and, therefore, does not degrade tryptophan. The whole cells or extracts of the cells are then used as biocatalysts in continuous flow, immobilized cell, or immobilized enzyme reactors to convert indole and serine to tryptophan.

Because tryptophan synthase occurs naturally in *E. coli,* the steps to clone the genes were fewer than those described for AGH. There was no need to isolate mRNA and make cDNA in this case. By means of restriction enzymes, the entire *trp* operon was cut out of the *E. coli* chromosome and inserted into a plasmid vector that was subsequently used to transform *E. coli*. Transformants were identified by complementation of *trp* operon mutants. Additional restric-

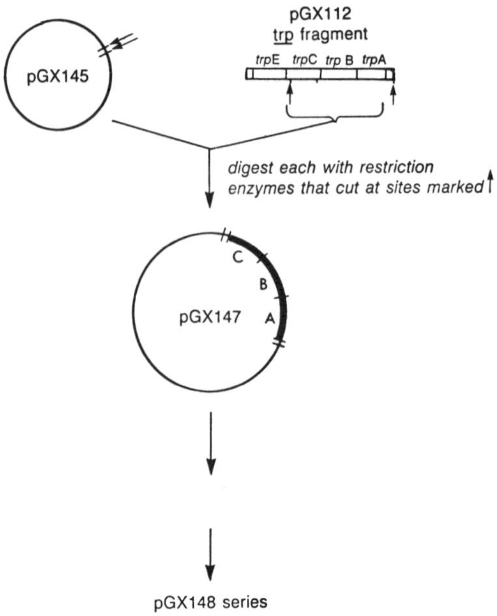

Figure 6.11. Construction of a vector for expression of high levels of tryptophan synthase.

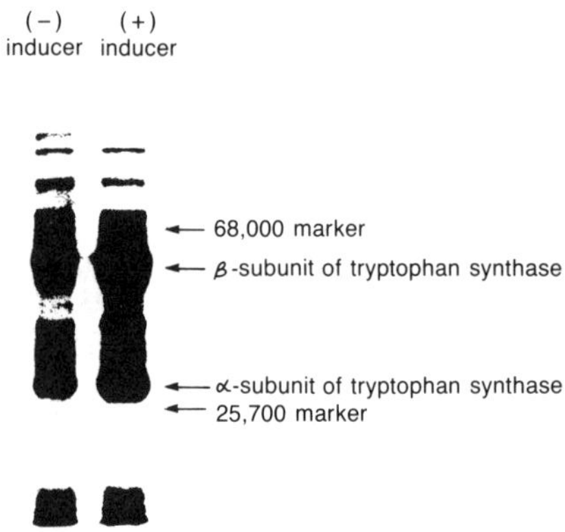

Figure 6.12. Polyacrylamide gel electrophoresis of lysates of pGX148-containing cells. The lane on the left represents a lysate of pGX148-containing *E. coli* cells which were grown in the absence of an inducer. The lane on the right shows that a lysate of the same cells contained even higher levels of the α and β subunits when the cells were grown in the presence of an inducer.

Table 6.1 Effect of Induction on Tryptophan Synthase Activity in *E. coli*

Plasmid	Units/mg protein	% Extractable protein as Tryptophan synthase
pGX112	19	0.73
pGX148-4 uninduced	80	3.08
pGX148-4 induced	468	18.0

tion enzyme digests followed by ligation yielded plasmid pGX112, which encodes the *trp* operon with an internal deletion for *trpD*.

Plasmids pGX112 and pGX145 were then cut with restriction enzymes in order to generate the same "sticky" ends (Fig. 6.11). The pGX112 fragment of interest contained the complete *trpA* and *trpB* genes, plus a portion of the *trpC* gene. This fragment was inserted into pGX145 to form plasmid pGX147. Through subsequent genetic manipulations, pGX147 gave rise to the pGX148 series of plasmids.

Cells transformed with plasmids of the pGX148 series have been found to express high levels of tryptophan synthase. Polyacrylamide gel electrophoresis of whole-cell protein extracts was used to assess the levels of intracellular tryptophan synthase in *E. coli* containing such plasmids (Fig. 6.12). The lane on the left in Figure 6.12 represents a lysate of pGX148-containing *E. coli* cells, which were grown in the absence of an inducer. The lane on the right shows that a lysate of the same cells contained even higher levels of the α and β enzyme subunits when the cells were grown in the presence of an inducer. In the uninduced cells, 3% of the total cell protein consisted of the tryptophan synthase subunits, while in the induced cells, 18% of the total cell protein consisted of the two subunits (Table 6.1).

Figure 6.13 illustrates the extent of tryptophan production in the presence of enzyme extracts from cells containing another high-expression vector for tryptophan, pGX141. The cells were fed indole gradually in order to avoid inhibition of tryptophan synthase activity. As a result, extremely high levels of tryptophan were obtained by continuing the reaction over prolonged periods of time. Levels of tryptophan over twice as great as those shown in Figure 6.13 have been achieved in this manner.

Utilization of bioreactors. As mentioned above, we are also developing an L-serine production process to interface with the tryptophan production process. Figure 6.14 shows the enzymatic reactions of these processes coupled. We believe that the most efficient method for production of tryptophan by means of these processes is to use immobilized cells in a multicolumn, packed-bed reactor system. The final production microorganisms are immobilized on beads and packed into the columns. Reactants are added to the columns, and the effluent is pumped to recovery operations. Figure 6.15 depicts a bioreactor

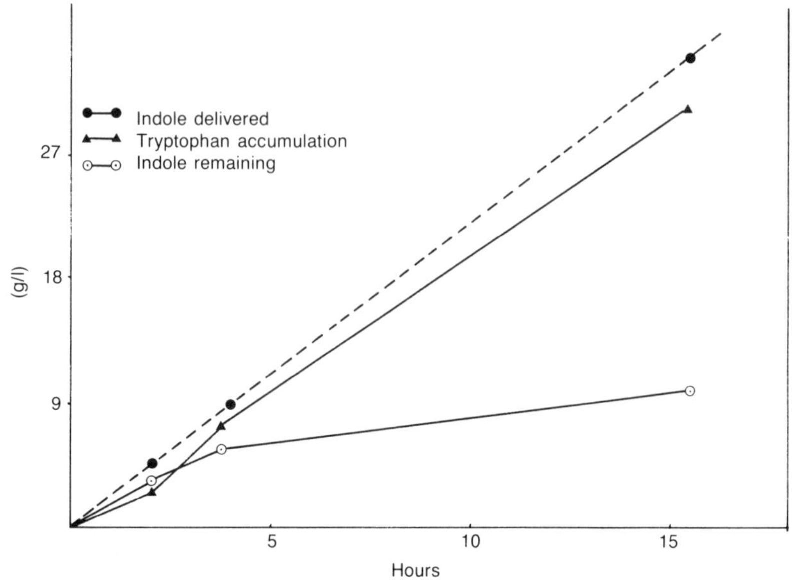

Figure 6.13. Levels of tryptophan production by *E. coli* with pGX141.

system in which the serine hydroxymethyl transferase and tryptophan synthase reactions take place in series in the columns shown. Indole, glycine, and formaldehyde are introduced at various ports, and the effluent containing the tryptophan is collected for recovery. There are three possible ways of setting up such a system:

1. A single production microorganism is genetically engineered to carry out both reactions.
2. Two different microorganisms, each genetically engineered to carry out a separate reaction, are combined in each column.
3. The two different microorganisms are added to separate columns.

An advantage of this system, regardless of its configuration, is that it enables one to produce either tryptophan, serine, or threonine as the end product, just by changing the mix of reactants (Table 6.2). If indole is omitted from the above mix, only serine will be produced. Moreover, because serine hydroxymethyl transferase also catalyzes the synthesis of threonine from glycine and acetaldehyde, the same bioreactors will produce threonine in the presence of the latter two reactants.

As a general rule, the capital investment for a batch fermentation system will amount to 150% to 200% of the capital investment for an immobilized cell process. Operating costs, such as labor and utilities, also tend to be higher for batch processes, given the same process conditions. We estimate that combin-

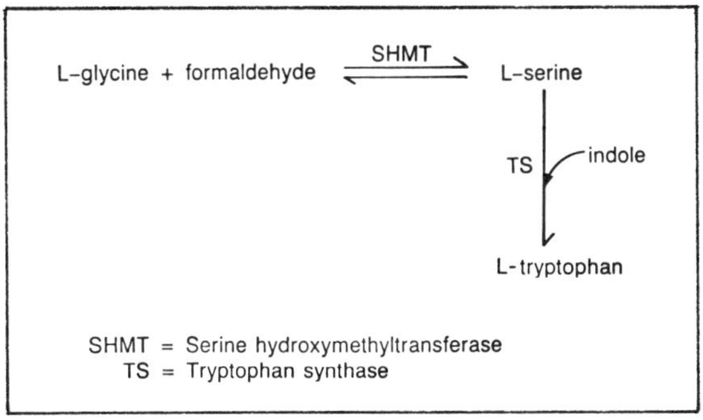

Figure 6.14. Pathway for tryptophan production.

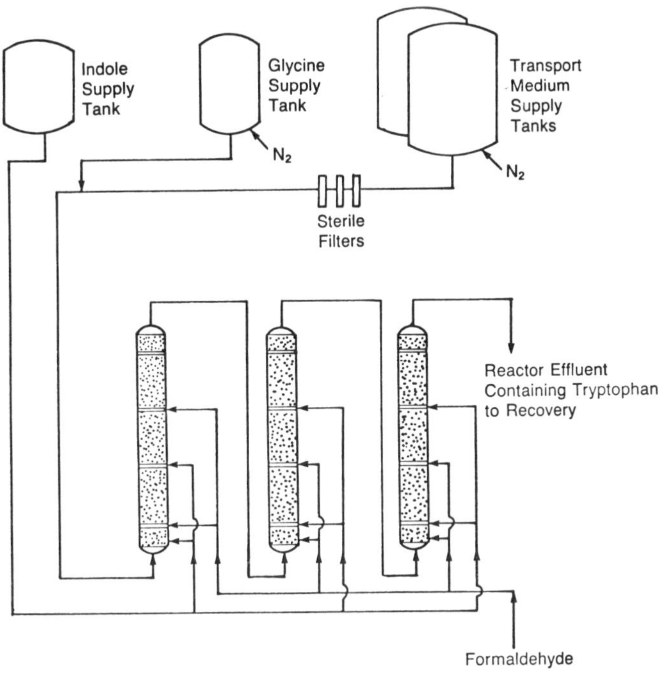

Figure 6.15. Use of bioreactors for production of tryptophan.

Table 6.2 Use of a Multipurpose Bioreactor System for the Production of Various Amino Acids

Products	Enzymes Involved		Addition of Feedstocks			
	Serine Hydroxymethyl Transferase	Tryptophan Synthase	Acetaldehyde	Glycine	Formaldehyde	Indole
Threonine	+		+	+		
Serine	+			+	+	
Tryptophan	+	+		+	+	+

ing genetic and biochemical engineering technologies for the production of tryptophan will result in a 50% to 90% reduction of current manufacturing costs.

LITERATURE CITED

Anonymous. 1982. *Elanco introduces growth promotant implant requiring no withdrawal time*. Feedstuffs 54:8.

Aviv, H., and P. Leder. 1972. *Purification of biologically active globin messenger RNA by chromatography on oligothymidylic acid-cellulose*. Proc. Nat. Acad. Sci. USA 69:1408–12.

Chang, L.M.S., and F.J. Bollum. 1971. *Deoxynucleotide-polymerizing enzymes of calf thymus gland. V. Homogeneous terminal deoxynucleotidyltransferase*. J. Biol. Chem. 246:909–16.

Crawford, I.P. 1980. *Comparative studies on the regulation of tryptophan synthesis*. Crit. Rev. Biochem. 8:175–89.

Dayhoff, M.O. 1976. *Atlas of protein sequences and structure*. National Biomedical Research Foundation, Vol. 5, Suppl. 2, pages 120–21.

Davis, B.D., and P. C. Tai. 1980. *The mechanism of protein secretion across membranes*. Nature (Lond.) 283:433–38.

Dicamelli, R.F., E. Balbinder, and J. Lebowitz. 1973. *Pressure effects on the association of the alpha subunit and beta-two subunit of tryptophan synthetase from Escherichia coli and Salmonella typhimurium*. Arch. Biochem. Biophys. 155:315–24.

Florent, J., and L. Ninet. 1979. *Vitamin B_{12}*. Pages 497–520 in H.J. Peppler and D. Perlman, eds., *Microbial Technology,* vol. 1. Academic Press, New York.

Grunstein, M., and J. Wallis. 1979. *Colony hybridization*. Pages 379–88 in R. Wu, ed., *Methods in Enzymology,* vol. 68. Academic Press, New York.

Irwin, M.R., D.R. Melendy, and D.P. Hutcheson. 1980. *Reduced morbidity associated with shipping fever pneumonia in levamisole phosphate-treated feedlot cattle*. The Southwestern Vet. 33:45–49.

Itoh, T., K. Toki, I. Chibata, and R. Yoshido. 1974. *Utilization of amino acids*. Pages 243–302 in T. Kaneko, Y. Izumi, I. Chibata, and T. Itoh, eds., *Synthetic Production and Utilization of Amino Acids*. Halsted Press, New York.

Kleid, D.G., D. Yansura, B. Small, D. Dowbenko, D.M. Moore, M.J. Grubman, P.D. McKercher, D.O. Morgan, B.H. Robertson, and H.L. Bachrach. 1981. *Cloned viral protein vaccine for foot-and-mouth disease: Responses in cattle and swine.* Science 214:1125–29.

Klenow, H., and I. Henningsen. 1970. *Selective elimination of the exonuclease activity of the deoxyribonucleic acid polymerase from Escherichia coli B by limited proteolysis.* Proc. Nat. Acad. Sci. USA 65:168–75.

Machlin, L.J. 1972. *Effect of porcine growth hormone on growth and carcass composition of the pig.* J. Anim. Sci. 35:794–800.

———.1973. *Effect of growth hormone on milk production and feed utilization in dairy cows.* J. Dairy Sci. 56:575–80.

Martial, J.A., R.A. Hallewell, J.D. Baxter, and H.M. Goodman. 1979. *Human growth hormone complementary DNA cloning and expression in bacteria.* Science 205:602–607.

Maxam, A.M., and W. Gilbert. 1977. *A new method for sequencing DNA.* Proc. Nat. Acad. Sci. USA 74:560–64.

Miller, W.L., J.A. Martial, and J.D. Baxter. 1980. *Molecular cloning of DNA complementary to bovine growth hormone messenger RNA.* J. Biol. Chem. 255:7521–24.

Moore, T.C. 1979. *Biochemistry and Physiology of Plant Hormones.* Pages 147–80. Springer Verlag, New York.

Nakayama, K., and H. Hagino. 1971. *Process for producing L-tryptophan.* U.S. Patent 3,594,279.

———. 1974. *Process for producing L-tryptophan.* U.S. Patent 3,849,251.

Peel, C.J., D.E. Bauman, R.C. Gorewit, and C.J. Sniffen. 1981. *Effect of exogenous growth hormone on lactational performance in high yielding dairy cows.* J. Nutr. 111:1662–71.

Pelham, H.R.B., and R.J. Jackson. 1976. *An efficient mRNA-dependent translation system from reticulocyte lysates.* Eur. J. Biochem. 67:247–56.

Perlman, D. 1979. *Microbial process for riboflavin production.* Pages 521–27 in H.J. Peppler and D. Perlman, eds., *Microbial Technology,* vol. 1. Academic Press, New York.

Platt, T. 1978. *Regulation of gene expression in the tryptophan operon of Escherichia coli.* Pages 263–302 in J.H. Miller and W.S. Reznikoff, eds., *The Operon.* Cold Spring Harbor Laboratory, New York.

Reynolds, I.P. 1980. *Correct use of anabolic agents in ruminants.* Vet. Rec. 107:367–69.

Sanger, F., S. Nicklen, and A.R. Coulson. 1977. *DNA sequencing with chain-terminating inhibitors.* Proc. Nat. Acad. Sci. USA 74:5463–67.

Schuuring, C. 1982. *New era vaccine.* Nature (Lond.) 296:792.

Seeburg, P.H., J. Shine, J.A. Martial, J.D. Baxter, and H.M. Goodman. 1977. *Nucleotide sequence and amplification in bacteria of structural gene for rat growth hormone.* Nature (Lond.) 260:486–94.

Shirai, T., K. Toi, I. Chibata, R. Yoshida, T. Kaneko, and K. Toki. 1974. *Synthetic methods for individual amino acids.* Pages 61–230 in T. Kaneko, Y. Izumi, I. Chibata, and T. Itoh, eds., *Synthetic Production and Utilization of Amino Acids.* Halsted Press, New York.

Turner, C.D. 1960. *The pituitary gland.* Pages 43–48 in C.D. Turner, ed., *General Endorcrinology,* W.B. Saunders Co., London.

Vogt, V.M. 1973, *Purification and further properties of single strand specific nuclease from Aspergillus oryzae.* Eur. J. Biochem. 33:192–200.

Wilhelmi, A.E., and J.B. Mills. 1972. *Studies on the primary structure of porcine growth hormone.* Pages 38–41 in A. Pecile and E. Muller, eds., *Growth and Growth Hormone.* Amsterdam: Excepta Medica Foundation Inter Cong. Ser. 244.

Windass, J.D., M.J. Worsey, E.M. Pioli, D. Pioli, P.T. Barth, K.T. Atherton, E.C. Dart, D. Byrom, K. Powell, and P.J. Senior. 1980. *Improved conversion of methanol to single-cell protein by Methylophilus methylotrophus.* Nature (Lond.) 287:396–401.

7] Production of a Vaccine for Foot-and-Mouth Disease Through Gene Cloning

by DOUGLAS M. MOORE*

ABSTRACT

Worldwide demand for foot-and-mouth disease (FMD) vaccine remains high. One approach to satisfy this demand is to produce by biosynthesis the immunogenic outer capsid polypeptide VP_3 for use in vaccines. Complementary transcripts of the FMD virus genome were cloned into *Escherichia coli,* the coding sequence of type A_{12} VP_{12} was identified, and a plasmid constructed which expressed VP_3 as an *E. coli*-VP_3 chimeric protein. The protein was produced in high yield and when used in a vaccine, protected cattle and swine against FMD virus challenge. Cloning of additional strains provided their respective VP_3 amino acid sequences, and distinct serotype and subtype variable zones were found. Stable expression in *E. coli* of VP_3 coding sequences from additional FMDV strains was engineered and immunogenicity demonstrated for types A_{24} and O_1. Comparison of amino acid sequences, antigenic analysis of VP_3 fragments, and capsid surface site studies located the position of an exposed capsid antigenic site within the primary structure of VP_3. A 32-amino acid polypeptide spanning the exposed region was expressed by subcloning into *E. coli.* This polypeptide produced high levels of neutralizing antibody when used as a vaccine in guinea pigs.

INTRODUCTION

Foot-and-mouth disease virus (FMDV) is a picornavirus of the genus *Aphthoviridae*. There are seven distinct serotypes of the virus (A, O, C, Southern African Territories (SAT) 1, SAT 2, SAT 3, and Asia 1) with 65 or more subtypes among them. The virus has a molecular size of about 7 x 10^6 daltons and a diameter of about 25 nm. The capsid contains four major

*Plum Island Animal Disease Center, Agricultural Research Service, U.S. Department of Agriculture, Greenport, New York, 11944.

polypeptides, VP_1, VP_2, VP_3 (about 25,000–30,000 daltons each), and VP_4 (about 10,000 daltons), and a plus-stranded RNA genome about 2.7×10^6 daltons or about 8,000 nucleotides in length (Bachrach 1977).

Foot-and-mouth disease (FMD) is a highly contagious disease primarily affecting cloven-hooved animals. It is an acute disease characterized by vesicular lesions of the mouth and feet as well as the snout and teats. Infected animals are febrile, frequently become lame, and may refuse food, resulting in a loss of production of meat and milk. Mortalities rarely occur in adults but can approach 50% in young animals. Infected animals release large quantities of virus through respiration, by excretion, and by rupture of vesicles. This release of virus results in nearly 100% infection within herds, and transmission to adjacent farms can occur by air or mechanical transmission.

Methods used to control FMD vary with the individual country. Disease-free countries have strict import and inspection regulations to prevent entry of the disease. If FMD is introduced, it is eradicated through slaughter of infected and exposed animals. Vaccine may be used until the disease can no longer be detected. Other countries having occasional outbreaks generally have vaccination programs that keep the incidence of disease low. Countries in which FMD is endemic regularly vaccinate against current strains found in the field, with varying degrees of success.

Vaccines for FMD are produced from virus grown in tissue culture cells or in explants of bovine tongue epithelium, and are chemically inactivated with formaldehyde or acetylethyenimine. Some attenuated live-virus vaccines are also in limited use. About 2.4 billion (monovalent equivalent) doses of inactivated vaccine are used annually worldwide (Bachrach 1978). Virus vaccines require refrigeration and have a limited shelf-life. Escape of virus from facilities producing virus for vaccine presents a risk to surrounding livestock populations, and incompletely inactivated vaccines have been linked to outbreaks of FMD (European Commission, FAO 1981; King et al. 1981). In countries successfully controlling FMD, the vaccine-related infection compares significantly with the risk of contracting the disease from other sources.

One of the capsid polypeptides of FMDV, designated here as VP_3, when isolated and used as a vaccine was shown to elicit neutralizing antibody responses (Laporte et al. 1973) and to protect swine and cattle from infection (Bachrach et al. 1975; Bachrach et al. 1982). In addition, certain fragments generated by cyanogen bromide or enzymatic cleavage spanned the middle portion of VP_3 and were also active, indicating that discrete segments might contain the antigenic determinant(s) responsible for the immunizing effect (Kaaden, Adam and Strohmaier 1977; Bachrach, Morgan, and Moore 1979; Bachrach et al. 1982).

As a vaccine, virus-derived VP_3 presents certain disadvantages. The amount of VP_3 required to elicit immunity has been greater than for inactivated whole-virus vaccines; lesser quantities of VP_3 vaccine than of conventional vaccine can be produced from a pool of virus harvest; the purification of VP_3 is tedious; and isolation of virus-derived VP_3 requires the propagation of infectious virus with the attendant risk of escape of FMDV from the laboratory. There are compensating advantages, however. Isolated VP_3 cannot by itself be infectious; it is stable to heat; and it can be stored in concentrated form indefinitely.

These advantages, together with the apparent simplicity of the immunogenic site(s) on VP_3 make production of the vaccine by organic synthesis or by microbial expression an attractive potential source of the antigen. This chapter will address the latter approach, which has been shown to be successful for efficient production of an effective biosynthetic vaccine for FMD.

CLONING THE FMDV GENOME

Biochemical map of the FMD genome. The approximate 8,000-nucleotide genome of FMDV has a small protein, VPg, covalently linked to its 5′-end and a polyadenylic acid (poly-A) tract about 40 to 100 nucleotides long at the 3′-end. A tract of poly-C 100 to 150 nucleotides long is located about 500 nucleotides from the 5′-end (Sangar 1979). Initiation of protein synthesis is believed to be to the 3′-side of the poly-C tract (Sangar et al. 1980). The proteins of the virus are produced as a precursor polyprotein, cleaved into primary cleavage products and subsequently into the structural proteins, VPg, a putative protease, the RNA polymerase, and other proteins of unidentified function (Fig. 7.1a). The structural proteins, translated in the sequence VP_4-VP_2-VP_1-VP_3, represent the 93,000-dalton precursor protein at the 5′-end of the genome, placing their location between nucleotides 700 and about the center of the genome (Sangar et al. 1977). This puts the location of the VP_3 protein coding information around 3000 to 3,500 nucleotides from the 5′-end. The N- and C-terminal amino acids of the capsid proteins are known as well as the N-terminus of the centrally located immunogenic cyanogen bromide (CNBr) cleavage product of VP_3 (Bachrach, Swaney, Van de Woude 1973; Bachrach et al. 1982).

Production of cloned FMDV genome segments. Descriptions of the procedures used to clone gene segments into *Escherichia coli* are given generally (Gilbert and Villa-Komaroff 1980) and in detail (Wu 1979). The following is a summary of the steps used to produce and identify genome inserts containing coding sequences for the immunogenic polypeptide VP_3, for FMDV type A_{12}. The FMDV RNA genome is itself message for synthesis of the viral proteins (through precursor polyproteins), so complementary DNA (cDNA) was produced by priming and reverse transcription of the intact genome RNA. During early experiments, the exact position of the VP_3 coding region was not known, and RNA sequence data was not available on the 3′-side of VP_3 for production of appropriate primers. Therefore, priming was accomplished by annealing oligo-dT to the poly-A tract at the 3′-end of the genome (Fig. 7.1b). cDNA was produced with reverse transcriptase, made double-stranded, and large transcripts were inserted into the Pst I site of the plasmid pBR322. This was accomplished by adding oligo-dC tails to the FMDV cDNA and oligo-dG tails to the cleaved plasmid, annealing the complementary tails together, and transforming the recombinant plasmids into *E. coli* K12 strain. Bacteria containing plasmids with cDNA inserts were selected for resistance to tetracycline and sensitivity to ampicillin.

Candidates for further analysis were surveyed for inserts containing 1,000

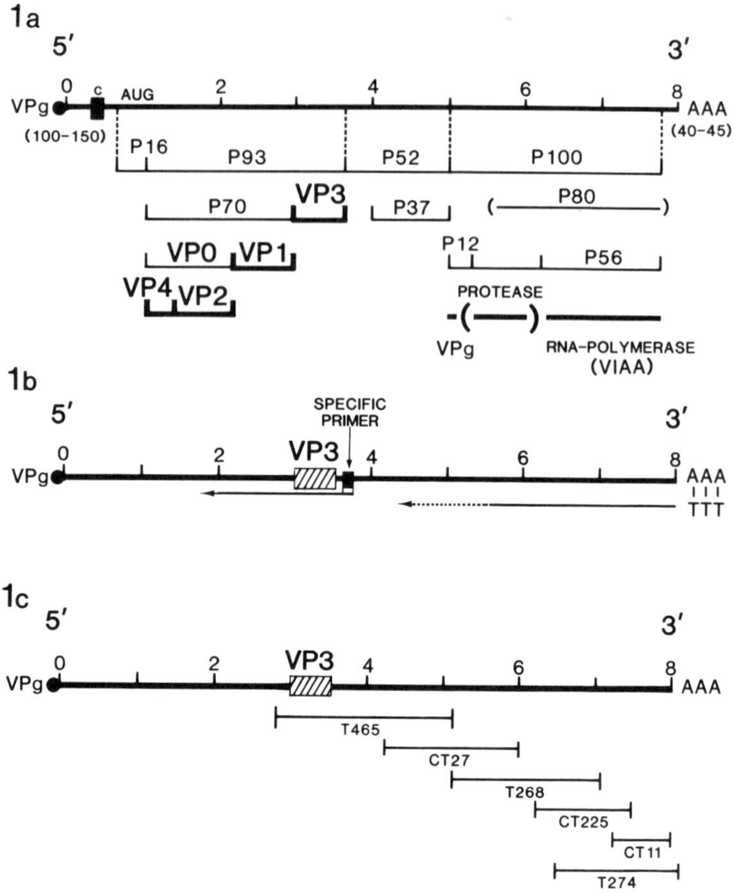

Figure 7.1. a. Physical and biochemical map of the FMDV genome; b. priming sites used to produce cDNA on the FMDV genome; c. a collection of plasmids and their approximate location on the FMDV genome determined by restriction endonuclease mapping. (D. Yansura, D. Dowbenko and B. Small, unpublished.)

base pairs (bp) or greater and were confirmed to contain FMDV complementary sequences through filter hybridization using ^{32}P-labeled FMDV RNA to probe the plasmid DNA preparations. A group of selected plasmids were organized into an overlapping series of inserts by "mapping" specific restriction endonuclease sites (Fig. 7.1c). Plasmids containing FMDV 3′-oriented inserts were identified by hybridization with a probe produced by brief oligo-dT-primed cDNA synthesis.

The map positions of plasmid inserts and the biochemical map of the FMDV genome were used to select a plasmid (T465), which had a large insert (2,500 bp) and which appeared to span the VP_3 coding region. Nucleotide sequence analysis confirmed that the full coding sequence of VP_3 was contained in the plasmid (Kleid et al. 1981a). Codons for previously determined N-terminal residues of VP_3 were located as well as those of the cyanogen bromide immunogenic fragment (Fig.7.2). The insert extended about 2,400 bp (800 amino acid codons) from the 5′-side of VP_3 toward the 3′-end of the genome.

Once data was available on the nucleotide sequence of the middle part of the FMDV genome, the process for obtaining plasmid inserts coding for VP_3 of additional strains of FMDV was streamlined (D. Yansura and D. Kleid, unpublished). Sequence homology of plasmid cDNA corresponding to the area just to the 3′-side of VP_3 was compared for FMDV types A_{12}, A_{27}, and C_3 (D. Yansura and D. Dowbenko, unpublished) and O_1 (Küpper et al. 1981). Three regions of 10 nucleotides were selected, about 200 through 350 nucleotides beyond the C-terminus of VP_3, which were each identical among the four strains. These oligonucleotides were chemically synthesized (Crea et al. 1978) and used to prime cDNA synthesis on genome RNA of additional FMDV strains. By starting near the VP_3 area, the probability of obtaining plasmids with VP_3 coding sequences was increased, and production of long transcripts from completely intact genome RNA was not required. Isolation of full-length FMDV RNA is hampered by its rapid degradation in the virus capsid and during purification (Denoya et al. 1978; Grubman, Baxt, and Bachrach 1979). Thus, plasmids containing cDNA coding for VP_3 were identified by hybridization, with a ^{32}P-labeled Pst I-Hind III segment of plasmid T465 representing the VP_3 N-terminus. The plasmids selected spanned the VP_3 coding region outside the C-terminus and extended to or beyond the N-terminus. The presence of VP_3 coding sequences was confirmed through location of previously identified restriction sites and by nucleotide sequencing. This method has been successful in isolating VP_3 coding inserts for five FMDV strains representing serotypes A, O, and C (D. Yansura and D. Dowbenko, unpublished). Derived amino acid sequences of VP_3 are shown for types A_{24} and O_1 in Figure 7.2.

Construction of VP_3 expression plasmids. To obtain expression of the A_{12} VP_3 polypeptide in *E. coli*, segments of plasmid T465 were inserted into a specially designed expression vector to produce VP_3 as a fusion protein (Kleid et al. 1981b). The expression plasmid is a modified pBR322 containing the *E. coli* tryptophan promoter-operator, which directs the expression of the *E. coli* protein trp Δ LE 1413 (Miozzari and Yanofsky 1978). This protein (LE′) consists of the first six amino acids of the trp leader peptide and the last third of the trp E polypeptide, resulting in a polypeptide 190 amino acids long. The LE′ polypeptide is insoluble in bacteria and generally stabilizes heterologous polypeptides attached to it against proteolytic degradation (Kleid et al. 1981a).

The construction of the expression plasmid pFM1 is shown in Figure 7.3a (Kleid et al. 1981a). The vector was cleaved with Eco RI and Bam HI. A segment of plasmid T465 coding for amino acids 8 to 211 of A_{12} VP_3 was obtained by cleavage with Pst I and Pvu II. The segment was joined to the LE′ coding segment by using a short DNA linker molecule containing an EcoRI and Pst I site spaced to join the two together in the correct VP_3 codon reading frame. The construction was completed by joining a 375-bp Eco RI to Bam HI

```
     1                                                                                                              30
A12 Thr Thr Ala Thr Gly Glu Ser Ala Asp Pro Val Thr Thr Thr Val Glu Asn Tyr Gly Gly Glu Thr Gln Val Gln Arg Arg His His Thr
A24 ... ... ... -   -   -   -   -   -   -   -   -   -   -   -   -   -   -   -   -   -   -   -   Ile -   -   -   -   -   -
O1   -   -  Ser Ala -   -   -   -   -   -   -   -   Ala -   -   -   -   -   -   -   -   -   -   Ile -   -   -   Gln -   -

     31                  36                                                                 54                      60
A12 Asp Val Ser Phe Ile MET Asp Arg Phe Val Lys Ile Lys Ser Leu Asn Pro Thr His Val Ile Asp Leu MET Gln Thr His Gln His Gly
A24  -  Ile -   Gly -   -   -   -   -   -   -   -   Gln -   -   Ser -   -   -   -   -   -   -   -   -   -   -   -   -   -
O1   -   -  -   -   -   -   -   -   -   -   -   Val Thr Pro Gln -   Gln Ile Asn Ile Leu -   -   -   -   Ile Pro Ser -   Thr

     61                                                                                                             90
A12 Leu Val Gly Ala Leu Leu Arg Ala Ala Thr Tyr Tyr Phe Ser Asp Leu Glu Ile Val Val Arg His Asp Gly Asn Leu Thr Trp Val Pro
A24  -   -  -   -   -   -   -   -   -   -   -   -   -   -   -   -   -   -   -   -   -   -   Glu -   -   -   -   -   -   -
O1   -   -  -   -   -   -   -   -   Ser -   -   -   -   -   -   -   -   -   Ala -   Lys -   Glu -   Asp -   -   -   -   -

     91                                                                                                             120
A12 Asn Gly Ala Pro Glu Ala Ala Leu Ser Asn Thr Gly Asn Pro Thr Ala Tyr Asn Lys Ala Pro Phe Thr Arg Leu Ala Leu Pro Tyr Thr
A24  -   -  -   -   -   Ser -   -   Leu -   -   Ser -   -   -   -   -   -   -   -   -   -   -   -   -   -   -   -   -   -
O1   -   -  -   -   -   Lys -   -   Asp -   -   Thr -   -   -   -   -   -   -   -   -   Leu -   -   -   -   -   -   -   -

     121                                                                                                        149
A12 Ala Pro His Arg Val Leu Ala Thr Val Tyr Asn Gly Thr Asn Lys Tyr Ser Ala Ser  Gly  Ser  Gly  Val Arg Gly Asp Phe Gly Ser
A24  -   -  -   -   -   Ser -   -   -   -   -   -   -   Ser -   -   Ala Val Gly  -    -    -    Arg -   -   -   MET -   Thr
O1   -   -  -   -   -   -   -   -   -   -   -   -   Glu Cys Arg -   Ser Arg Asn Ala Val Pro Asn Val -   -   -   Leu Gln Val
   (121)                                                                                                           (150)

     150                                                                                                            179
A12 Leu Ala Pro Arg Val Ala Arg Gln Leu Pro Ala Ser Phe Asn Tyr Gly Ala Ile Lys Ala Glu Thr Ile His Glu Leu Leu Val Arg MET
A24  -   -  Ala -   -   Val Lys -   -   -   -   -   -   -   -   -   -   -   -   -   Asp Ala -   -   -   -   -   -   -   -
O1   -   -  Gln Lys -   -   -   Thr -   -   Thr -   -   -   -   -   -   -   -   -   Thr Arg Val Thr -   -   -   Tyr -   -
   (151)                                                                                                           (180)

     180                                                                                                            209
A12 Lys Arg Ala Glu Leu Tyr Cys Pro Arg Pro Leu Leu Ala Ile Glu Val  Ser  Ser  Gln  Asp Arg His Lys Gln Lys Ile Ile Ala Pro Gly
A24  -   -  -   -   -   -   -   -   -   -   -   -   -   -   -   -    -    -    -    -   -   -   -   -   -   -   -   -   -   Ala
O1   -   -  -   -   Thr -   -   -   -   -   -   -   -   -   His  Pro  Thr  Glu  Ala  -   -   -   -   -   -   Val -   -   Val
   (181)                                                                                                           (209)

     210         212
A12 Lys Gln     Leu
A24  -   -  Leu  -
O1   -   -  Thr  -
   (210)       (213)
```

Figure 7.2. The amino acid sequence of the VP_3 polypeptides of types A_{12}, A_{24}, and O_1 derived from the nucleotide sequence. Abbreviations for the amino acid residues are: Ala, Alanine; Arg, arginine; Asp, aspartic acid; Asn, asparagine; Cys, cysteine; Glu, glutamic acid; Gln, glutamine; Gly, glycine; His, histidine; Ile, isoleucine; Leu, leucine; Lys, lysine; Met, methionine; Phe, phenylalanine; Pro, proline; Ser, serine; Thr, threonine; Trp, tryptophan; Tyr, tyrosine, Val, valine. Gaps in the sequence (A_{12}: 139–143, 211–212; O_1: 195–199) indicate a deleted residue, "–" indicates the same residue as for type A_{12} directly above, and ". . ." indicates undetermined residues. Residue numbers in parentheses beneath the lines refer to the 0_1 sequence positions.

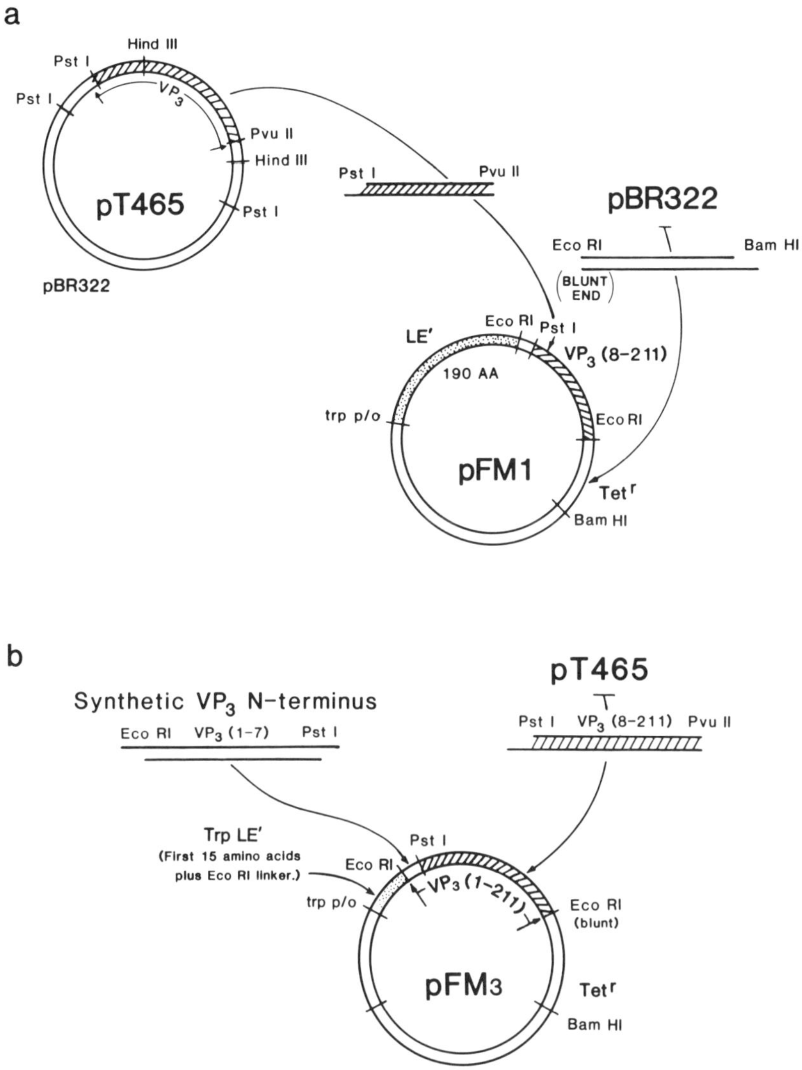

Figure 7.3. Construction of expression plasmids designed to produce type A_{12} VP_3 as a fusion protein under the control of the tryptophan promoter-operator. a. Long LE′ – VP_3 fusion expression plasmid pFM1; b. short LE′ – VP_3 fusion expression plasmid pFM3.

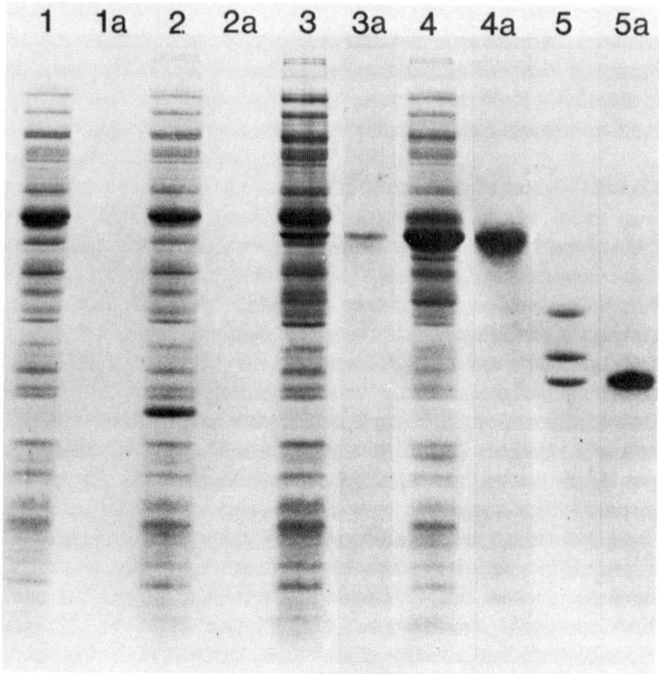

Figure 7.4. Polyacrylamide gel analysis of LE′ $-VP_3$ fusion protein. Each sample is shown in pairs. The first lane of the pair shows the total bacterial or viral protein stained with Coomassie brilliant blue. The second lane of the pair is an autoradiogram of protein that has been transferred to CNBr-treated paper and exposed to ^{125}I-labeled antibody to VP_3. Lanes 1 and 1a, *E. coli* 294; lanes 2 and 2a, *E. coli* 294/pLE control in which only the LE protein was expressed; lanes 3 and 3a, *E. coli* 294/pFM_1 partially induced for fusion protein expression (tryptophan in the growth media was not completely depleted); lanes 4 and 4a, *E. coli* 294/pFM_1 fully induced; lanes 5 and 5a, FMDV A_{12} 119ab VP_1, VP_2, and VP_3 isolated from FMDV (VP_4 is not visible). (From Kleid et al. 1981a, used with permission. See Table 7.1.)

fragment of pBR322 to the blunt Pvu II site of the VP_3 segment and the Bam HI site of the vector. The last segment provided the tetracycline promoter and first part of the tetracycline resistance gene of pBR322, which served as a marker for selection of transformed *E. coli*. The plasmid as constructed would direct the expression of an *E. coli*-FMDV fusion protein of 405 amino acids containing 190 trp LE′, 6 linker-coded, 205 FMDV VP_3 and 4 pBR322-coded amino acids. This molecule is designated long LE′-VP_3 protein.

Another construction of a plasmid designed to express the VP_3 polypeptide is designated pFM3 and is shown in Figure 7.3b. This construction is similar to pFM1, with the following improvements: the trp LE′protein is greatly reduced in size to 17 amino acids by cleavage at a Bgl II site and addition of a synthetic

DNA linker with an Eco RI site. A synthetic DNA segment was produced to insert a methionine and restore the first seven N-terminal amino acids of VP_3. It was bounded by an Eco RI and Pst I site to join the LE′ and VP_3 coding segments. The T465 VP_3 coding segment (codons 8–211) was as above (pFM1). The pFM3 construction would express a polypeptide, short LE′-VP_3, containing 232 amino acids: 17 LE′, 211 VP_3, and 4 pBR322 (B. Small and D. Kleid, unpublished).

Expression of FMDV VP_3 in E. coli. The *E. coli* bacteria containing plasmid pFM1 (long LE′-fusion) were grown and transferred to tryptophan-depleted medium to induce the fusion protein. The bacteria were lysed, and the insoluble material was pelleted by centrifugation. Proteins in the pellets were analyzed by sodium dodecyl sulfate (SDS) acrylamide gel electrophoresis (Fig. 7.4). Normal *E. coli* and *E. coli* with a plasmid containing only the trp LE′ insert were also examined. The pFM1-transformed *E. coli* contained a new protein not found in normal or trp LE′-plasmid–transformed *E. coli*. Its molecular size was about 45,000 daltons and corresponded to the predicted size of the 405-amino acid fusion protein. The fusion protein represented about 38% of the crude *E. coli* lysate protein content, or about 17% of total *E. coli* protein (Kleid et al. 1981a). This represented 1–2 million molecules per bacterium. In a blotting test for antigenic activity, only the fusion protein LE′-VP_3 and VP_3 from FMDV were reactive with ^{125}I-labeled VP_3 antibody (Fig. 7.4, "a" lanes). Induction of expression of plasmid pFM3 similarly yielded large quantities of the short-fusion LE′-VP_3 polypeptide. Its apparent size on acrylamide gels corresponded to the molecular size of about 25,000 daltons expected for the predicted 232-amino acid polypeptide.

The fusion proteins with either short or long LE′-segments were purified by electrophoresis on two successive preparative SDS acrylamide slab gels containing 8 M urea (Kleid et al. 1981a). The eluted polypeptides were quantitated by analytical scanning and compared with FMDV VP_3 in radioimmunoassay for their ability to compete with ^{125}I-virus for antibody specific for VP_3 sites. Both the long- and short-fusion LE′-VP_3 polypeptides competed and could not be distinguished from FMDV VP_3 (Fig. 7.5).

Expression of VP_3 for additional FMDV strains. A number of methods have been used to accomplish expression of VP_3 of additional FMDV strains. Several FMDV strains with available restriction sites at or near the VP_3 N- and C-terminal coding regions of the plasmid inserts were substituted for the equivalent A_{12} segment in the expression plasmids pFM1 or pFM3. This was accomplished by a direct switch (type A24) or by blunt ligation to join dissimilar sites (type O_1 (D. Yansura and D. Dowbenko, unpublished). Expression plasmids for the A_{24} and O_1 FMDV strains were transformed into *E. coli*, and both were found to direct the expression of a stable fusion protein in large quantity. Analysis by immunological staining showed reactivity of the fusion proteins with anti-VP_3 immunoglobulin as shown for type A_{12} in Figure 7.4 (data not shown).

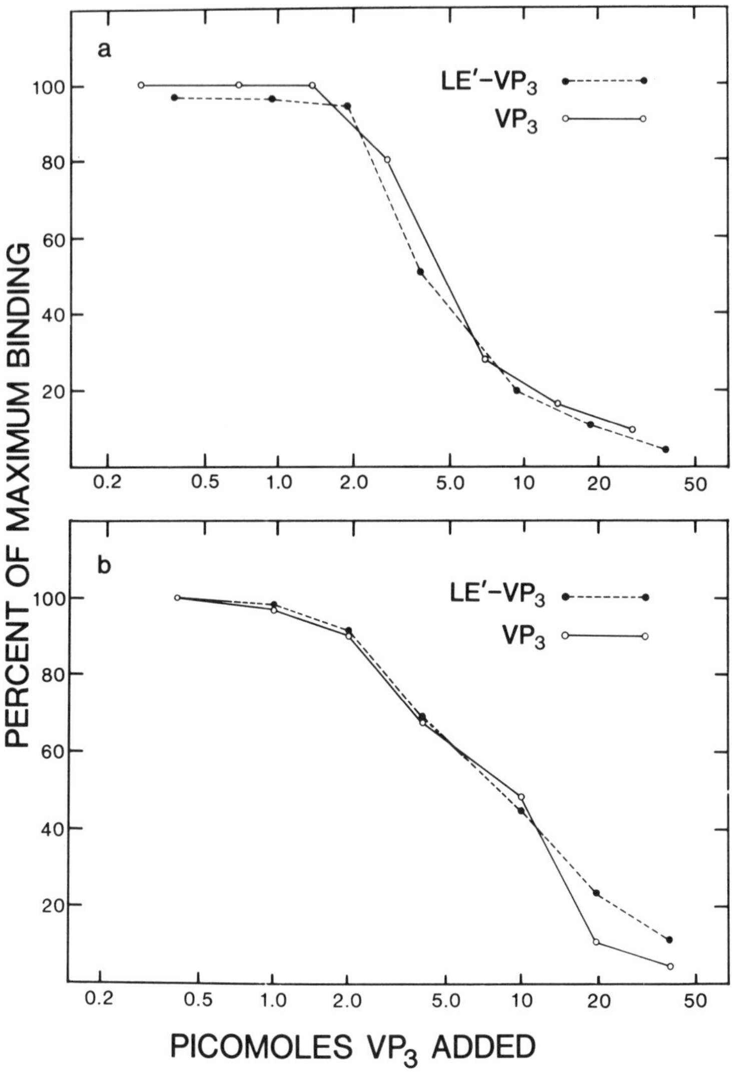

Figure 7.5. Comparison of A_{12} LE′ – VP_3 fusion proteins with FMDV VP_3 in a FMDV ^{125}I-virion-aVP_3 radioimmunoassay (Kleid et al. 1981a). a. Long LE′ – VP_3 *vs* FMDV VP_3; b. short LE′ – VP_3 *vs* FMDV VP_3.

IMMUNE RESPONSES TO CLONED VP_3

Vaccination with type A_{12} LE′-VP_3. A preparation of A_{12} long-fusion LE-VP_3 was tested as a vaccine in steers and swine. The vaccine was prepared by pulverizing a slice of the acrylamide gel containing the protein and emulsifying in incomplete Freund oil adjuvant. The vaccine was given to six steers and two swine in two 250-μg doses. The animals showed high levels of immunity to challenge with FMDV (Kleid et al. 1981a). The results of the experiment are summarized in Table 7.1. One steer (No. 5) vaccinated with LE′-VP_3 developed a single small foot lesion, but did not develop VIA antibody (Cowan and Graves 1966) or show a rise in neutralizing antibody titer, indicating that systemic infection had not occurred. The experiment demonstrated that bacterially synthesized VP_3, like FMDV VP_3, can immunize livestock against FMDV infection.

Antibody responses to various LE′-VP_3 preparations. A series of experiments were performed to determine the immune responses of guinea pigs to FMDV VP_3, A_{12} LE′-VP_3 with long and short LE′-proteins, and short LE′-VP_3 from several field strains of FMDV. For these tests, FMDV VP_3 and the fusion proteins were purified by a modification of an ion-exchange chromatography method previously shown effective for FMDV VP_3 (S. Shire, unpublished; Bernard et al. 1974). The VP_3 or LE′-VP_3 were maintained in solution with 6 M urea or 0.1% SDS. In one test, LE′-VP_3 was isolated as an acrylamide gel slurry from a preparative SDS acrylamide gel as previously used for cattle and swine.

The responses to a single injection (except Groups 7 and 8) of antigen are shown in Table 7.2. The long-leader A_{12} fusion protein induced titers equivalent to those of the natural FMDV VP_3. As little as 4 μg of the fusion protein gave good antibody responses, but a fivefold lower dose (0.8 μg) generally failed to induce a response when given in a single vaccine dose (D. Moore, unpublished). The short-leader A_{12} LE′-VP_3 was immunogenic, although a lower response was obtained at the 4-μg dose. The same protein injected in an acrylamide gel slurry induced lower levels of antibody at less than a 100-μg dose.

Fusion proteins were expressed for A_{24} and O_1 short LE′-VP_3 and purified as above. Purified short-fusion LE′-VP_3 for A_{24} and O_1 induced neutralizing antibody responses (Table 7.2, Groups 5, 6, and 7). The titers induced by type O_1 LE′-VP_3 were lower, however. On revaccination (Group 7), a 500-μg dose induced a high neutralizing titer. Two 100-μg doses did not markedly increase the titer over that obtained with a single 100-μg dose.

The results indicate that both long- and short-leader LE′-VP_3 induce high levels of FMDV neutralizing antibody in guinea pigs. Use of LE′-VP_3 in an acrylamide gel slurry as originally used in steers and swine may reduce the ability of VP_3 to induce an immune response. Fusion protein from the A_{24} and

Table 7.1 Neutralizing Antibody and Immune Responses of Cattle and Swine Vaccinated with LE′-VP_3 Fusion Protein or VP_3 Isolated from Virus

		Neutralizing antibody day after vaccination[a]								Challenge of immunity[b]		
Subject	Antigen	7	14	21	28	35	42	46	56	Mouth lesions	Foot lesions	VIA antibody[c]
Cattle												
1	LE′-VP_3	< .3	.5	.3	1.2	2.3	2.7		2.9	0	0	–
2	LE′-VP_3	.2	.9	1.3	2.0	2.4	2.7		2.9	0	0	–
3	LE′-VP_3	.2	.4	1.1	1.7	1.9	2.1		2.9	0	0	–
4	LE′-VP_3	.2	.7	.7	1.1	2.2	2.6		2.9	0	0	–
5	LE′-VP_3	< .3	.9	1.0	1.1	2.6	2.5		2.6	0	+	–
6	LE′-VP_3	< .3	1.0	1.2	1.3	2.1	2.1		2.3	0	0	–
7	VP_3	.1	1.0	2.1	1.9	2.2	2.6		2.1	0	0	–
8	VP_3	< .3	.4	.4	.7	1.7	1.6		4.1	+++	+++	+
9	None[b]								3.4	+++	+++	+
10	None								3.5	+++	+++	+
11	None								3.3	+++	+++	+
12	None								3.3	+++	+++	+
Swine												
13	LE′-VP_3	< .3	< .3	.5	1.5				3.0	0	0	–
14	LE′-VP_3	< .3	.3	.6	1.4				3.0	0	0	–
15	None[b]								3.5	+++	+++	+
16	None								3.4	+++	+++	+

[a] Revaccinations on day 24 for swine and day 28 for cattle; titer is $-\log_{10}$ of serum dilution that protects 50% of suckling mice (Skinner 1953). The 46- and 56-day serums were collected 14 days after challenge.

[b] Challenged on day 32 for swine and day 42 for cattle. Non-vaccinated animals 9 to 12 and 15 and 16 constituted the challenge groups; half were inoculated with virulent type A_{12}119ab virus and half were contact transmission controls; +, single small lesion, not generalized disease; +++, numerous lesions and generalized infection.

[c] Presence, +, of virus-infection-associated antigen (VIA) antibody in serums collected 14 days after challenge indicated animal experienced FMD (Cowan and Graves 1966); –, VIA antibody absent.

Source: From Kleid et al. 1981a. Copyright 1981 by the American Association for the Advancement of Science, used with permission.

Table 7.2 Neutralizing Antibody Responses of Guinea Pigs Using LE′-VP_3 Fusion Proteins or FMDV VP_3

Group	Antigen	Buffer[b]	Neutralizing antibody response at given dose (μg)[a] 4.0	20	100	500
1	FMDV $A_{12}VP_3$	Urea-Tris	3.5	3.9	3.6	–
2	Long-leader A_{12} LE′-VP_3	Urea-Tris	3.2	3.9	4.8	–
3	Short-leader A_{12} LE′-VP_3	Urea-Tris	2.3	3.6	3.6	–
4	Short-leader A_{12} LE′-VP_3	Gel slurry SDS-Tris	< .3	1.7	4.2	–
5	Short-leader A_{24} LE′-VP_3	Urea-Tris	–	–	3.2	3.6
6	Short-leader O_1 LE′-VP_3	Urea-Tris	–	–	1.5	2.2
7	Short-leader O_1 LE′-VP_3	Urea-Tris	–	–	1.8[c]	3.2[c]
8	Long-leader A_{12} LE′-VP_3 tandem peptide amino acids #137-167	Urea-Tris	–	–	> 3.8[d]	–

[a]Neutralizing antibody titers determined by the suckling mouse serum neutralization test (Skinner 1952) of guinea pig sera, taken 28–35 days after vaccination. Values are the $-\log_{10}$ serum dilution 50% endpoint of a pool of 5 sera, protecting mice from 100 LD_{50} doses of the homologous virus. For vaccination appropriate doses in a 2 ml aqueous-Freund incomplete oil adjuvant emulsion were given subcutaneously in the skin of the neck. "–" indicates not tested.

[b]Indicates the aqueous phase of the vaccines: Urea-Tris = 6M urea, 0.05 M β-mercapto-ethanol, 0.014 M Tris-HCl, pH 8.6; gel slurry SDS-Tris = pulverized polyacrylamide preparative gel slice diluted in 0.1% SDS, 0.05 M Tris-glycine, pH 8.1.

[c]Neutralizing antibody titer 21 days after revaccination at 35 days with the same vaccine preparation.

[d]Neutralizing antibody titer on day 28, revaccinated on day 14 with same preparation.

O_1 field strains induced neutralizing antibody, but apparently there was a difference between the A-types and the O_1 antigen. However, when sufficient amounts of O_1 antigen were used, high levels of antibody were obtained. Careful titration of immunizing potency will be required for additional types cloned to establish appropriate vaccine doses. Important differences in protein folding and protein-protein interaction between the types may affect the activity of the antigenic sites. Also, antigenic site studies (see next section) will show whether fundamental differences exist among the serotypes regarding location, number, and potency of immunizing sites.

EXPOSED VP_3 SITES IN THE VIRUS CAPSID

Studies using enzymes and radioiodination have shown that VP_3 is situated at the surface of the virus capsid (Laporte and Lenoir 1973; Bachrach et al. 1975; Moore and Cowan 1978). In these studies only VP_3 was accessible to cleavage by trypsin and chymotrypsin or for labeling in tyrosine residues with ^{125}iodine. Furthermore, a centrally located cyanogen bromide fragment (residues 55–179) and enzyme digest fragments containing the N-terminus through residues ca 140–144 of VP_3 all contain an immunogenic site (Bachrach et al. 1982; Robertson et al. 1982). Comparison of amino acids derived from nucleotide sequences for several types of VP_3 shows several areas in which amino acids vary considerably (Fig. 7.2). The area of greatest variation between serotypes as well as within subtypes is between residues 133 and 154; another area of serotype variation occurs at residues 42–60 and 193–199. The results show that certain areas of VP_3 are more exposed on the viral surface, and further suggest that the variation in amino acid sequence in these areas specifies the antigenic sites associated with serotype and subtype antigenic variation.

Virus capsids were radioiodinated and the labeled VP_3 isolated by SDS acrylamide gel electrophoresis. Also, VP_3 labeled in all tyrosine residues was obtained by first isolating VP_3 and radioiodinating it in free solution. The VP_3 preparations were subjected to trypsin digestion and ^{125}I-tyrosine-containing peptides analyzed by ion-exchange chromatography (Fig. 7.6). With in situ-labeling, a single peptide was principally labeled (Fig. 6b), while with isolated VP_3 many sites became labeled (Fig.6a). A mixing experiment showed the in situ-labeled peptide to be identical to one found on trypsinized isolated VP_3 (Fig. 6c). The size of the exposed site-labeled peptide was determined to be 900–1000 daltons by gel filtration; manual Edman degradation identified ^{125}I-tyrosine as the N-terminal residue (Robertson et al. 1982). The tryptic peptide with these characteristics is found at residues 136–144 in the centrally located variable region of VP_3. This peptide was subsequently shown to react with VP_3 antibody specific for FMDV capsids, but failed to react with a VP_3 reactive antiserum from which anti-capsid specific antibody had been absorbed (Moore and Robertson 1982). The results show that the variable region (residues 136–154) is exposed on the surface of virus capsid and that a portion of the segment is specifically reactive with antiviral antibody.

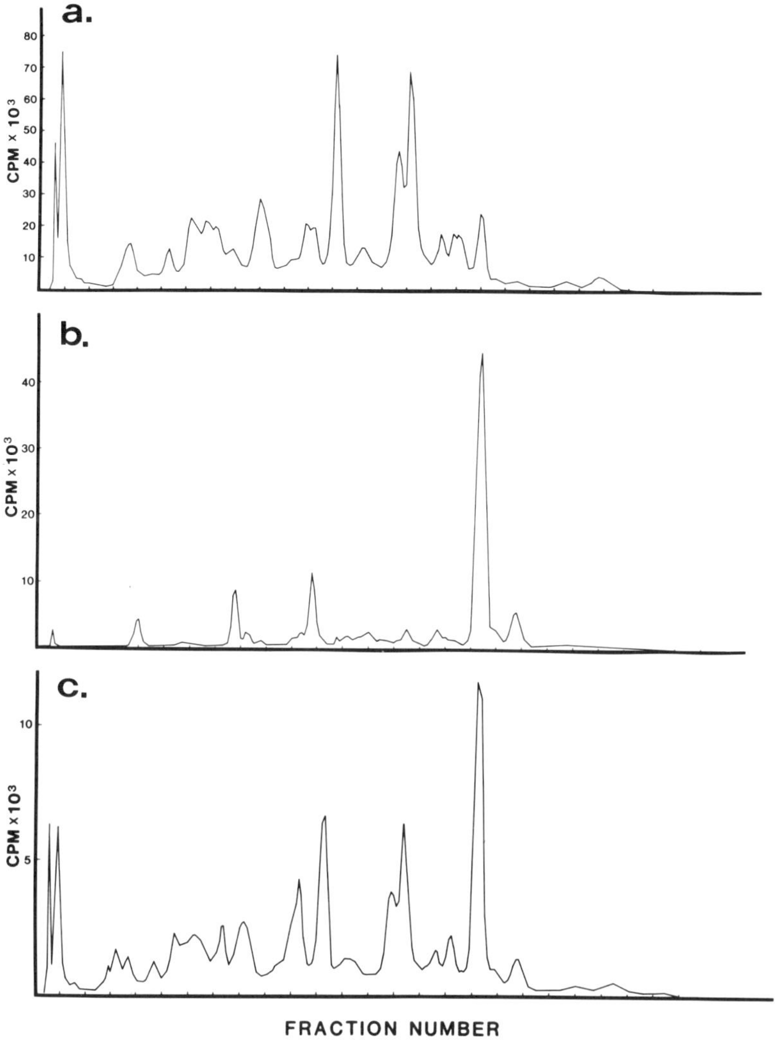

Figure 7.6. Analysis of ^{125}I-tryosine containing tryptic peptides by cation-exchange chromatography. Peptides were separated on a Phoenix 7% cross-linked sulfonated styrene divinylbenzene column, using a 0.05 to 2.0 M pyridine acetate buffer gradient, pH 3.5 to 5.7. a. Tyrosine-containing tryptic peptides of VP_3 labeled after isolation from the capsid; b. the same as a., but VP_3 labeled in situ prior to isolation from the capsid; c. a mixture of a. and b. identifying the peak in fully labeled VP_3 to be the exposed site in virion V_3.

A segment of A_{12} plasmid T465 corresponding to VP_3 codons 137–167 was subcloned into a long LE′-expression vector. The construction resulted in the insertion of two tandem VP_3 gene coding segments for residues 137–167 (D. Dowbenko, unpublished). The cleavage site altered the tyrosine-136 codon so that the effective expression of the VP_3 segment was for 31 codons (137–167). A fusion protein of about 29,000 daltons (appropriate for 190 LE′ plus 2 × 31 VP_3 plus 13 vector amino acids) was induced, isolated from *E. coli,* and purified (S. Shire, unpublished). Guinea pigs were vaccinated with 100 μg of the protein in incomplete Freund's adjuvant. A repeat dose was given at 14 days, and the animals were bled at 28 days. A very high neutralizing antibody titer was induced (Table 7.2, Group 8). The titer obtained suggests that the short VP_3 segment was as effective as the whole VP_3 molecule. The tandem insert peptide spans nearly the entire variable region (residues 133–154, see Fig. 7.2). It contains the antigenically reactive tryptic peptide (136–144) and begins at one residue past the tyrosine exposed at the surface of the virus capsid. Experiments are underway to identify the exact position and the size of the immunogenic site within this polypeptide.

CONCLUSION

Cloning the genome of FMDV into *E. coli* has been directly applied to the expression of the immunogenic polypeptide VP_3. This has provided capability for production of large quantities of this protein, and vaccine produced from it was effective in protecting livestock from FMDV infection. This methodology eliminates the risk of escape of FMDV from production facilities or persistence of infectious virus in conventional vaccines.

An additional benefit of the cloning experiments has been to identify the primary structure of the FMDV polypeptides. Comparison of amino acid sequences among the serotypes and subtypes will provide clues to the location of sites responsible for immunogenic and other functions of the capsid polypeptides. This, coupled with studies on exposed sites, antigenicity, and conformation of virus polypeptides should provide a description of immunogenic sites on the virus capsid. Improvements may be made on potency and stability of conventional vaccines; methods for diagnosis and detection of FMDV; and for design of effective polypeptide vaccines produced by microbial expression or by direct organic synthesis.

ACKNOWLEDGMENTS

The production of a cloned polypeptide vaccine for FMDV reported in this chapter was a result of research under a cooperative agreement between the Plum Island Animal Disease Center and Genentech, Inc., South San Francisco, California. Contributing from Plum Island were Douglas M. Moore, Marvin J. Grubman, Peter D. McKercher, Donald O. Morgan, Betty H.

Robertson, and Howard L. Bachrach; and from Genentech, Dennis G. Kleid, Daniel Yansura, Barbara Small, Donald Dowbenko, Maureen Hoatlin, Gregory Weddell, Steven Shire, Lawrence Bock, Stewart Builder, and John Ogez.

LITERATURE CITED

Bachrach, H. L. 1977. *Foot-and-mouth disease virus: Properties, molecular biology and immunogenicity.* Pages 3–32 in J. A. Romberger, ed., *Beltsville Symposia in Agricultural Research I. Virology in Agriculture.* Allanheld, Osmun, Montclair, N. J.

——— 1978. *Foot-and-mouth disease: World-wide impact and control measures.* Pages 229–310 in K. Maramorosch and E. Kurstak, eds., *Viruses and Environment.* Academic Press, New York.

Bachrach, H. L.; D. M. Moore; P. D. McKercher; and J. Polatnick. 1975. *Immune and antibody responses to an isolated capsid protein of foot-and-mouth disease virus.* J. Immunol. 115:1636–41.

Bachrach, H. L.; D. O. Morgan; P. D. McKercher; D. M. Moore, and B. H. Robertson. 1982. *Foot-and-mouth disease virus: Immunogenicity and structure of fragments derived from capsid protein* VP_3 *and of virus containing cleaved* VP_3. Vet. Microbiol. 7:85–96.

Bachrach, H. L., D. O. Morgan, and D. M. Moore. 1979. *Foot-and-mouth disease virus immunogenic capsid protein* VP_T*: N-terminal sequences and immunogenic peptides obtained by CNBr and tryptic cleavages.* Intervirology 12:65–72.

Bachrach, H. L., J. B. Swaney, and G. F. Van de Woude. 1973. *Isolation of the structural polypeptides of foot-and-mouth disease virus and analysis of their C-terminal sequences.* Virology 52:520–28.

Bernard, S., J. Wantyghem, J. Grosclaude, and J. Laporte. 1974. *Chromatographic preparation of purified structural proteins from foot-and-mouth disease virus.* Biochem. Biophys. Res. Commun. 58:624–32.

Cowan, K. M., and J. H. Graves. 1966. *A third antigenic component associated with foot-and-mouth infection.* Virology 30:528–40.

Crea, R., A. Kraszewski, R. Hirose, and K. Itakura. 1978. *Chemical synthesis of genes for human insulin.* Proc. Nat. Acad. Sci. USA 75:5765–69.

Denoya, C., E. A. Scodeller, B. M. Gimenez, C. Vasquez, and J. L. LaTorre. 1978. *Foot-and-mouth disease virus: I. Stability of its ribonucleic acid.* Virology 84:230–35.

European Commission for the Control of Foot-and-Mouth Disease, Report of the 24th session (FAO Rome, April 1981).

Gilbert, W., and L. Villa-Komaroff. 1980. *Useful proteins from recombinant bacteria.* Sci. Amer. 242 (4):74–94.

Grubman, M. J., B. Baxt, and H. L. Bachrach. 1979. *Foot-and-mouth virion RNA: Studies on the relation between the length of its 3′*-poly(A) segment and infectivity. Virology 97:22–31.

Kaaden, O. R., K. H, Adam, and K. Strohmaier. 1977. *Induction of neutralizing antibodies and immunity in vaccinated guinea pigs by cyanogen bromide-peptides of* VP_3 *of foot-and-mouth disease virus.* J. Gen. Virol. 34:397–400.

King, A. M. Q., B. O. Underwood, D. McCahon, J. W. I. Newman, and F. Brown. 1981. *Biochemical identification of viruses causing 1981 outbreaks of foot-and-mouth disease in the UK.* Nature (Lond.) 293:479–80.

Kleid, D. G., D. Yansura, B. Small, D. Dowbenko; and D. M. Moore, M. J. Grubman, P. D. McKercher, D. O. Morgan, B. H. Robertson, H. L. Bachrach. 1981a. *Cloned viral protein vaccine for foot-and-mouth disease: Responses in cattle and swine.* Science 214:1125–29.

Kleid, D. G., D. Yansura, G. Miozzari, M. Heyneker. 1981b. European Patent Application Publ. No. 0036776, 30 September 1981.

Kupper, H., W. Keller, C. Kurz, S. Forss, M. Schaller, R. Franze, K. Strohmaier, O. Marquardt, V. G. Zaslavsky, and P. M. Hofschneider. 1981. *Cloning of cDNA of major antigen of foot-and-mouth disease virus and expression in E. coli.* Nature (Lond.) 289:555–59.

Laporte, J., J. Grosclaude, J. Wantyghem, S. Bernard, and P. Rouze. 1973. *Neutralization en culture cellulaire du pouvoir infectieux du virus de la fievre aphteuse par des serums provenant de porcs immunises a l'aide d'une proteine virale purifiee.* C. R. Acad. Sci. 276:3399–3401.

Laporte, J., and G. Lenoir. 1973. *Structural proteins of foot-and-mouth disease virus.* J. Gen. Virol. 20:161–68.

Miozzari, G. F., and C. Yanofsky. 1978. *Translation of the leader region of the Escherichia coli tryptophan operon.* J. Bacteriol. 133:1457–66.

Moore, D. M., and K. M. Cowan. 1978. *Effect of trypsin and chymotrypsin on the polypeptides of large and small plaque variants of foot-and-mouth disease virus: Relationship to specific antigenicity and infectivity.* J. Gen. Virol. 41:549–62.

Moore, D. M., and B. H. Robertson. 1982. *Foot-and-mouth disease virus antigenic sites specified by capsid polypeptide VP_3.* Abstr., Ann. Mtg. Amer. Soc. for Microbiol. Page 264.

Robertson, B. H., D. M. Moore, M. J. Grubman, and D. G. Kleid. 1982. *Localization of the exposed immunogenic region in the outer capsid polypeptide from foot-and-mouth disease virus.* Abstr. Ann. Mtg. Amer. Soc. Microbiol. Page 264.

Sangar, D. V. 1979. *The replication of picornaviruses.* J. Gen. Virol. 45:1–13.

Sangar, D. V., D. N. Black, D. J. Rowlands, and F. Brown. 1977. *Biochemical mapping of the foot-and-mouth disease virus genome.* J. Gen. Virol. 35:281–97.

Sangar, D. V., D. N. Black, D. J. Rowlands, T. J. R. Harris, and F. Brown. 1980. *Location of the initiation site for protein synthesis of foot-and-mouth disease virus RNA by in vitro translation of defined fragments of the RNA.* J. Virol. 33:59–68.

Skinner, M. M. 1953. *One week old white mice as test animals in foot-and-mouth research.* Proc. 15th Int. Vet. Congr., Stockholm, vol. 1. Part 1, pp. 195–99.

Wu, R., editor. 1979. *Recombinant DNA. Methods in Enzymology,* vol. 68. Academic Press, New York.

8] Hybridoma Technology and Its Application to Problems in Veterinary Research

by RICHARD A. GOLDSBY,* S. SRIKUMARAN,† ALBERT J. GUIDRY,** STEVEN J. NICKERSON,†† and RONA P. SHAPIRO†

ABSTRACT

The demonstration that the techniques of somatic cell hybridization can be used to construct continuous cell lines that produce monoclonal antibody has revolutionized immunology. Though a recently introduced technology, the production of antibody-secreting cell lines, known as hybridomas, obtained by fusing activated lymphocytes with plasmacytomas, has already become a standard technique in many areas of basic and applied biology. Monoclonal antibodies have followed (and, in some cases, displaced) their antecedent conventional antisera into areas as diverse as neurobiology, parasitology, immunogenetics, tumor immunology, virology, developmental biology, biochemistry and, recently, the veterinary and agricultural manifestations of these subdisciplines. In this chapter the development and methodology of hybridoma technology are reviewed, and some examples are presented of its application to veterinary problems.

INTRODUCTION

The extremely useful methodologies of recombinant DNA technology evolved from fundamental studies of bacteria and their viruses. Hybridoma technology also has roots in basic biology. It is the utilitarian capstone of years of basic studies of the nature and behavior of somatic cell hybrids. A perspective on

*Department of Biology, Amherst College, Amherst, Massachusetts 01002; †Department of Dairy Science, University of Maryland, College Park, Maryland 20742; **Laboratory of Milk Secretion and Mastitis Research, Agricultural Research Service, United States Department of Agriculture, Beltsville, Maryland 20705; ††Mastitis Research Laboratory, North Louisiana Hill Farm, Homer, Louisiana 70140.

GENETIC ENGINEERING: APPLICATIONS TO AGRICULTURE (Beltsville Symposium 7—Lowell D. Owens, ed.) Rowman & Allanheld, Totowa

Table 8.1 History of Somatic Cell Hybridization (from an immunological perspective)

1960	Barski, Sorieul, and Cornefert discover cell hybrids produced in culture by the spontaneous fusion of two mouse sarcoma-producing cell lines.
1964	Littlefield introduces the use of mutant cell lines and selective media for the isolation of hybrid cells.
1966	Yerganian and Nell demonstrate that sendai virus can be used as a fusion agent to greatly increase the yield of hybrid cells.
1967	Weiss and Green discover the preferential elimination of human chromosomes in human-mouse hybrids and demonstrate that this phenomenon can be used to assign genes to specific chromosomes.
1973	Schwaber and Cohen produce human lymphocyte–mouse myeloma hybrids which secrete immunoglobulin.
1975	Pontecorvo demonstrates that PEG is an efficient fusing agent for producing hybrids in animal cell cultures.
1975	Kohler and Milstein produce myeloma-spleen cell hybrids secreting monoclonal antibodies of predefined specificity.

this rapidly spreading technology for the "immortalization" of clones of differentiated cells that secrete useful products can be obtained by an examination of its intellectual pedigree.

The landmark developments in somatic cell hybridization (at least from the immunologists' point of view) are summarized in Table 8.1. Although specialized examples of cell fusion (gametic fusion and the fusion of myoblasts) are widespread and have been recognized for 50 years, the discovery that other somatic cells would undergo fusion in vitro is a relatively recent one. The field of somatic cell hybridization is generally conceded to have been launched with the demonstration by Barski, Sorieul, and Cornefert (1961) that the cocultivation of cell cultures of two mouse sarcoma-producing lines resulted in the occasional spontaneous formation of hybrid cells. This approach, while pioneering, suffered from the limitation that the hybrids could be identified and selected only on the basis of subjective morphological criteria. In 1964, Littlefield introduced a rational strategy for the selection of somatic cell hybrids that employed the HAT (*h*ypoxanthine, *a*minopterin, and *t*hymidine) selective system and established mutant cell lines lacking one or both of the salvage enzymes HGPT (*h*ypoxanthine *g*uanine *p*hosphoribosyl *t*ransferase) and thymidine kinase (Littlefield 1964). The demonstration by Yerganian and Nell (1966), that UV-inactivated Sendai virus increased the incidence of cell hybrids by inducing fusion between neighboring cells, was also an extremely important methodological development. Using Sendai to promote fusion and the HAT system to select the hybrids, it was possible to obtain large numbers of hybrids from the parasexual matings of an extremely wide variety of cell types. Ten years later, the introduction from plant protoplast research of fusions mediated by inexpensive, stable, easily prepared polyethylene glycol

(PEG) preparations into animal cell systems has again simplified the task of making hybrids (Pontecorvo 1976). It has also made it possible to fuse combinations of cells even when one or both partners lack receptors for Sendai virus. This has opened the way for the fusion of cells across profound evolutionary distances, and barriers have been breached between phyla and even between kingdoms (human Hela cells X tobacco protoplasts).

In reviewing, however briefly, the history of somatic cell hybridization, it is critical to emphasize the landmark and pivotal observation of Weiss and Green (1967). These workers discovered that in human X mouse hybrids there is preferential loss of human chromosomes. Furthermore, they pointed out that by concordance analysis of interspecific hybrids in which the chromosomes of one of the participating species is preferentially lost, it is possible to assign particular genes to particular chromosomes. This observation provided a general approach to the problem of assigning genes to chromosomes, thereby providing an alternative to assignment by classical genetic methods which involve the deliberate crossing of individuals and the subsequent analysis of their progeny to determine linkage by recombination analysis. These latter techniques, which are the stock in trade of mouse and *Drosophila* geneticists, represent an approach to gene mapping which is unethical in the human and, because of the generation time and expense of rearing, impractical in many species of veterinary interest.

The first antibody-producing hybridomas were made in 1973 by Schwaber and Cohen (1973), who used inactivated Sendai virus to fuse human lymphocytes from normal donors to a mouse plasmacytoma. These workers were the first to demonstrate that one could immortalize the antibody production of a *normal cell of the B lymphocyte lineage* by fusing it with a plasma cell tumor. It should be noted that these workers found that some of their hybrids produced immunoglobulin of human as well as mouse origin. Their report was therefore the first to establish the feasibility of fusing mouse myeloma cells with normal immunocytes of another species as an approach to the production of non-murine monoclonal antibodies. Other workers have used this approach to produce human monoclonal antibodies (Nowinski et al. 1980), rat monoclonal antibodies (Springer et al. 1978) and, as reported in this paper, bovine monoclonal antibodies.

In 1975, Kohler and Milstein (1975) devised and demonstrated a strategy for the deliberate and rational construction of continuous cell lines which secrete monoclonal antibodies of a desired specificity. They fused a HAT-selectable mouse myeloma line with spleen cells from mice which had been *previously immunized* with sheep red blood cells. They then *screened* the resulting hybrids for the production of antibody specific for sheep red blood cell, the *immunizing antigen*. Their success, which has been widely reproduced, revolutionized serology, provided tools for the examination of basic immunological mechanisms, and created an industry. Table 8.2 provides a partial list of the antigens that have been successfully targeted by monoclonal antibodies. It is clear that in the short time since its inception, monoclonal antibody technology has influenced an extraordinary range of pure and applied biological problems.

Table 8.2 Monoclonal Antibodies Have Been Produced with Specificity for:

Viruses
Bacteria and bacterial products
Parasites
Hormones
Drugs
Tumor-associated antigens
Soluble antigens useful in tumor detection
Subpopulations of Neuronal cells
Classes and subpopulations of leukocytes
Major histocompatibility complex (MHC)-determined antigens
DNA (single-stranded and double-stranded)
RNA
Numerous soluble proteins from immunoglobulins to tubulin
Glycolipids
Carbohydrates

Contrasts between monoclonal and polyclonal antibody production and properties. Most antigens of practical interest bear many distinct antigenic determinants (epitopes). This is obviously true of the manifestly complex "forests" of antigens presented by the surface of microorganisms and eucaryotic cells. It is also true that a protein molecule or a complex carbohydrate displays a number of epitopes, some of which are sequence determinants while others are conformational determinants. Sequence epitopes arise from the local variations in the nature and order of linkage of a macromolecule's monomeric units. Conformational epitopes are antigenic determinants that arise from the tertiary (or quarternary) structure of the native macromolecule.

The consequences of immunizing an animal with a multideterminant antigen are illustrated in Figure 8.1. Each epitope has the potential to trigger the activation of one or more B cell clones to divide and differentiate, thereby producing a number of distinct, antibody-secreting, clonal populations of plasma cells. The monoclonal antibodies characteristic of each activated B cell clone pool in the body fluids and, consequently, the serum harvested from the animal is an intimate polyclonal mixture of different antibody molecules. Even in those cases where it is feasible to subject polyclonal serums to affinity chromatography on immobilized antigen columns, one only succeeds in purifying a mixture of different immunoglobulin (Ig) molecules that share a capacity to bind to some determinants of the antigen. Antibodies thus isolated will differ in amino acid sequence and probably in Ig class and subclass. Furthermore, the composition of this polyclonal mixture will change from day to day and from animal to animal. In most practical programs of antiserum production, different batches of antiserum, even though raised by well-standarized protocols, will differ from batch to batch.

Fortunately, a very different state of affairs results if one harvests spleen cells from recently immunized animals, constructs hybrids by fusion with a suitable plasmacytoma line, and then identifies those hybridoma clones that secrete antibody to the immunizing antigen. As illustrated in Figure 8.1, each of the selected hybridomas produces a monoclonal Ig that recognizes only a particular epitope of the antigen. Use of the hybridoma technology allows the investigator an alternative to the shotgun approach of polyclonal antiserum production and offers the following advantages:

1. Once stabilized, a hybridoma provides a "perpetual" source of well-defined homogeneous antibody.
2. Large amounts (hundreds of milligrams or even grams) of a particular monoclonal antibody can be obtained with a relatively modest investment of resources and personnel.
3. Monoclonal antibodies highly specific for a particular target antigen can be obtained even when the antigen is grossly impure or present only in trace amounts.
4. Since monoclonal antibodies react with determinants in an "all or none" fashion, there is no need (in fact it is infeasible) to resort to absorption to improve specificity.
5. The ability of monoclonal antibodies to examine and compare the cross reactivities of individual epitopes within an complex antigen between different species is a tool of great power for examining evolutionary relationships.

While noting the advantages of monoclonal antibodies, it is useful to be aware of some quirks of monoclonal serology that contrast with polyclonal serology. As we noted earlier, polyclonal antisera are mixtures of immunoglobulins and typically contain antibodies of many different affinities, specificities, antibody classes, and subclasses. A monoclonal antibody is uniform with respect to all of these parameters. These contrasts between monoclonal and polyclonal reagents are important because different serological assays depend on different properties and classes of antibody. Assays involving complement fixation will require IgM or only certain IgG subclasses (IgG_{2a} and IgG_{2b} in the mouse). Precipitation assays such as Ouchterlony or radial immunodiffusion (RID) require a sufficiently multivalent interaction between antigen and antibody to form a latticework. Radioimmunossay or ELISA of compounds present in trace amounts ($<10^{-5}$M) are dependent upon antibodies with high affinities for the antigen of interest. A polyclonal serum will often contain a subpopulation of antibody molecules with properties that are appropriate to any one of the assay strategies described above. A monoclonal antibody may lack the properties necessary to function in one or more of these commonly used assays, however. The importance of these considerations in selecting a suitable monoclonal antibody for a given type of assay cannot be overemphasized.

With these considerations in mind let us examine, specifically, how one proceeds to construct and cryopreserve hybridomas. We shall also describe methods for passaging hybridomas as tumors and discuss a system for handling

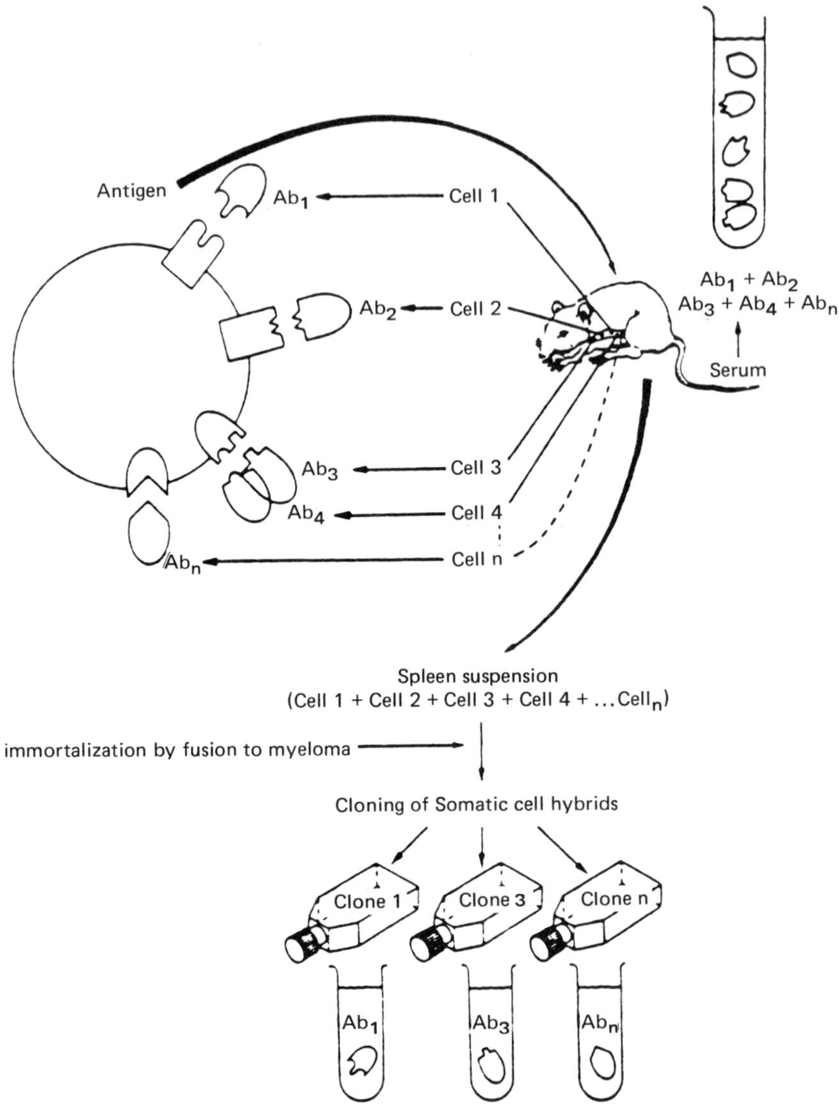

Figure 8.1. Monoclonal and polyclonal antibody production contrasted. (Adapted from D. Secher.)

the rather large number of samples routinely encountered in hybridoma work. We will then present examples which illustrate the use of hybridoma technology to solve the following problems:

1. the development of reagents for the analysis of bovine immunoglobulins;
2. the development of reagents that recognize the differentiation antigens of a specific group of bovine immunocytes—the PMNs;
3. the production of monoclonal bovine Ig; and
4. the production of specific antibody without specific immunization.

METHODS FOR THE CONSTRUCTION OF ANTIBODY-PRODUCING HYBRIDOMAS

There are many procedures for the successful production of antibody-producing hybridomas. The one detailed here is used in our laboratory.

Materials and Equipment

1. A suitable plasmacytoma cell line.
2. RPMI 1640, serum free.
   RPMI 1640 containing 2mM L-glutamine, 1mM sodium pyruvate, 15% fetal calf serum, 100 units/ml penicillin, 100 units/ml streptomycin or, because of its stability in cell culture medium and broad spectrum, 50 μg/ml of Gentamycin is the antibiotic of choice. This is called complete RPMI 1640 in the fusion protocol.
3. Complete RPMI 1640 containing 1×10^{-4}M hypoxanthine, 4×10^{-7}M aminopterin and 1.6×10^{-5}M thymidine (complete RPMI 1640 + HAT).
4. Complete RPMI 1640 containing 1×10^{-4}M hypoxanthine and 1.6×10^{-5}M thymidine (complete RPMI 1640 + HT).
5. Sterile, frosted microscope slides.
6. 50 ml and 15 ml sterile, plastic, centrifuge tubes.
7. 96-well, flat-bottomed, microtiter dishes.
8. 50% polyethylene glycol 1500 or 4000 (PEG 1500 or PEG 4000) in RPMI 1640, serum free.
9. 24-well, cluster plates (Costar or Linbro).
10. Several gross of sterile-cotton plugged Pasteur pipettes and bulbs.
11. 100 mm tissue culture dishes.
12. Complete RPMI 1640 + 10% dimethyl sulfoxide (DMSO).
13. Freezing Vials (Nunc #N 1076-1).
14. Deep Freeze (−70°C).
15. Liquid Nitrogen Freezer.
16. Water-jacketed CO_2 incubator.
17. Inverted microscope.
18. Freezing medium (mix 20% v/v fetal calf serum, 10% v/v dimethyl sulfoxide and 70% v/v Dulbecco's PBS, adjust to pH 7.4, filter sterilize and store at 2° − 8°C).

Immunization. Two immunizations (a priming injection and a boost 3–4 weeks later) are usually quite sufficient to insure the presence of immune spleen cells.

While many feel that a boost is helpful, it is not an absolute requirement. The literature contains reports of hybrids being derived after only a single priming injection of antigen. We suggest the following guidelines for immunization:

1. Prime more mice (two or three times the anticipated need) than required for a hybridization. In case a mishap or a failure to obtain an adequate number of hybrids makes it necessary to repeat the hybridization, a supply of antigen-primed mice can save time.
2. Suggested procedure for soluble antigens: prime mice by intraperitoneal injection of 100 μg of the antigen emulsified in complete Freund's adjuvant. Boost after an interval of 21 days or more by intravenial or intraperitoneal (i.p.) injection of 0.2 ml of an aqueous solution containing 10–50 μg of the antigen. The suggestion that one prime with 100 μg of antigen should be viewed only as a rough guideline. Many workers have consistent success with 50-μg or even 20-μg doses. Some workers have succeeded with doses of 1 μg and even less. When such small quantities of antigen are employed, however, a route of immunization other than i.p. is recommended. One might, for example, inject antigen emulsified in CFA into the foot pad and subsequently harvest and hybridize lymphocytes from the inquinal lymph nodes.
3. Suggested procedure for cellular antigens: harvest cells and wash them twice in phosphate-buffered saline. Prime by injecting 1-2 $\times 10^7$ cells i.p. (the cells may be emulsified in complete Freund's adjuvant, but this is not necessary in many cases). Boost after 21 days by i.p. injection of 1×10^7 cells.
4. If hyperimmune mice are to be used as spleen cell donors, the animals should be rested 30 days before receiving the final boost.

Source of spleen and plasmacytoma (myeloma) cells

Spleen cells: Timing of animal sacrifice is important because it is the actively dividing lymphoblastoid population that gives rise to most of the hybridomas. Three days post immunization has been found to be an optimal time to harvest spleen cells for a number of antigens. It is useful to exsanquinate the mice by decapitation and to harvest and save the serum to aid in evaluating the success of the immunization protocol and for use as a positive control in assay procedures. While the appearance of antibody in the serum three days after boosting is proof that one has activated lymphocytes, a negative result or a low serum titer does not mean the activation has failed. It is sometimes advisable to proceed with the fusion of spleen cells from immunized animals that present negative sera.

Following sacrifice and serum collection, totally immerse the carcasses in 70% ethanol and transfer to a laminar flow hood. Remove the spleens sterilely and place them in 15 ml of serum-free RPMI 1640 and tease them apart with sterile needles or by grinding between the frosted faces of sterile microscope slides. Transfer the suspension to a 15-ml centrifuge tube and allow the larger tissue fragments to settle (about 2 min). After sedimentation, transfer the spleen cell suspension to a second 15 ml tube and harvest them by sedimentation for 10 minutes at 400xg. Resuspend the cells in 15 ml of serum-free RPMI 1640. Dilute 0.5 ml of this suspension in 2% acetic acid (in order to lyse red cells) and immediately count on a hemacytometer. A single spleen from a boosted animal will yield $1.5–2.0 \times 10^8$ spleen cells. It is recommended that the spleen cells for hybridization be taken from a pool of at least two spleens in order to increase the potential diversity of hybridomas.

Plasmacytoma (myeloma) cells: Maintain the cell line in log phase until it is harvested for use in a fusion. This requires that the cells be subcultured at densities of at least 2×10^4 cells/ml. Most workers report a drastic decrease of viability if the density of the culture reaches 5×10^5 cells/ml. Many have found it useful to initiate cultures at 5×10^4 cells/ml three days prior to their use for hybridization and then split them 1→3 each day. It is good practice to maintain the myeloma line in selective medium (since the lines used for hybridoma production are HGPRT-7; this means in 2.5 μg/ml 6TG or 25 μg/ml 8-azaguanine) continuously or at least for the three days prior to their use as fusion partners in a hybridization. Immediately prior to their use in a hybridization, the plasmacytoma cells are spun out of selective medium (400xg for 10 min) and washed 2× in serum-free RPMI 1640.

Fusion. The following protocol is recommended:

1. In a 37°C water bath, warm complete RPMI 1640 to which the HAT selective system has been added, serum-free RPMI 1640, and the PEG solution.
2. Make a 37°C water bath for use in the laminar flow hood by placing one glass beaker containing 37°C water within a second larger beaker also containing 37°C water. Maintain this improvised water bath at 37°C until it is needed.
3. Mix 2×10^8 pooled spleen cells with 2×10^8 plasmacytoma cells and copellet the mixture by centrifugation at 400xg for 10 minutes at room temperature. It should be noted that the myeloma/spleen cell ratio may be varied from 1 to 0.1 and the total number of cells can be routinely reduced to between 2×10^7 and 1×10^8. With appropriate scaledown of volumes and vessels, hybridomas have been obtained from as few as 10^4 activated lymphocytes.
4. Completely remove all supernatant from the pellet by aspiration with a pipette and bulb.
5. Loosen and reslurry the pellet by repeatedly tapping the bottom of the tube on the hard metal surface of the hood.
6. Place the tube containing the pellet in the water bath and slowly (dropwise) add 2.0 mls of warm PEG over course of 45 seconds. Gently mix the PEG and cell slurry by agitation of the tube during the period of addition. Continue to incubate the PEG-cell slurry for an additional 20 seconds.
7. Using a 2 or 5 ml pipette stir in 2 ml of 37°C serum-free RPMI 1640 over a period of 1 minute.
8. Wait 1 minute and then repeat Step 7. Wait an additional minute and again repeat Step 7.
9. Wait 2 minutes and stir in 6 ml of 37°C serum-free RPMI 1640. Wait 2 minutes and then slowly stir in 37°C serum-free RPMI 1640 until the volume has reached 50 ml. This step should take 1 to 2 minutes.

10. Pellet the cell suspension by centrifugation at 400xg for 10 minutes.

11. Discard the supernatant and by gentle pipetting, resuspend the cells in 50 ml of complete RPMI 1640 containing the HAT selective system.

12. Use 30 ml of this suspension as is and dilute one 15-ml portion with 30 ml of complete RPMI 1640 + HAT and the remaining 5 ml with 25 ml of complete RPMI 1640 + HAT. Add 0.1 ml of the undiluted suspension (4×10^5 spleen cell + 4×10^5 myeloma cell input) to each of the 96 wells of microtiter dishes which have been previously seeded with 0.1 ml of a feeder cell suspension in complete RPMI 1640 + HAT. The 3-fold and 6-fold dilutions of the cell suspensions are also seeded into respective sets of 3 feeder-layer–containing microtiter dishes.

Feeder cells may be mouse or rat thymocytes, mouse macrophages, or irradiated spleen feeder cells. In 96-well, microtiter dishes use 1×10^5 thymocytes/well, 1×10^4 macrophages/well, or 1×10^5 irradiated mouse spleenocytes/well. If spleen cells are used they should be irradiated with 1300–2000 rads or a sufficient dose of UV light (cells must be in PBS during the UV irradiation, since growth medium contains a number of UV-absorbing constituents) to kill 90% of a plasmacytoma cell culture. If unirradiated spleen cells are used, overgrowth of well cultures by fibroblasts is a likely outcome.

13. The plates are incubated in a 37°C humidified incubator containing a 5% CO_2—95% air atmosphere.

14. After 3 days, approximately one half the medium is removed from each well and 0.1 ml of complete RPMI 1640 + HT is added. This process should be repeated every 5 days.

15. When clones have reached a diameter of 1–2 mm (usually 9–14 days) a sample of their culture fluid should be tested to determine if they are secreting antibody of the desired specificity. It should be pointed out that not all hybridomas develop at the same rate. Wells should be maintained for at least 3 or even 4 weeks before concluding that they do not contain viable hybridomas. Replace whatever medium is removed for testing with 0.1 ml of complete RPMI 1640 + HT.

16. Within 24 hours of sampling, transfer the positive clones to the wells of 24-well dishes. Prior to transfer of hybrid clones the wells of these dishes should be charged with 1 ml of a feeder cell suspension (1×10^6 thymocytes or irradiated spleen cells, but 1×10^5 macrophages will suffice) in complete RPMI 1640 + HT.

17. Two days after transfer add 1 ml of complete RPMI 1640 + HT.

18. Depending on the rate of cell growth, remove the medium as completely as possible 24 to 48 hours after Step 17 and replace this medium with 2 ml of complete RPMI + HT. When the cells are nearly confluent, test the supernatant for the presence of the desired antibody.

19. Immediately clone the positive cultures by limiting dilution on feeder layers in complete RPMI 1640 + HT in 96-well, microtiter dishes. Clone at the following cell densities; 24 wells at 1 cell/well, 24 wells at 2 cells/well, 24 wells at 5 cells/well, and 24 wells at 10 cells/well.

20. Expand the balance of the culture by dilution to 5 ml with complete

medium + HT and transfer to a 60-mm tissue culture dish or 25-sq cm culture flask.

21. When these cultures become dense (about 1 × 10^5 cells/ml) transfer to a 100-mm dish and add 5 ml of complete RPMI 1640 + HT. With rapidly growing hybrids it will be necessary to split these cultures 1 to 2 within 24 to 48 hours.

22. After two 100-mm cultures have been obtained, the hybridoma cells should be frozen as a protection against accidental loss. Also, a portion of each hybridoma should be innoculated into a mouse so that significant quantities of potentially valuable antibodies begin to accrue even before the first cycle of cloning prescribed in Step 19 has produced a batch of clones.

23. When positive clones of interest have been identified from Step 19, two subclones of each of the positive cultures cloned in Step 19 should be expanded, frozen, and injected into mice as indicated in Steps 20, 21, and 22.

24. Use the following simple protocol for freezing cells: (a) Select healthy logarithmically growing cultures for cryopreservation. (b) Centrifuge the cells out of the growth medium (400–600xg for 10 minutes) at room temperature. (c) Resuspend in freezing medium. (d) Put 1 ml of the cell suspension into as many 2-ml Nunc freezing vials as necessary. Immediately place the ampules in a mailing tube or its equivalent and place in a −70°C or −80°C freezer and leave overnight. The next day quickly transfer the ampules to liquid nitrogen. (e) To return frozen cultures to growth in culture, quickly thaw the ampule in a 37°C water bath (wear heavy gloves and eye protection—ampules sometimes contain liquid nitrogen and explode when it assumes the vapor phase). (f) Immediately dilute the contents of the ampules to 10 ml with growth medium containing 20% serum and centrifuge at 400–600xg for 10 minutes. Remove the supernatant and resuspend the cells in 2 ml of 20% serum-containing growth medium. Determine the viable cell count and adjust the cell number to 5×10^5 cells per ml and plate in flasks or dishes. NOTE OF CAUTION: On thawing, some cells do not immediately flourish when cultured in large volumes of liquid. They will often recover, however, if cultured in 0.2 ml aliquots on suitable feeder layers in the wells of microtiter dishes. Consequently, it is prudent to establish four or five 0.2-ml cultures of a thawed culture at the same time one initiates the mass culture.

25. Growth of hybridomas as tumors:

a. For solid tumors:

1. Harvest cells from growth medium by centrifugation.
2. Resuspend to a cell concentration of $2\text{-}5 \times 10^7$/ml.
3. Inject 2 or more mice with 0.2 ml of the cell suspension subcutaneously just anterior to the right or left rear leg.
4. Tail-bleed the mouse when the tumor attains the size of a garden pea (2–4 weeks) and at 3- to 4-day intervals thereafter until the tumor becomes life-threatening. At this point the mouse should be exsanguinated and the tumor immediately excised, placed in 10 mls of serum-free growth medium and minced. Aliquots of mince (0.2 ml/mouse) may be reinjected into a number of mice. The balance of the tumor should be harvested by centrifugation, resuspending in freezing me-

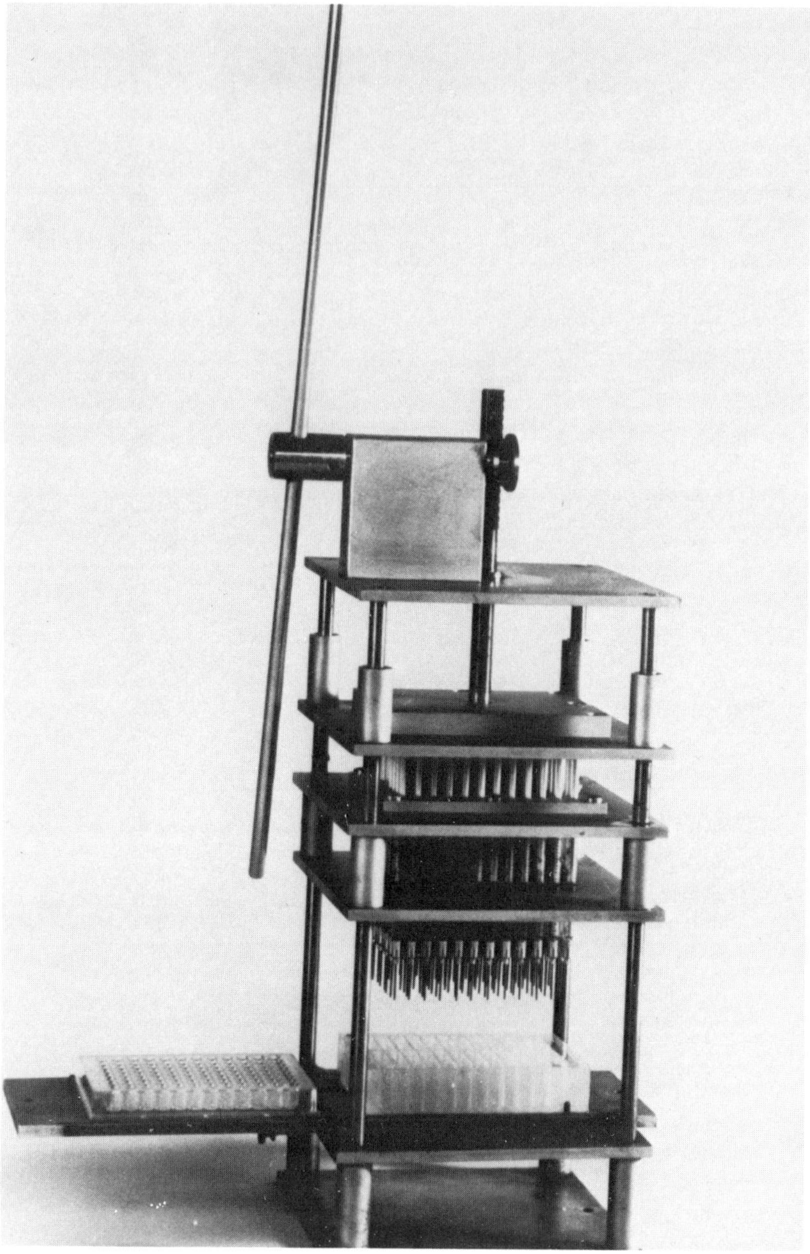

Figure 8.2. Syringe replicator with a chambered trough and 96-well, microtiter dish on the ways.

dium and frozen according to the protocol for cryopreservation of cells outlined in Step 24.

b. For ascities:

1. Inject the mice intraperitoneally with 0.5 ml of pristane (Aldrich Chemical Company).

2. One to eight weeks after pristane-priming, inject (use a 20-gauge needle) 5×10^6-1×10^7 logarithmically growing hybridoma cells in a volume of 0.2–0.4 ml of serum-free medium or PBS.

3. Tap the ascities fluid when the abdomen becomes visually swollen by insertion of an 18-gauge (no syringe) needle. Allow the fluid to drip into a suitable collection vessel (a 15-ml centrifuge tube).

4. Remove the cells by centrifugation (500xg for 10 min). Pipette off the antibody-containing supernatant for assay and freezing. Resuspend the cells to a concentration of 5×10^7/ml in PBS and innoculate as many primed mice as desired with 0.2 ml of the cell suspension. The balance of the cells may be spun down, resuspended in freezing medium, and stored in liquid nitrogen as outlined in Step 24.

An apparatus to facilitate the handling and transfer of cells and fluids. While clone transfer and feeding can be accomplished with Pasteur pipettes or hand-held multichannel pipettors, a considerable saving in time and effort can be realized by use of a syringe replicator. Figure 8.2 is a photograph of a syringe replicator. It is a rigid matrix of 96 plastic, disposable, 1-ml syringes outfitted with 20-gauge disposable needles (from which the bevels have been filed), which are mounted concentrically with the 96 wells of a microtiter dish. This apparatus, which can be readily fabricated by a competent machine shop, is well suited to such tasks as the seeding of cells, cloning at limiting dilution, feeding, replica plating, washing particulates, and the precise (100 μl ± 7%) collection and transfer of measured aliquots of culture fields for screening procedures. For a detailed discussion of the use of complementary matrices for cell and fluid handling and replica plating, see Goldsby and Zipser (1969) and Goldsby and Mandell (1973).

THE DEVELOPMENT OF REAGENTS FOR THE ANALYSIS OF BOVINE IMMUNOGLOBULINS

Highly specific antisera to recognized classes and subclasses of bovine immunoglobulins (Igs) are useful reagents for the analysis of the humoral immune response in this economically important animal. Antisera prepared by conventional methods often lack class and subclass specificity, however, and require removal of antibodies of unwanted specificity by adsorption onto appropriate immobilized antigens. Even after multiple adsorption procedures, the resulting antisera may still have some degree of cross-reactivity with closely related antigens. Such cross-reactivity is often encountered in preparing polyclonal antisera to particular subclasses of Ig because of the extensive sharing of antigenic determinants among Ig classes and subclasses. The construction of

Table 8.3 Reactivity of Anti-bovine IgG Monoclonal Antibodies

Monoclonal antibody	IgG_1	IgG_2	IgM	BSA
DAS 1	Strong	Strong	Weak	No
DAS 2	No	Strong	No	No
DAS 3	Weak	Strong	Weak	No
DAS 4	Weak	Strong	Weak	No
DAS 5	Weak	Strong	Weak	No
DAS 6	No	No	Strong	No

hybridomas producing monoclonal antibodies specific for the Ig determinants uniquely borne by a particular Ig subclass circumvents this problem.

Hybrids were derived by fusing the hybrid plasmacytoma line SP2/0 with spleen cells from two 4-month old female Balb C/J mice which had been primed by i.p. injectin of 100 μg of alum-precipitated bovine Ig emulsified in complete Freund's Adjuvant and 4 weeks later boosted with a second i.p. injection of 50 μg alum-precipitated bovine Ig. Spleen cells were harvested and fused with SP 2/0 after boosting. Of 88 hybrid clones examined, 16 produced antibody to bovine IgG. After three passages in culture, five of the hybrids retained secretory capacity and were grown to mass cultures, passed as tumors, and cryopreserved. Their specificities, along with that of a sixth monoclonal antibody, DAS 6, which was derived later, are summarized in Table 8.3.

Our experience in purifying and characterizing one of these hybridomas, DAS 2, illustrates the pattern of results obtained with monocolonal antibodies. A monoclonal antibody produced from a hybridoma made by fusing spleen cells with a nonsecreting plasmacytoma line will be of a single Ig subclass and will have a single species of light chain. Ouchterlony analysis of DAS 2 revealed that the monoclonal antibody has a IgG_1 heavy chain and a kappa light chain (Fig. 8.3). Isoelectric focusing of DAS 2 resulted in a multiband pattern (Fig. 8.4), which was centered around an isoelectric point of pH 6.2. Multiple closely spaced bands instead of a single band are a consequence of posttranslational modification of the monoclonal antibody. Pure monoclonal antibody for the studies just mentioned was readily obtained by affinity chromatography. The elution pattern shown in Figure 8.5 was obtained by passing sera from mice bearing the DAS 2 hybridoma over a sepharose 4B-bovine IgG-coupled column.

The anti-bovine Ig class and subclass monoclonal antibodies, because of their specificity, reproducibility, and high titer, will be of value in studies in which a specific antibody probe for a particular subclass or class is needed. One might expect to see these reagents find the following applications: (a) assaying particular bovine Ig species in serum, milk, or colostrum; (b) prepara-

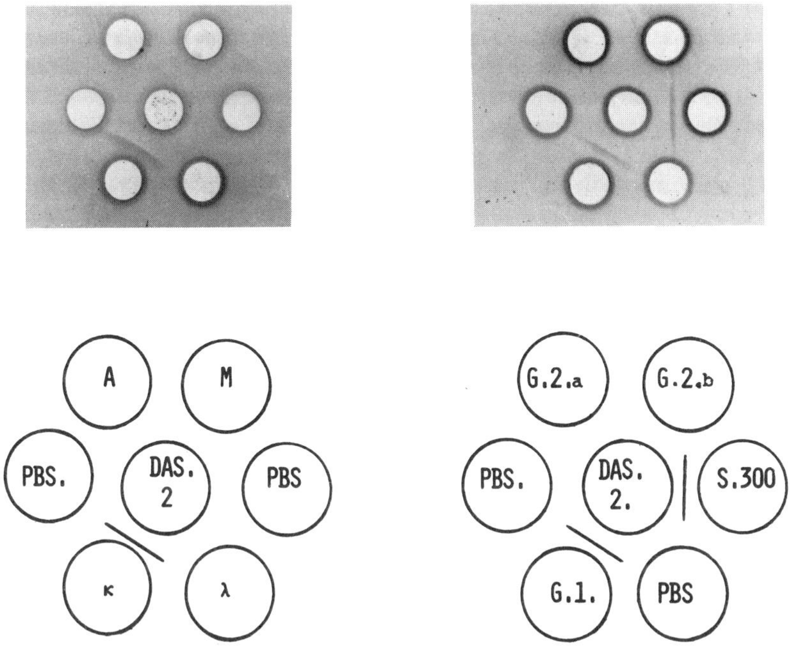

Figure 8.3. Ouchterlony analysis of affinity purified DAS-2. The wells labeled G_1, G_{2a}, G_{2b}, A, M, kappa and lambda represent rabbit anti-mouse I_gG_1, I_gG_{2a}, I_gG_{2b}, I_gA, I_gM, kappa and lambda light chains respectively. S 300 is a polyclonal rabbit anti-mouse I_g which is included as a positive control. DAS-2 is shown by this analysis to possess I_gG_1 heavy chains and kappa light chains.

tion of large quantities of highly purified bovine Ig classes and subclasses; (c) identification of functional domains of bovine Ig heavy chains and the isolation of those domains after the molecule's dissection by an appropriate proteolytic enzyme.

THE DEVELOPMENT OF REAGENTS WHICH RECOGNIZE THE DIFFERENTIATION ANTIGENS OF A SPECIFIC GROUP OF BOVINE IMMUNOCYTES

The immune system is a complex, interacting network of many different cell types that collaborate to produce an immune response. The surface properties of immunocytes are of great interest because they identify populations of cells that conduct different functions of the immune system. It is well established that B lymphocytes of all species synthesize and display membrane-bound Ig (Froland and Natvig 1972; Fu, Winchester and Kunkel 1974). A variety of surface antigens are found on the T lymphocytes of the mouse (Cantor and Boyse 1975a, b) and the human (Reinherz and Schlossman 1980). Further-

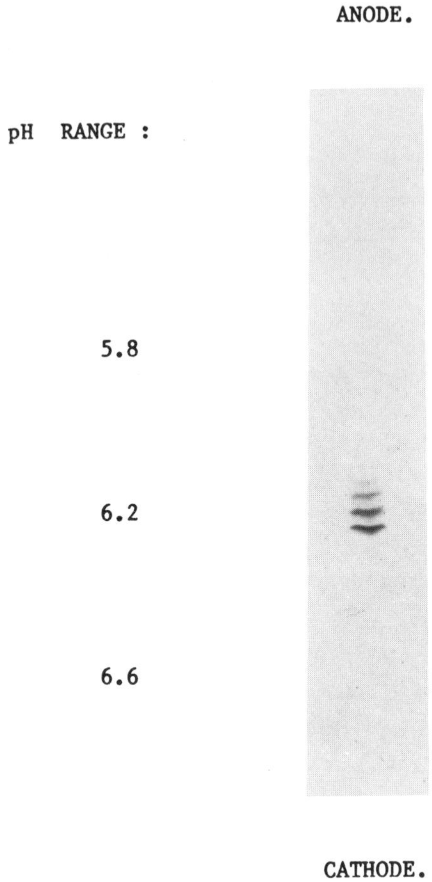

Figure 8.4. Isoelectric focusing of intact DAS-2 in an agarose matrix.

more, in the mouse and in the human, discrete functional (amplifier, suppressor, cytotoxic) subpopulations can be identified and isolated by taking advantage of their display of distinctive and identifying surface antigens. These considerations make it apparent that highly specific, high-titer antibody preparations to particular bovine immunocyte populations and subpopulations would assist greatly in extending our ability to detect changes in immunocyte populations.

Until the advent of successful approaches to the construction of monoclonal antibody-producing hybrid cell lines, one of two standard methods of obtaining antibodies to cell-surface antigens was employed. One of these, immunization with cells bearing the target antigens across xenogeneic or allogeneic barriers, required an absorption protocol deemed sufficient to remove unwanted activities and render the antiserum specific. The other approach, a technically

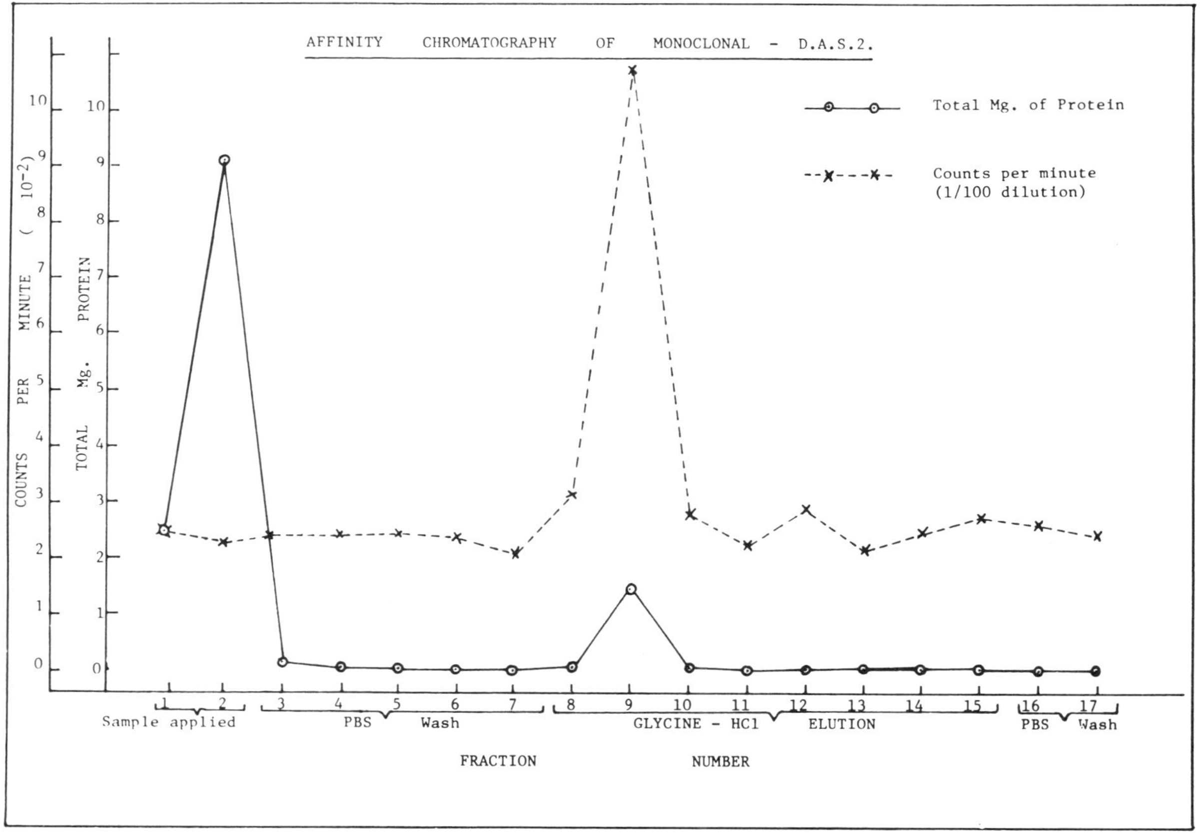

Figure 8.5. One-step purification of DAS-2 directly from the serum of animals in which the hybridoma grew as a tumor.

formidable one, involved simplification of the immunogen by isolation and purification of the identifying differentiation antigen and the subsequent use of the purified antigen in a program of immunization.

We chose to develop a library of monoclonal antibodies specific for neutrophilic polymorphonuclear leukocytes (PMN), a population of immunocytes which constitute an essential defense mechanism against infection. In the bovine system it is well established (Paape et al. 1979) that PMN constitute a major immunologic barrier to invading mastitis pathogens. Using monoclonal antibodies, binding sites on PMN, such as those involved in recognition of chemotactic agents, antigen-antibody complexes, phagocytosis, and intracellular kill could eventually be studied, and peripheral blood subpopulations identified that infiltrate the udder in response to infection. The importance of understanding these processes and factors to influencing the course of mastitic infections is clear.

Monoclonal antibodies were derived by hybridizing spleen cells, from Balb C/J mice hyperimmunized with bovine peripheral blood leukocytes (PBLs) (priming 2×10^7 PBLs in CFA and three subsequent boosts of 2×10^7 cells in PBS, i.p. 24, 54, and 120 days post priming). The hybridization yielded 261 clones, 82 of which secreted antibody which bound to bovine PBLs but not to bovine erythrocytes, bovine gamma globulin, or whole bovine serum (Fig. 8.6). On continued culture and differential screening, 16 of the 82 hybrids initially shown to secrete antibody binding to PBLs were determined to produce monoclonal antibody specific for PMNs and were unreactive toward bovine lymphocytes, erythrocytes, Igs, or serum components (Figure 8.7). Preliminary studies using fluoresence methods have demonstrated that some of the monoclonal antibodies described in Figure 8.4 react only with a subset of the total PMN population. Such reagents are extremely interesting because they provide a means of establishing heterogeneity in the neutrophil population and, furthermore, they provide an approach to the separation of neutrophil subpopulations.

It is appropriate that a great deal of effort be devoted to the development of reagents and methods that permit the recognition and separation of neutrophil subpopulations. The importance of the problem lies in the realization that just as with other immunocytes, such as T lymphocytes, different PMN subpopulations may perform different roles, be present at different stages of differentiation or have different tissue tropisms. The conduct of a rigorous and detailed analysis of the PMN population depends critically on the ability to recognize and isolate different PMN subsets. The initial results indicate that the construction of antibody-producing hybrids for the production of highly specific, high-titer, PMN-specific antibody preparations will be of determinative importance in the study of this important group of immunocytes.

THE PRODUCTION OF MONOCLONAL BOVINE IMMUNOGLOBULIN

The availability of monoclonal Igs, first from myelomas and lately from hybridomas, has been of critical importance to the development of our current

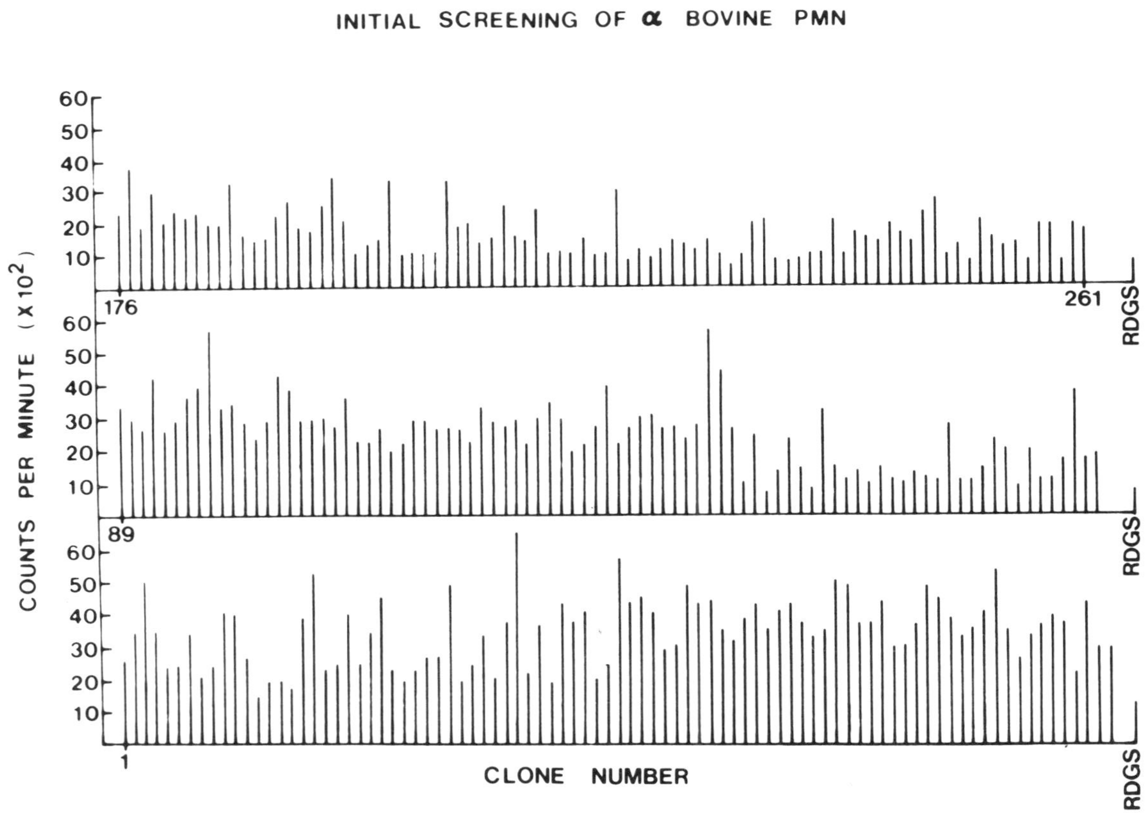

Figure 8.6. Initial screening of hybridomas produced by fusion of spleen cells from mice immunized with bovine PBLs.

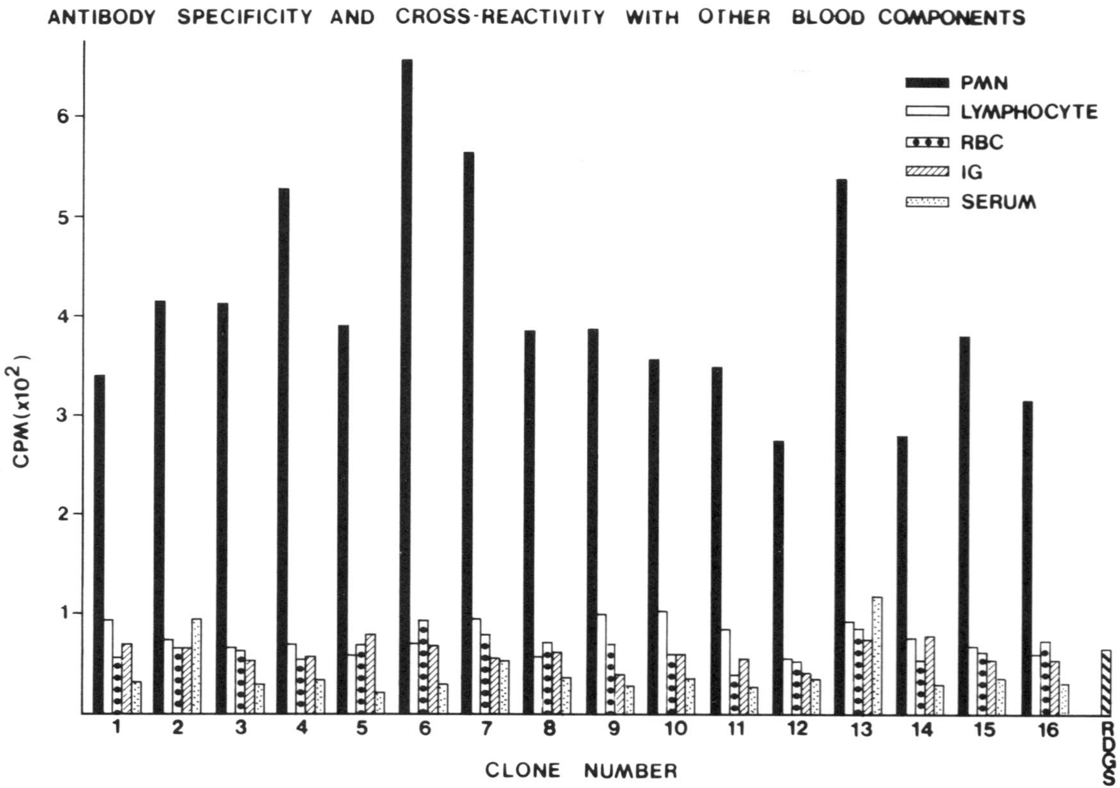

Figure 8.7. Specificity of 16 of the hybrids described in Figure 8.6 for bovine PMNs. RBC = red blood cell; I_g = bovine immunoglobulin; RDGS = growth medium negative control.

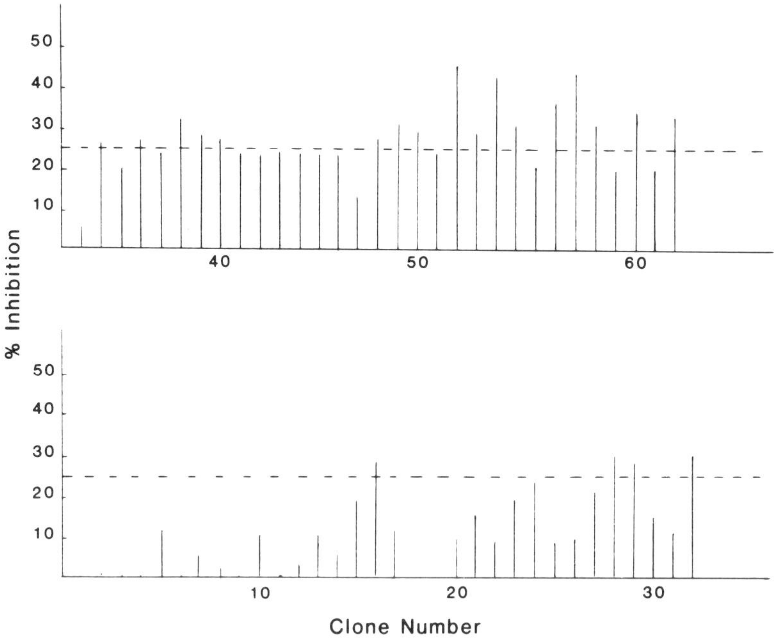

Figure 8.8. Legend for initial screening of BXM. Assay of supernatants from BXM hybrids. Flexible 96-well, microtiter dishes were coated with rabbit anti-bovine λ light chain by adding 20 ml of the antibody (50 μg/ml in PBS) to each of the wells, incubating at room temperature for 1 hour and then washing 3 times with PBS containing 1% bovine serum albumin (PBS-BSA). Supernatants were tested for the presence of bovine I_g by adding 20 ml of the supernatant and 20 ml of ^{125}I-labeled bovine I_gG (10 μCi/μg) to the antibody-coated wells. Those wells which received 20 ml of growth medium which had no contact with the hybrid cells and 20 ml of ^{125}I-labeled bovine I_gG served as controls. After 1 hour the wells were washed 3 times with PBS-BSA and counted in a γ counter. The count is inversely proportional to the bovine I_g content of the supernatant.

picture of the structure of human and mouse Ig and their genes. Antibody-secreting hybridomas have been produced by intraspecific fusion of myeloma cells with normal cells of the B lymphocyte lineage. Nevertheless, a general approach to the derivation of cell lines secreting monoclonal bovine Ig using normal cells of the B lymphocyte lineage has not been demonstrated in the bovine system. Hence the field of veterinary immunology has been without a source of monoclonal immunoglobulins for sequence studies and for the preparation and testing of class-specific antisera. Hybridoma technology has enabled us to solve this problem by the interspecific fusion of the mouse plasmacytoma line SP 2/0 with normal bovine spleen cells to produce stable hybrid cell lines that secrete monoclonal bovine Ig molecules. Interspecific hybridomas were produced because the unavailability of HAT-sensitive bovine myeloma cell lines made the more conventional intraspecific fusion infeasible.

Bovine spleen cells for fusion were obtained by mincing six randomly

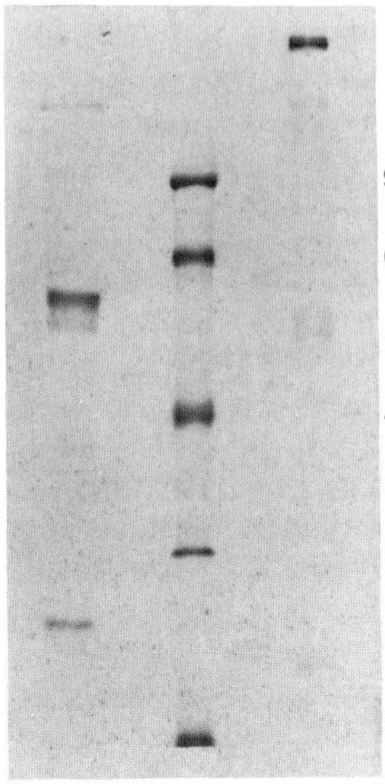

Figure 8.9. Autoradiograph of the labeled product secreted by the hybrid cell line and analyzed by SDS-PAGE. Hybrid cells (1×10^7) were cultured in 2 ml of methionine-free DME medium containing 1% fetal calf serum and 0.5 mCi ^{35}S-methoinine for 6 hours. The supernatant was applied on 7% polyacrylamide gel slabs before reduction (lane C) and after reduction (lane A). Lane B shows (methyl-^{14}C) methylated molecular weight standards.

selected cubes (~2 cm^2) of tissue taken from the spleen of an adult female Holstein cow. After washing the bovine lymphocytes twice in serum-free RPMI 1640, they were mixed in a 1:1 ratio with 1×10^8 SP 2/0 cells and co-pelleted by centrifugation at 400xg for 15 minutes at room temperature. The fusion was conducted along the outlines of the hybridization protocol detailed earlier in this paper. Between 14 and 21 days post-fusion, 63 single hybrid clones were obtained. Initial screening of supernatants from these hybrids revealed 21 clones secreting bovine Ig (Fig. 8.8).

After three cycles of subcloning, three cell lines continued to secrete bovine Ig in quantity (one line secretes 5–10 μg IgG_1/ml/5×10^5 cells/24 hours). A panel of monoclonal antibodies to bovine Ig classes was used to determine that, of the three lines stabilized, one secretes bovine IgG_1, one secretes bovine IgG_2, and the third is a bovine IgM secretor. Analysis of the products secreted by these bovine x mouse ($B \times M$) hybridomas by SDS-PAGE confirmed the

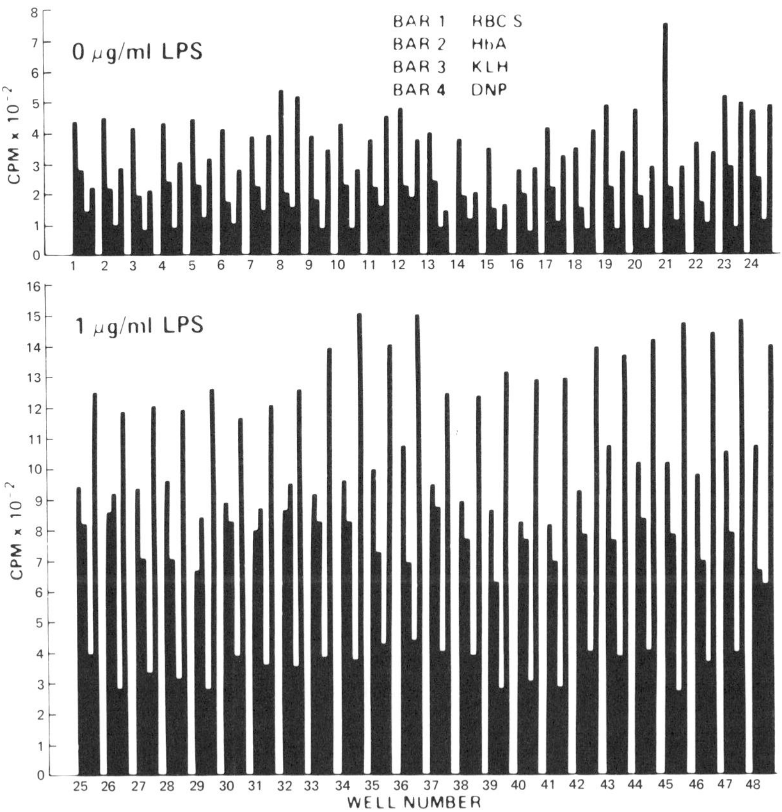

Figure 8.10. Polyclonal response of (Balb/c XSJL)F_1 spleen cells. After 96 hours, the fluids from each of 48 cultures stimulated by the indicated concentrations of LPS were harvested and tested against the indicated panel of antigens.

serological assays. All of the hybridomas produced Igs, which ran as single bands under nonreducing conditions and yielded light and heavy chains on reduction. Figure 8.9 illustrates the pattern obtained when the IgG_1 produced by a B × M hybridoma is analyzed by SDS-PAGE.

It is our hope that continuous cultures of B × M hybrid cell lines will supply homogeneous bovine Ig in virtually unlimited quantities. Such monoclonal bovine immunoglobulins will be useful as serological standards and as antigens of choice for the production of polyclonal and monoclonal antisera to bovine Ig isotypes for use in diagnostic assays. Monoclonal antibody-secreting hybrid cell lines can be used to provide material for the examination of problems heretofore not experimentally accessible. They can be expected to produce: (a) monoclonal Ig for sequencing studies; (b) homogeneous Ig for the precise serological and structural delineation of bovine Ig isotypes and allotypes; and (c) mRNA for the production of cDNA probes for the cloning of bovine Ig genes and the systematic analysis of Ig genes in the bovine genome.

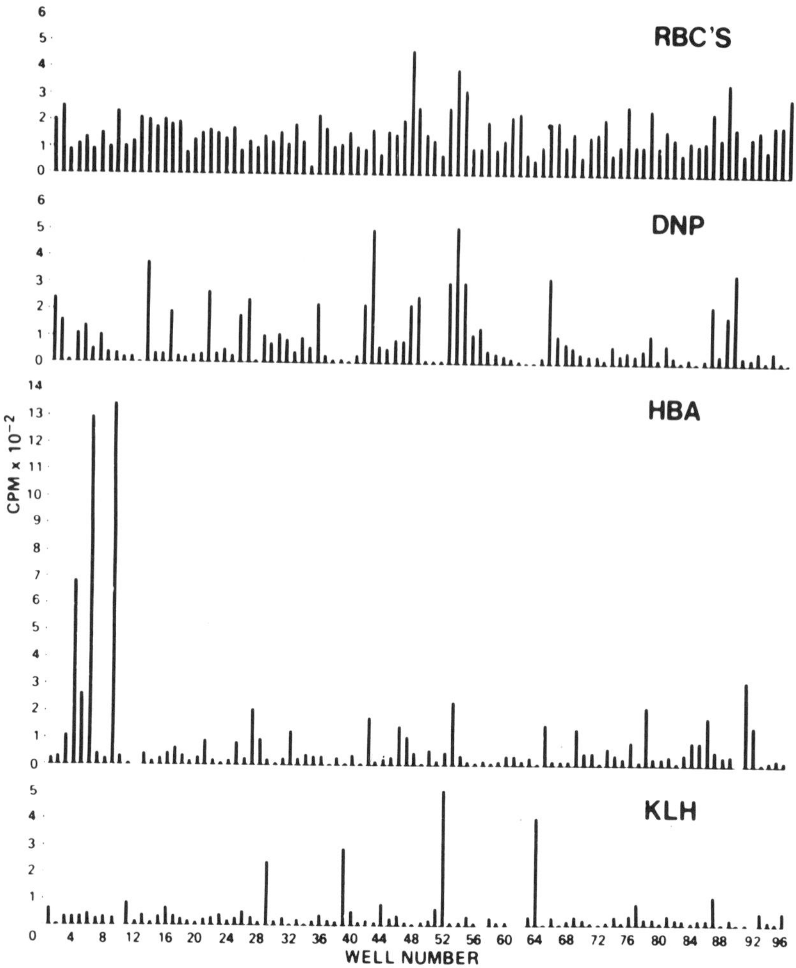

Figure 8.11. Preservation of the polyclonal response by hybridization. The fluids from 96 hybrid populations, each of which was made up of several (>6) clones, were assayed and the amount of antibody to the indicated antigen produced in each culture was determined.

THE PRODUCTION OF SPECIFIC ANTIBODY WITHOUT SPECIFIC IMMUNIZATION

The stimulation of B lymphocytes by bacterial lipopolysaccharides (LPS) results in their division and differentiation to yield cells that synthesize and secrete Ig (Anderson, Sojberg, and Moller 1972; Kearney and Lawton 1975a, b). Each mitogen-reactive B lymphocyte divides to produce a clone of cells that secretes Ig molecules containing a single set of variable regions and, therefore, are specific only for antigens of a given configuration. Because a

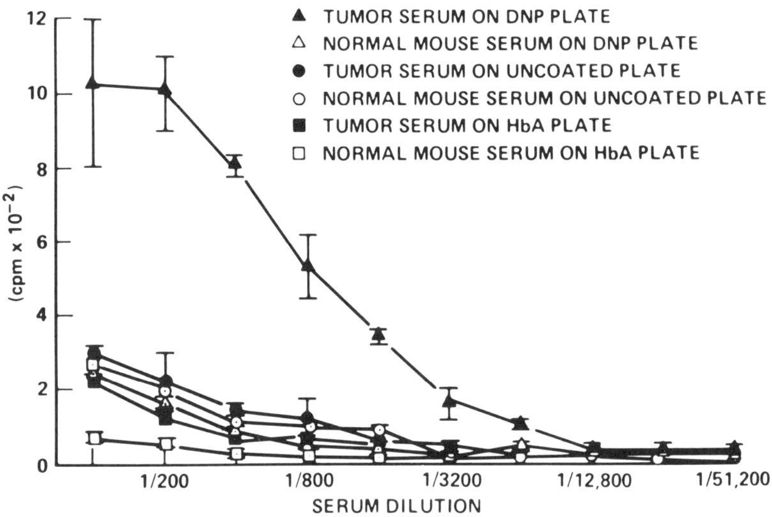

Figure 8.12. Antigen specificity of serum from an animal bearing an anti-DNP–secreting tumor. The antibody-containing serum was examined for antibody to DNP or human hemoglobin A in a plate-binding radioimmunoassay. Only plates coated with DNP-conjugated BSA and treated with dilutions of serum from tumor-bearing animals bound significant amounts of mouse I_g.

substantial fraction of a B lymphocyte population exposed to LPS is triggered to division (Andersson et al. 1977a, b), LPS-stimulated cultures are polyclonal. Furthermore, since each Ig-secreting clone of cells secretes Ig of a particular specificity, mitogen-activated cultures contain antibodies with an impressive variety of specificities for a broad range of antigens (See Fig. 8.10). Indeed, under appropriate conditions, the fraction of B lymphocytes stimulated in a population of mouse spleen cells may be large enough to constitute a representative sample of the total repertory of antibody specificities that reside in the entire set of B lymphocytes.

Due to cell death and other factors, however, the polyclonal response of LPS-stimulated cultures declines with the passage of time, and after 6–9 days ceases altogether. The transitory nature of polyclonal response prevents its direct exploitation as a strategy for the production of useful amounts of antibodies to particular antigens. It occurred to us that hybridoma technology might provide a means of preserving the polyclonal response for an indefinite period of time. The data of Figure 8.11 demonstrate that this is, in fact, the case. Furthermore, by resort to cloning it was possible to obtain monoclonal antibody production out of the polyclonal production of a population of LPS-stimulated spleen cell X myeloma hybrid clones. We chose to search out and isolate an anti-DNP producing hybridoma clone. The growth of this clone in mice as an antibody-producing tumor allowed us to obtain useful titers of specific antibody (see Fig. 8.12) without resort to any program of immunization whatsoever.

Clearly, the hybridization of LPS-stimulated B cells, in principle, permits the production of each B cell's characteristic Ig in quantity. Furthermore, it is important to emphasize that this approach allows one to immortalize and examine antibodies, such as autoantibodies, that are profoundly suppressed and cannot be elicited by immunization. Put another way, the hybridization of LPS-stimulated cultures suggests an avenue for the production, in large quantities if desired, of monoclonal antibodies to antigens that may not themselves be immunogenic in the system under investigation.

CONCLUSIONS

The examples presented here and the already enormous monoclonal antibody literature provide an ample demonstration of the impact hybrid hybridoma technology has had on experimental biology. As monoclonal serology becomes increasingly widely adopted, the useful but undefined conventional polyclonal antisera will be replaced by precisely defined monoclonal reagents. We can confidently expect that hybridoma technology will provide standard analytical and reference reagents for the fields of food technology, veterinary medicine, and agriculture. Already, the use of monoclonal probes has greatly sharpened our ability to ask questions in the fields of veterinary immunology, animal parasitology, and plant virology.

LITERATURE CITED

Andersson, J., O. Sojberg, and G. Moller. 1972. *Induction of immunoglobulin and antibody synthesis "in vitro" by lipopolysaccharides*. Eur. J. Immunol. 2:349–53.

Andersson, J., A. Coutinho, and F. Melchers. 1977a. *Frequencies of mitogen-reactive B cells in the mouse. I. Distribution in different lymphoid organs from different inbred strains of mice at different ages*. J. Exp. Med. 145:1511–19.

———. 1977b. *Frequencies of mitogen-reactive B cells in the mouse. II. Frequencies of B cells producing antibodies which lyse sheep or horse erythrocytes and trinitrophenylated or nitroiodophenylated sheep erythrocytes*. J. Exp. Med. 145:1520–30.

Barski, G., S. Sorieul, and F. Cornefert. 1961. *"Hybrid" type cell in combined cultures of two different mammalian cell strains*. J. Nat. Cancer Inst. 26:1269–77.

Cantor, H., and E. A. Boyse. 1975a. *Functional subclasses of T lymphocytes bearing different Ly antigens. I. The generation of functionally distinct T cell subclasses is a differentiative process independent of antigen*. J. Exp. Med. 141:1376–89.

———. 1975b. *Functional subclasses of T lymphocytes bearing different Ly antigens. II. Cooperation between subclasses of Ly+ cells in the generation of killer activity*. J. Exp. Med. 141:1390–99.

Froland, S. S., and J. B. Natvig. 1972. *Class, subclass and allelic exclusion of membrane-bound lg of human B-lymphocytes*. J. Exp. Med. 136:409–14.

Fu, S. M., R. J. Winchester, and H. G. Kunkel. 1974. *Occurrence of surface IgM, IgD and free light chains of human lymphocytes*. J. Exp. Med. 139:451–56.

Goldsby, R. A., and E. Zipser. 1969. *The isolation and replica plating of mammalian cell clones*. Exp. Cell Res. 54:271–75.

Goldsby, R. A., and N. Mandell. 1973. *The isolation and replica plating of cell clones*. Pages 261–68 in D. M. Prescott, ed., *Methods in Cell Biology*, vol. 7. Academic Press, New York.

Kearney, J. F., and A. R. Lawton. 1975a. *B-lymphocyte differentiation induced by lipopolysaccharide. I. Generation of cells synthesizing four major immunoglobulin classes*. J. Immunol. 115:671–76.

———. 1975b. *B-lymphocyte differentiation induced by lipopolysaccharide. II. Response of fetal lymphocytes*. J. Immunol. 115:677–81.

Kohler, G., and C. Milstein. 1975. *Continuous cultures of fused cells secreting antibody of predefined specificity*. Nature (Lond.) 256:495–97.

Littlefield, J-W. 1964. *Selection of hybrids from matings of fibroblasts in vitro and their presumed recombinants*. Science 145:709–10.

Nowinski, R., C. Berglund, J. Lane, M. Lostrom, I. Bernstein, W. Young, S. Hakemori, L. Hill, and M. Cooney. 1980. *Human monoclonal antibody against Forssman Antigen*. Science 210:537–39.

Paape, M. J., W. P. Wergin, A. J. Guidry, and R. E. Pearson. 1979. *Leukocytes—second line of defense against invading mastitis pathogens*. J. Dairy Sci. 62:135–53.

Pontecorvo, G. 1976. *Production of mammalian somatic cell hybrids by means of polyethylene glycol (PEG) treatment*. Somatic Cell Genet. 1:397–400.

Reinherz, E. L., and S. F. Schlossman. 1980. *The differentiation and functions of human T lymphocytes*. Cell 19:821–27.

Schwaber, J., and S. P. Cohen. 1973. *Human x mouse somatic cell hybrid clone secreting immunoglobulins of both parental types*. Nature (Lond.) 244:444–47.

Springer, T., G. Galfre, D. Secher, and C. Milstein. 1978. *Monoclonal xenogeneic antibodies to mouse leukocyte antigens: Identification of macrophage-specific and other differentiation antigens*. Curr. Top. Microbiol. Immunol. 81:45–50.

Weiss, M.C., and H. Green. 1967. *Human-mouse hybrid cell lines containing partial complements of human chromosomes and functioning human genes*. Proc. Nat. Acad. Sci. USA 58:1104–11.

Yerganian, G., and M. B. Nell. 1966. *Hybridization of dwarf hamster cells by UV-inactivated Sendai virus*. Proc. Nat. Acad. Sci. USA 55:1066–73.

four
DNA CLONING

9] Organization and Expression of Soybean Seed Protein Genes

by ROBERT B. GOLDBERG*

ABSTRACT

My laboratory has been investigating soybean seed protein gene organization and expression in order to understand the molecular events that regulate gene activity during plant development. Clones representing glycinin, β-conglycinin, Kunitz trypsin inhibitor, lectin, and 15-kilodalton (kD) protein mRNAs were selected from an embryo cDNA library. Solution and DNA gel blot hybridization studies showed that each seed protein mRNA is encoded by a small gene family that is not selectively amplified during development. DNA/RNA hybridization experiments demonstrated that seed protein gene expression is under strict developmental control. Seed protein genes are expressed during defined embryonic periods and are inactive in the organ systems of the mature plant at both the nuclear and cytoplasmic levels. Hybridization experiments with "runoff" RNA synthesized in vitro by isolated nuclei suggested that seed protein gene expression is regulated primarily at the transcriptional level. Seed protein genes were selected from libraries of soybean genomic DNA constructed in phage lambda. R-loop and S-1 nuclease analyses showed that the Kunitz trypsin inihibitor, lectin, and 15-kD protein genes have no detectable introns. On the other hand, glycinin genes have at least one, 0.7-kilobase pair (kbp) intron, while β-conglycinin genes have several, small (< 0.1 kbp), introns. Preliminary "walking" studies have shown that intrafamily linkage of glycinin, β-conglycinin, and Kunitz trypsin inhibitor genes does occur. Although members of different seed protein gene families are not dispersed among each other, several seed protein genes are closely linked to unidentified genes that are regulated differently during the life cycle. To begin to identify sequences necessary for seed protein gene expression, the molecular basis of a mutation which blocks lectin gene expression was studied. Our experiments showed that the mutant lectin gene is not transcribed at detectable levels, and that it possesses a 3.5-kbp insertion sequence which resembles a transposable element.

*Department of Biology, University of California, Los Angeles, California 90024.

GENETIC ENGINEERING: APPLICATIONS TO AGRICULTURE (Beltsville Symposium 7—Lowell D. Owens, ed.) Rowman & Allanheld, Totowa

Table 9.1 Characteristics and Identity of Seed Protein mRNA Clones

cDNA clone[a]	Insert size[a] (kbp)	Size of[a] complementary mRNA (kb)	Size of[b] polypeptides synthesized from hybrid-selected mRNA (kD)	Antibody which[b] precipitates polypeptides synthesized in vitro
A-16	0.68	2.5 (major) 1.7 (minor)	83 (major) 53 (minor)	β-conglycinin
A-28	0.72	2.1	60 56	glycinin
A-37	0.75	0.90	23	Kunitz trypsin inhibitor
L-9	0.30	1.1	33.5	seed lectin
A-36	0.69	0.85	15.5	?

[a]Taken from Goldberg et al. 1981a; and from R. B. Goldberg, G. Hoschek, and L. O. Vodkin, unpublished.

[b]Unpublished results of M. L. Crouch, L. O. Vodkin, and R. B. Goldberg. cDNA clone identities were confirmed independently by direct sequencing (N.E. Nielsen, personal communication; L. O. Vodkin, P. Rhodes, R. B. Goldberg, unpublished), from the results of others (Tumer, Thanh, and Nielsen 1981; Beachy, Barton, and Jarvis 1981; Vodkin 1981; Barton et al. 1982), and/or by mRNA gel blot studies with mutant soybean lines (R. B. Goldberg and L. O. Vodkin, unpublished).

INTRODUCTION

This chapter will review briefly the research that has been carried out in my laboratory on the organization and expression of soybean seed protein genes. These genes constitute only a tiny fraction of the 20,000 genes which are expressed in soybean cells (Goldberg et al. 1981b), but they direct the synthesis of several, superprevalent mRNA classes which collectively represent approximately half the embryo mRNA mass at a given developmental stage (Goldberg et al. 1981a, 1981b). The genes I will focus on encode the storage proteins glycinin and β-conglycinin, the Kunitz trypsin inhibitor, seed lectin, and an unidentified 15,500-dalton protein. The storage proteins are rapidly degraded after seed germination and serve as a "food" supply for the rapidly growing seedling (Larkins 1981). The physiological functions of seed lectin and the Kunitz trypsin inhibitor are not well understood, but they have been implicated in several host/pathogen related processes (Lis and Sharon 1981; Ryan 1981).

My laboratory has been investigating seed protein genes in order to understand the molecular events that control gene expression during plant development. If molecular genetic engineering procedures are to be used to improve crop plants, it is important to introduce genes into plant chromosomes so that they are expressed in the correct developmental context. Deciphering the regulatory circuitry which controls plant gene expression is a crucial first step in this process.

SEED PROTEIN GENE EXPRESSION

Cloning seed protein mRNAs. To study the organization and expression of specific seed protein genes, a complementary DNA (cDNA) library of mid-maturation stage embryo mRNA was constructed in the plasmid pBR322 (Goldberg et al. 1981a). At this developmental stage, seed protein mRNAs are highly prevalent (Goldberg et al. 1981b), and the average seed protein accumulation rates are high (Hill and Briedenbach 1974; Sengupta et al. 1981). An abundant cDNA fraction was used to select seed protein mRNA clones, and a combination of restriction endonuclease analyses, mRNA gel blot studies, and hybrid-selected translation experiments were used to correlate the selected cDNA clones with specific seed protein mRNAs (Goldberg et al. 1981a). The characteristics of cDNA clones representing glycinin, β-conglycinin, Kunitz trypsin inhibitor, seed lectin, and 15-kilodalton (kD) protein mRNAs are presented in Table 9.1. With the exception of A-16, each cDNA clone represents a single, unique-sized embryo mRNA class (Goldberg et al. 1981a). A-16, a β-conglycinin cDNA clone, is complementary to two, distinct-sized mRNA classes. As I will discuss below, the reason for this is that β-conglycinin is coded for by a complex gene family that directs the synthesis of homologous, discrete-sized mRNAs.

Temporal regulation of seed protein gene expression during embryogenesis. What is the temporal program of seed protein gene expression during embryogenesis? To answer this question mRNAs were isolated from different embryonic stages, and the prevalence of each seed protein message was estimated by cDNA excess DNA/RNA titration experiments in solution (Goldberg et al. 1981a). The results, summarized in Figure 9.1 and Table 9.2, show that there is striking temporal control of seed protein gene expression at the mRNA level. Each seed protein mRNA class accumulates during early-maturation, and then decays during late-maturation when dehydration occurs. At dormancy, and in post-germination cotyledons, seed protein mRNAs have diminished to the level of only a few molecules per cell (Table 9.2). Since the accumulation and diminution of seed protein mRNAs correspond roughly with changes in the average seed protein deposition rates (Hill and Briedenbach 1974), the rate-limiting step in seed protein synthesis appears to be the cytoplasmic availability of seed protein mRNA—at least as a first approximation. These data do not rule out, however, the possibility that translational-level controls are superimposed upon newly synthesized seed protein mRNAs.

Regulation of seed protein gene expression with respect to embryonic cell type. Is seed protein gene expression regulated with respect to embryo cell

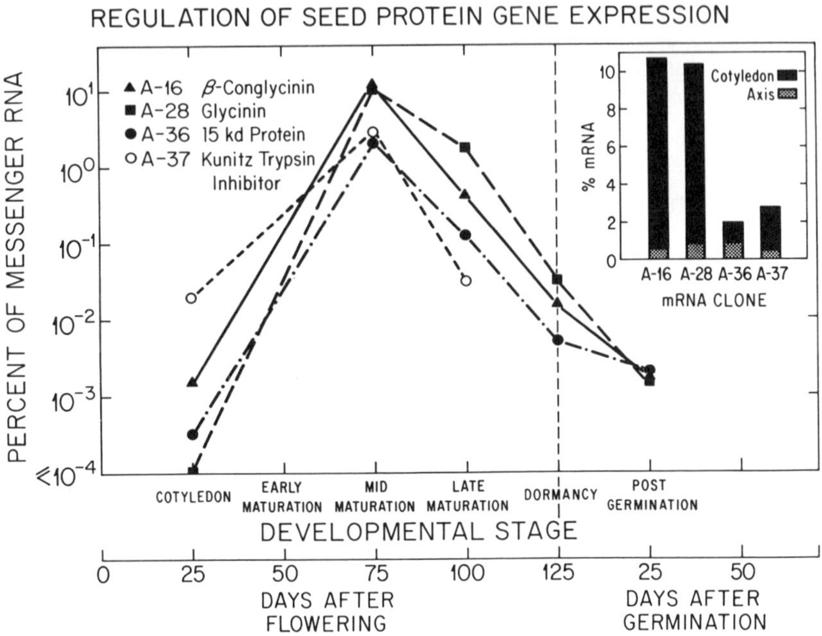

Figure 9.1. Developmental regulation of seed protein gene expression. (Adapted from Goldberg et al. 1981a.)

type? Soybean embryos possess two developmentally distinct organ systems. The cotyledons, which contain 80% of the three million embryo cells (Goldberg et al. 1981b), are responsible primarily for food reserve synthesis during embryogenesis and senesce after seed germination. The axis, on the other hand, possesses the shoot and root meristems and develops into the mature plant at the end of dormancy. Table 9.2, and the inset to Figure 9.1, compares seed protein mRNA prevalences in axis and cotyledon cells at the mid-maturation stage of development. These data show that there are striking differences. While glycinin, β-conglycinin, Kunitz trypsin inhibitor, and 15-kD protein mRNAs comprise approximately 25% of the cotyledon mRNA mass, they constitute less than 3% of the axis message population (Goldberg et al. 1981a). Thus, not only is seed protein gene expression regulated temporally in development, but it is regulated with respect to embryo cell type as well.

Seed protein gene expression in the mature plant. Are seed protein genes expressed in the cells of the mature plant? To answer this question, mRNAs were isolated from leaf, root, and stem polysomes, and the representation of seed protein gene transcripts was measured by (a) hybridization of seed protein cDNA plasmids to RNA gel blots (Goldberg et al. 1981a), (b) plasmid DNA excess, solution titration studies (Goldberg et al. 1981a), and (c) hybridization of labeled, random-primed leaf, root, and stem cDNAs with DNA gel

Table 9.2 Representation of Seed Protein Gene Transcripts in Embryo and Leaf RNA Populations

	Number of Molecules per Cell				
RNA	Glycinin	β-Conglycinin	Kunitz Trypsin Inhibitor	15-kD Protein	Lectin[a]
Cotyledon stage mRNA	< 0.30[b]	4	125	2	—
Mid-maturation stage					
a. Whole embryo mRNA	27,000	23,000	16,000	12,000	3,000
b. Axis mRNA	525	300	450	1,400	—
c. Whole embryo nuclear RNA	2,900	2,500	200	1,500	950
Late-maturation mRNA	1,560	325	70	275	—
Dry seed poly(A) RNA	10	5	—	4	—
Post-dormancy mRNA	0.5	0.5	—	2	—
Leaf					
a. mRNA	< 0.005[b]	0.05	< 0.005[b]	< 0.005[b]	—
b. Nuclear RNA	< 0.20[b]	< 0.20[b]	< 0.20[b]	< 0.20[b]	—

[a]Unpublished results of R. B. Goldberg, G. Hoschek, and L. O. Vodkin. mRNA gel blot studies showed that lectin mRNA accumulates and decays in the same developmental framework as other seed protein messages.
[b]No transcripts were detected. Represents the upper limit of mRNA sequences which would have been detected in each experiment.

Source: Taken from the data shown in Figure 9.1, and Goldberg et al. 1981a. Developmental stages were described in Goldberg et al. 1981b.

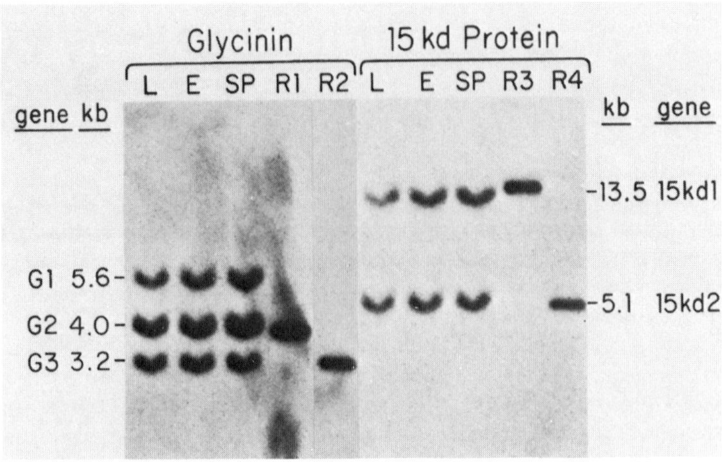

Figure 9.2. Hybridization of glycinin and 15-kD protein cDNA clones with DNA blots. L, E, and SP refer to leaf, embryo, and single-plant leaf DNA lanes, respectively. R lanes contain a single-copy equivalent of each restriction endonuclease fragment. (Adapted from Fischer and Goldberg 1982).

blots containing seed protein genome clones (Fischer and Goldberg 1982; J. K. Okamuro and R. B. Goldberg, unpublished). No physiologically meaningful levels of seed protein mRNAs were observed, even though transcripts present at less than 1 molecule per 10 cells would have been detected easily (Goldberg et al. 1981a; Fischer and Goldberg 1982). In addition, seed protein mRNAs were not detected in a nonpolysomal cytoplasmic compartment (Goldberg et al. 1981b). Together, these data show that seed protein genes are not expressed at the cytoplasmic level in the cells of the mature plant, and appear to be expressed only at defined embryonic stages (see Fig. 9.1).

Are seed protein genes expressed at the nuclear level in the organ systems of the mature plant? To address this question leaf nuclear RNA was isolated, and the presence of seed protein gene transcripts was measured by DNA-excess, solution titration studies with seed protein cDNA plasmids (Goldberg et al. 1981a). The results, presented in Table 9.2, indicate that the leaf steady-state nuclear RNA population contains no detectable seed protein gene transcripts.

Seed protein gene expression is controlled primarily at the transcriptional level. The experimental findings outlined above strongly suggest that seed protein genes are activated during early embryogenesis, are repressed during late embryogenesis, and remain inactive during the post-germination phase of the life cycle. To test this "on-off" model it is necessary to measure the actual rates of seed protein mRNA synthesis during embryogenesis, rather than simply monitoring the accumulation and diminution of cytoplasmic mRNAs. Since embryos develop in seed pods of the mature plant, the pulse-labeling experiments required to measure transcription rates cannot be performed readily.

To circumvent this problem a surrogate approach was used. Leaf nuclei were isolated, and preinitiated RNA chains were allowed to "run off" in the presence of ^{32}P-UTP (McKnight and Palmiter 1979; Luthe and Quatrano 1980a, b; Groudine, Peretz, and Weintraub 1981). The labeled nuclear "runoff" RNA was isolated, and then hybridized to filter-bound cDNAs (cloned in plasmids) of seed protein mRNAs and/or DNA gel blots containing seed protein genomic clones (L. Walling and R. B. Goldberg, unpublished). The rationale for this approach is that a transcribed gene should have RNA polymerase molecules bound to it. When placed in an in vitro system the RNA polymerases should cause chain elongation to occur, releasing labeled, nascent RNA molecules into the nucleoplasm. On the other hand, a "repressed" gene should have no bound RNA polymerases, and should not contribute detectably to the labeled "runoff" nuclear RNA population. No significant hybridization signals were observed with any seed protein gene sequence (L. Walling and R. B. Goldberg, unpublished). This finding, and that obtained with leaf steady-state nuclear RNA (Table 9.2), strongly suggests that seed protein genes are "off" in leaf cells, and that seed protein gene expression is controlled primarily at the transcriptional level.

To determine the developmental periods in which seed protein genes are activated and repressed, labeled nuclear "runoff" RNAs from embryos at different embryonic periods were hybridized with filter-bound, seed protein plasmid DNAs. Preliminary findings indicated that glycinin genes are switched "on" during early-maturation, and that lectin genes are turned "off" during mid-maturation (L. Walling and R. B. Goldberg, unpublished). In general, the relative seed protein RNA synthesis rates increase during pre- to early-maturation, and then decrease during mid- to late-maturation (L. Walling and R. B. Goldberg, unpublished). Assuming that in vitro nuclear "runoff" transcription studies reflect actual in vivo events, these findings strongly support the proposition that seed protein genes are activated and repressed during embryogenesis, and that they remain repressed after dormancy ends.

SEED PROTEIN GENE ORGANIZATION

Seed proteins are encoded by small gene families. How many copies of glycinin, β-conglycinin, seed lectin, Kunitz trypsin inhibitor, and 15-kD protein genes are there in soybean chromosomes? To address this question two different experimental approaches were used. First, labeled, single-stranded probes representing the anticoding strand of each plasmid cDNA insert were hybridized with excess, unlabeled genomic DNA, and the hybridization kinetics were measured (Goldberg et al. 1981a). Second, labeled plasmids carrying cDNAs of seed protein mRNAs were reacted with DNA gel blots containing restriction endonuclease digested genomic DNA (Fischer and Goldberg 1982; R. B. Goldberg et al., unpublished). A representative example of the latter strategy is shown in Figure 9.2, and the results of both approaches are summarized in Table 9.3. It can be seen that each seed protein gene is represented a small number of times in soybean chromosomes. In addition,

Table 9.3 Reiteration Frequency of Soybean Seed Protein Genes

Gene	DNA	Copy Number[a]	
		Kinetic method[b]	DNA blot method[c]
Glycinin	Leaf	4	3
	Embryo	—	3
β-Conglycinin	Leaf	5	10
	Embryo	5	10
Kunitz trypsin inhibitor	Leaf	5	2
	Embryo	—	2
Seed lectin	Leaf	2	2
	Embryo	2	2
15-kD protein	Leaf	2	2
	Embryo	—	2

[a]May represent minimal estimates since sequences > 30% divergent from a given cDNA probe would not have been detected by the hybridization conditions employed.
[b]Taken from Goldberg et al. 1981a; and from R. B. Goldberg and L. O. Vodkin, unpublished.
[c]Taken from Fischer and Goldberg 1982; and from R. B. Goldberg, J. J. Harada, D. Jofuku, G. Hoschek, and L. O. Vodkin, unpublished.

similar results were obtained with both leaf and mid-maturation–stage embryo DNAs. Together, these findings show that seed proteins are coded for by small, nonhomologous gene families, and that these families are not selectively amplified or altered during embryogenesis.

Seed protein genes have relatively simple intragenic structures. To characterize seed protein gene structures, two λ Charon 4 libraries of soybean DNA were constructed. One library contains DNA fragments prepared by partial Eco RI digestion (Fischer and Goldberg 1982), while the other contains partially digested Hae III/Alu I DNA fragments inserted into the vector with Eco RI linkers (D. Jofuku and R. B. Goldberg, unpublished). Genomic clones containing seed protein genes were identified by plaque hybridization with seed protein cDNA plasmids (Table 9.1). Seed protein gene structures were then visualized in the electron microscope by R-loop analysis (Fischer and Goldberg 1982; R. B. Goldberg et al. unpublished).

A collage of different seed protein genes is presented in Figure 9.3. No introns are visible in the Kunitz trypsin inhibitor, seed lectin, and 15-kD protein genes. S-1 nuclease mapping studies with the 15-kD protein and seed lectin genes confirmed these observations (Fischer and Goldberg 1982; R. B. Goldberg et al., unpublished), indicating that if introns are present they must be very short (< 0.05 kbp) and located near the gene ends. The top panel of Figure 9.3 shows that three different glycinin genes, G1, G2, and G3 (see Fig. 9.2), each have one, prominent intron. While the intron position appears to be

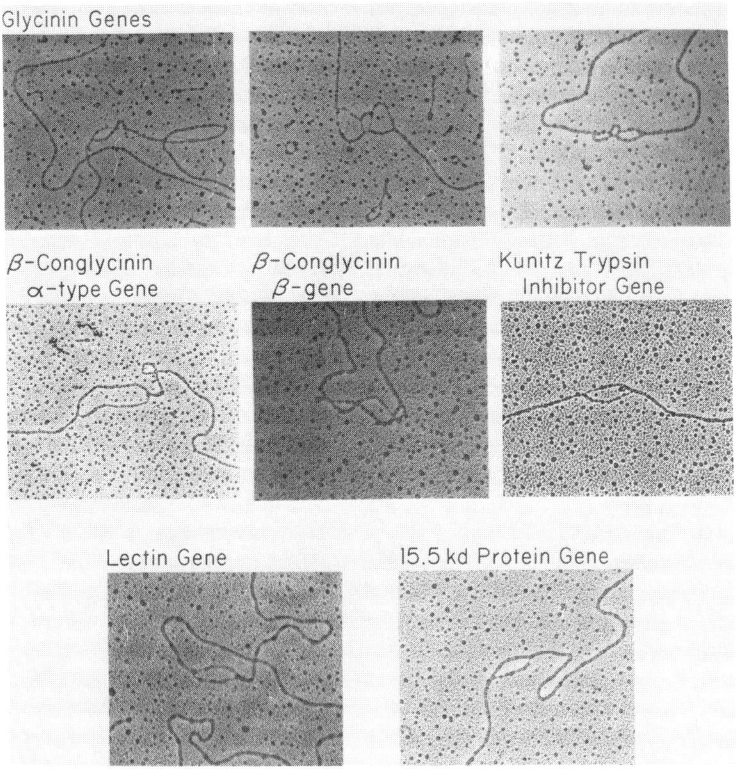

Figure 9.3. Visualization of seed protein gene structures.

similar in each gene, intron sizes vary from 0.5 kbp to 0.7 kbp (Fischer and Goldberg 1982; N. C. Nielsen, personal communication). In addition two smaller introns are also present in these genes (R. Fischer and R. Goldberg, unpublished).

In contrast to the relative structural uniformity of glycinin genes, β-conglycinin storage protein genes differ from each other in size and structural organization (J. J. Harada and R. B. Goldberg, unpublished). The α-type β-conglycinin gene displayed in Figure 9.3 shows strong homology with the 2.5-kilobase (kb) β-conglycinin mRNA size class (see Table 9.1). This gene is approximately 3 kbp in length and has no visible introns; however, the presence of short intervening sequences (~0.1 kbp) has been inferred from restriction endonuclease mapping studies. The β-type gene shown in Figure 9.3 is most closely related to the 1.7-kb β-conglycinin mRNA size class (see Table 9.1). This gene is approximately 2 kbp in length and has a few, very small introns. These data show that β-conglycinin gene structures vary, and that the 2.5-kb and 1.7-kb β-conglycinin mRNAs are synthesized from different genes, rather than from differential processing of a common pre-mRNA molecule.

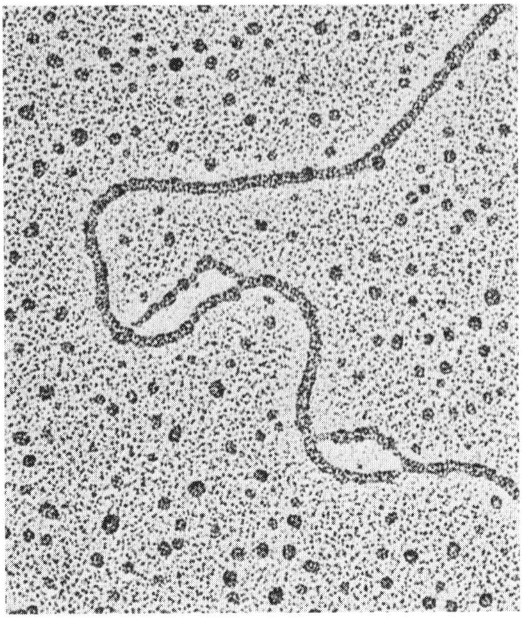

Figure 9.4. Linkage of two Kunitz trypsin inhibitor genes.

Organization of seed protein gene families. How are seed protein gene families organized in soybean chromosomes? To answer this question a series of chromosome "walks" are currently being carried out. Figure 9.4 shows a genomic clone which contains two Kunitz trypsin inhibitor genes separated by about 1 kbp, indicating that at least two Kunitz trypsin inhibitor family members are clustered (D. Jofuku and R. B. Goldberg, unpublished). Similarly, a genomic clone has been isolated which contains the G1 and G2 glycinin genes linked within 5 kbp (R. L. Fischer and R. B. Goldberg, unpublished), and three different genomic clones have been characterized which contain two β-conglycinin genes each (J. J. Harada and R. B. Goldberg, unpublished). These data indicate that intrafamily linkage of some Kunitz trypsin inhibitor, glycinin, and β-conglycinin genes does occur. Whether all members of each family are clustered, and whether more than one cluster exists as a consequence of ancient tetraploidization events, remains to be established.

Differentially regulated genes are contiguous to seed protein gene regions. Irrespective of the exact organizational pattern of each seed protein family, members of different families are not dispersed among each other (Fischer and Goldberg 1982; R. B. Goldberg et al., unpublished)—at least not at the distances investigated to date (20–50 kbp). Are there nonseed protein genes in the vicinity of highly regulated seed protein gene regions? To answer this question, labeled cDNAs were prepared from random-primed leaf, root, and

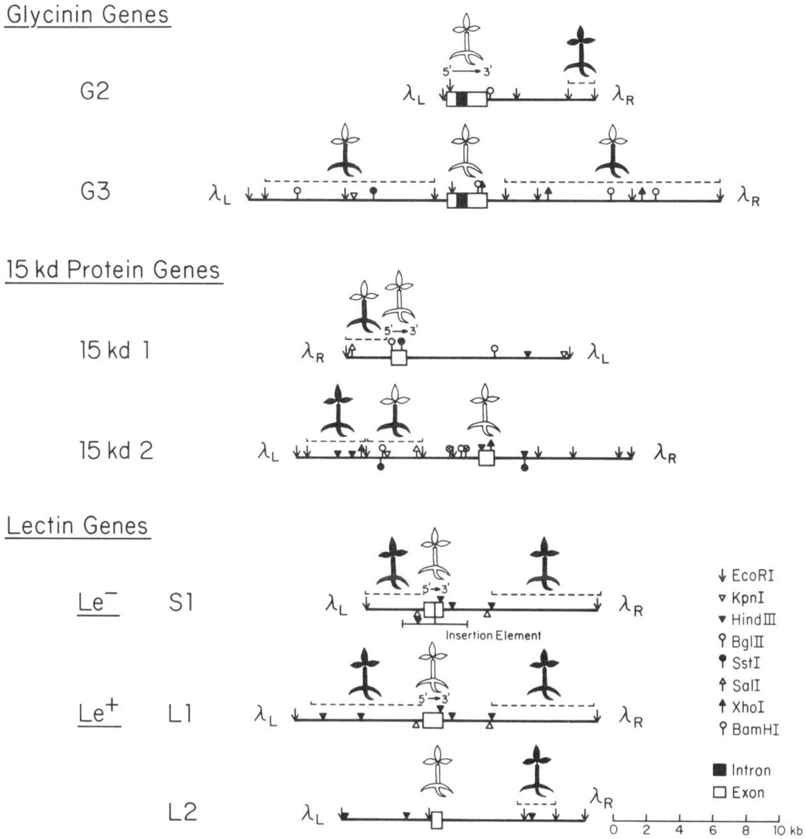

Figure 9.5. Nonseed protein genes linked to seed protein gene regions. Plants summarize the expression pattern for each gene region. Dark areas indicate that the gene is expressed in that organ system, while light areas indicate that the gene is inactive at the mRNA level.

stem mRNAs and hybridized with DNA gel blots containing seed protein genomic clones (Fischer and Goldberg 1982; J. K. Okamuro and R. B. Goldberg, unpublished). The relevant findings are portrayed in Figure 9.5, and show that genes expressed at the mRNA level in the mature plant flank several different seed protein gene regions. Gel blot studies were carried out to measure the sizes and prevalences of the nonseed protein mRNAs. Representative mRNA blots, shown in Figure 9.6, indicate that each nonseed protein mRNA has a unique size, and that it is present only a few times per cell. Since seed protein genes are not expressed in the mature plant, these data show that there is a short range clustering of genes that is expressed at distinct periods, and at quantitatively different levels within the soybean life cycle. This finding implies that each gene of the cluster possesses a tightly linked sequence that is responsible for regulating its expression during development.

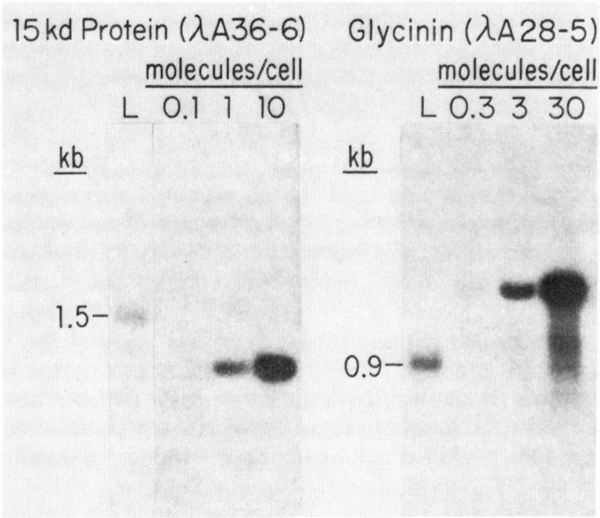

Figure 9.6. Hybridization of glycinin and 15-kD protein genomic clones with leaf mRNA gel blots. The reconstruction lanes contain known quantities of mid-maturation stage embryo mRNA. (Adapted from Fischer and Goldberg 1982.)

MOLECULAR BASIS OF A SEED PROTEIN GENE MUTATION

The central question is, of course, what are the sequences responsible for the cell-specific expression of plant genes? While the results discussed in the previous section strongly suggest that these sequences are closely linked to their contiguous structural genes, the problem is to identify them. To approach this question directly it is necessary to perform transformation studies with in vitro mutated, cloned gene regions (for example, see Palmiter, Chen, and Birnster 1982; Robins et al. 1982). Since widely applicable transformation procedures have not yet been developed for plant cells, a different approach was employed—an investigation of the structure and expression of a mutant seed lectin gene found naturally in soybean populations (Pull et al. 1978).

Several soybean lines have been identified that lack detectable amounts of seed lectin, both in the dry seed (Pull et al. 1978; Su, Pueppke and Friedman 1980) and in the mature plant (Pueppke et al. 1978). What is the molecular basis for the absence of seed lectin in these lines? Several different experimental approaches were used to answer this question, utilizing both lectin cDNA and genomic clones (R.B. Goldberg, G. Hoschek, and L.O. Vodkin, unpublished). The results showed that (a) seed lectin mRNA represents approximately 0.6% of the Le^+ mid-maturation stage mRNA population, and is 1.1 kb in length (see Tables 9.1 and 9.2); (b) seed lectin mRNA is detectable only at defined embryonic stages in Le^+ plants, and seed lectin gene expression appears to be controlled at the transcriptional level; (c) no physiologically meaningful levels of seed lectin gene transcripts are found in the nucleus and cytoplasm of Le$^-$ cells, suggesting that the molecular defect is either at the level of transcription or posttranscriptional processing; (d) Le$^+$ and Le$^-$ plants possess two related lectin genes—one gene is the "functional" seed lectin gene, while the other is either an inactive pseudogene or is expressed under physiological conditions yet to be defined (Figure 9.5); the latter gene is similar in both Le$^+$ and Le$^-$ plants; (e) the Le$^-$ "functional" seed lectin gene contains a 3.5-kbp insertion sequence that divides the gene in half (Figure 9.5); (f) sequencing studies suggest that this insertion resembles a transposable element (L. O. Vodkin, P. Rhodes, and R. B. Goldberg, unpublished).

CONCLUSIONS

Soybean seed protein genes represent an excellent model for the study of plant gene regulation. These structurally "simple" genes are organized into discrete gene families, and are under strict developmental control. The sequences that control the expression of these genes, and the molecules that interact with the regulatory regions, remain to be identified.

ACKNOWLEDGMENTS

I express deep appreciation to my technician Gisela Hoschek, my students Diane Jofuku and Jack Okamuro, and my former and current postdoctoral associates Gary Ditta, Marti Crouch, Bob Fischer, John Harada, Linda Walling, and Tom Sims for their time, labor, and insight into the questions addressed in this paper. These studies were supported by grants from the National Science Foundation and the U.S. Department of Agriculture.

LITERATURE CITED

Barton, K. A., J. F. Thompson, J. T. Madison, R. Rosenthal, N. P. Jarvis, and R. N. Beachy. 1982. *The biosynthesis and processing of high molecular weight precursors of glycinin subunits*. J. Biol. Chem. 257: 6089–95.

Beachy, R. N., N. P. Jarvis, and K. A. Barton. 1981. *Biosynthesis of the soybean 7S protein*. J. Mol. Appl. Genet. 1:19–27.

Fischer, R. L., and R. B. Goldberg. 1982. *Structure and flanking regions of soybean seed protein genes*. Cell 29:651–60.

Goldberg, R. B., G. Hoschek, G. S. Ditta, and R. W. Breidenbach. 1981a. *Developmental regulation of cloned superabundant embryo mRNAs in soybean*. Dev. Biol. 83:218–31.

Goldberg, R. B., G. Hoschek, S. H. Tam, G. S. Ditta, and R. W. Breidenbach. 1981b. *Abundance, diversity, and regulation of mRNA sequence sets in soybean embryogenesis*. Dev. Biol. 83: 201–17.

Groudine, M., M. Peretz, and H. Weintraub. 1981. *Transcriptional regulation of hemoglobin switching in chicken embryos*. Mol. Cell. Biol. 1:281–88.

Hill, J. E., and R. W. Breidenbach. 1974. *Proteins of soybean seeds. II. Accumulation of the major protein components during seed development*. Plant Physiol. 53:747–51.

Larkins, B.A. 1981. *Seed storage proteins: Characterization and biosynthesis*. Pages 449–89 *in* A. Marcus, ed., *Biochemistry of Plants*, vol. 6. Academic Press, New York.

Lis, H., and Sharon, N. 1981. *Lectins in higher plants*. Pages 371–433 in A. Marcus, ed., *Biochemistry of Plants*, vol. 6. Academic Press, New York.

Luthe, D. S., and Quatrano, R. S. 1980a. *Transcription in isolated wheat nuclei I. Isolation of nuclei and elimination of endogenous ribonuclease activity*. Plant Physiol. 65:305–308.

———. 1980b. *Transcription in isolated wheat nuclei II. Characterization of RNA synthesized in vitro*. Plant Physiol. 65:309–13.

McKnight, G. S., and R. D. Palmiter. 1979. *Transcriptional regulation of the ovalbumin and conalbumin genes by steroid hormones in chick oviduct*, J. Biol. Chem. 254:9050–58.

Palmiter, R. D., H. Y. Chen, and R. L. Birnster. 1982. *Differential regulation of metallothionein-thymidine kinase fusion genes in transgenic mice and their offspring*. Cell 29:701–10.

Pueppke, S. G., W. D. Bauer, K. Keegstra, and A. L. Ferguson. 1978. *Role of lectin in plant-microorganism interactions II. Distribution of soybean lectin in tissues of Glycine max*. Plant Physiol. 61:779–84.

Pull, S. P., S. G. Pueppke, T. Hymowitz, and J. H. Orf. 1978. *Soybean lines lacking the 120,000 dalton seed lectin*. Science 200:1277–79.

Robins, D. M., I. Paek, P. H. Seeburg, and R. Axel. 1982. *Regulated expression of human growth hormone genes in mouse cells*. Cell 29:623–31.

Ryan, C. A. 1981. *Plant proteinases*. Pages 351–69 in A. Marcus, ed., *Biochemistry of Plants*, vol. b. Academic Press, New York.

Sengupta, G., V. DeLuca, D. Bailey, and D. P. S. Verma. 1981. *Biosynthesis of soybean storage proteins: Post-translational processing of 7S and 11S components*. Plant Mol. Biol. 1:19–34.

Su, L. C., S. G. Pueppke, and H. P. Friedman. 1980. *Lectins and the soybean-Rhizobium symbiosis I. Immunological investigation of soybean lines, the seeds of which have been reported to lack the 120,000 dalton soybean lectin*. Biochim. Biophys. Acta 629:292–304.

Tumer, N. E., V. K. Thanh, and N. C. Nielsen. 1981. *Purification and characterization of mRNA from soybean seeds. Identification of glycinin and β-conglycinin precursors*. J. Biol. Chem. 256:8756–60.

Vodkin, L. O. 1981. *Isolation and characterization of mRNAs for seed lectin and Kunitz trypsin inhibitor in soybean*. Plant Physiol. 68:766–71.

10] Expression of α-Amylase Genes in Barley Aleurone Cells

by S. MUTHUKRISHNAN* and G. R. CHANDRA†

ABSTRACT

The induction of α-amylase by the plant hormone gibberellic acid (GA) in barley aleurones has been studied using a cloned α-amylase DNA probe. The cDNA clone was prepared by reverse transcription of an enriched α-amylase mRNA preparation, followed by insertion of the cDNA into pBR322 DNA by the standard "G-C tailing" procedures. Changes in levels of α-amylase mRNA in aleurones have been followed by probing RNA blots with the cloned α-amylase probe. α-Amylase mRNA becomes detectable in aleurones within an hour after addition of the hormone and continues to increase with time. Addition of 30 μM cycloheximide along with GA at the beginning of the incubation results in nearly complete inhibition of the appearance of α-amylase mRNA. Delaying the addition of cycloheximide results in a partial recovery of synthesis of α-amylase mRNA. GA appears to be required for the induction of α-amylase mRNA in isolated aleurones. The mechanism by which GA brings about an increase in levels of α-amylase in this tissue appears to be mediated through the synthesis of new mRNA for α-amylase.

INTRODUCTION

During germination of barley seeds, gibberellic acid (GA) synthesized by the embryo and scutellum diffuses to the aleurone layer cells, where it induces the synthesis of a variety of enzymes including α-amylase (Yomo and Varner 1971; Jones and Jacobsen 1978). This enzyme, which is responsible for the breakdown of the starch reserves of the endosperm, is synthesized *de novo* from precursor amino acids and accounts for a large percentage of the proteins secreted into the medium (Mozer 1980). Following addition of GA, there is an

*Department of Biochemistry, Kansas State University, Manhattan, Kansas 66506, and †Seed Research Laboratory, U.S. Department of Agriculture, Beltsville, Maryland 20705.

GENETIC ENGINEERING: APPLICATIONS TO AGRICULTURE (Beltsville Symposium 7—Lowell D. Owens, ed.) Rowman & Allanheld, Totowa

increase in labeling of polyadenylated (poly (A^+)) RNA in barley aleurones (Jacobsen and Zwar 1974; Ho and Varner 1976). In vitro translation experiments have revealed that treatment of aleurones with GA results in an increase in recovery of translatable α-amylase mRNA. Studies with cordycepin, an inhibitor of RNA synthesis, have also supported the notion that GA induces the synthesis of new mRNA for α-amylase (Ho and Varner 1976; Muthukrishnan, Chandra, and Maxwell 1979). By the use of cDNA prepared from an enriched α-amylase mRNA fraction, it has been shown that GA treatment results in an increase in the amount of complementary mRNA in the aleurones (Bernal-Lugo, Beachy, and Varner 1981). We have recently prepared and characterized a recombinant clone containing α-amylase cDNA sequences inserted into the PstI site of plasmid vector pBR322 DNA (Muthukrishnan, Chandra, and Maxwell 1983). The availability of a molecular probe for α-amylase mRNA allowed us to investigate some aspects of the mode of action of the hormone, GA, in barley aleurones at the molecular level. This chapter outlines some of these results.

RESULTS

Cloning of α-amylase cDNA sequences. The appearance of α-amylase following the addition of GA has been followed in earlier studies by assaying for this enzyme in the incubation medium (Yomo and Varner 1971). However, a time lag can occur between the appearance of mRNA and its translation to yield a protein. Also, additional sites of regulation may exist at the level of processing of a primary RNA transcript or of the protein molecule. A more accurate measure of the effect of GA on α-amylase gene expression may require a determination of the changes in the level of α-amylase mRNA (or its precursors), rather than the changes in the level of the protein per se. This problem has been approached by isolating mRNA fractions from aleurones treated with GA for different time periods and then translating them using an in vitro protein-synthesizing system from wheat germ. Functional mRNA for α-amylase has been detected in barley aleurones as early as 4 hrs after GA addition by this technique (Higgins, Zwar, and Jacobsen 1976; Muthukrishnan, Chandra, and Maxwell 1979). This assay was not sensitive enough to detect α-amylase mRNA at nanogram levels, however. Further, the possibility that inactive precursor forms of α-amylase mRNA were present in aleurones prior to GA treatment could not be ruled out. A molecular probe that would hybridize exclusively to α-amylase mRNA sequences would be useful to detect and quantitate low levels of this mRNA in aleurones at early times after addition of the hormone. Therefore, we have cloned the DNA sequences for α-amylase starting from poly (A^+) RNA from GA-treated aleurones. The details of this cloning procedure have been described in detail elsewhere (Muthukrishnan, Chandra, and Maxwell 1983). In brief, an enriched α-amylase mRNA preparation was reverse-transcribed and the cDNA product was cloned into the PstI site of plasmid vector pBR322 by the "G-C tailing" technique (Villa-Komaroff et al. 1978). From among the recombinant clones,

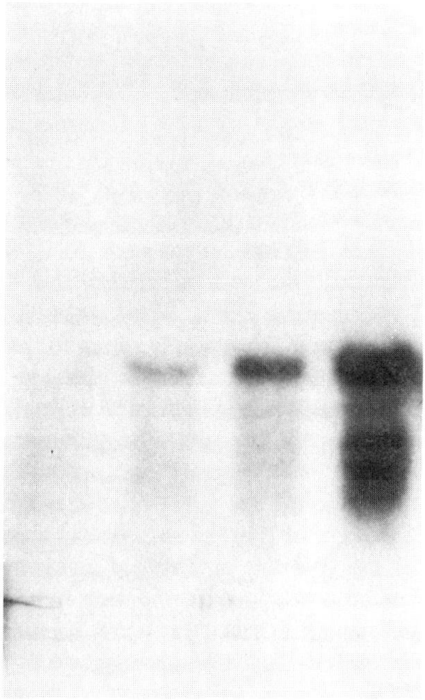

Figure 10.1. Time course of appearance of α-amylase mRNA in barley aleurones. Aleurones from embryo-free half-seeds that have been soaked on moist sand for three days were incubated with 1 μM GA under standard conditions as described in an earlier report (Muthukrishnan *et al.* 1979). Aleurones from 180 half-seeds were used for extraction of RNA for each time sample. 20 μg of RNA from each sample was treated with 1 M glyoxal in 50% DMSO for 1 hr at 50°C and analyzed by electrophoresis on 1% agarose gels. After staining with acridine orange, the RNA in the gel was blotted onto nitrocellulose paper according to the procedure of Thomas (1980). The nitrocellulose paper was hybridized with 20×10^6 cpm of 'nick-translated' α-amylase clone 103 DNA (4×10^8 cpm/μg) under standard conditions (Muthukrishnan *et al.* 1983). After washing, the blot was exposed to X-ray films for 24 hours using intensifier screens.
Lanes: 1, 2, 3 and 4, RNA from aleurones treated with 1 μM GA for 1, 2, 3 and 4 hr, respectively.

those containing α-amylase sequences were identified by in vitro translation of hybrid-selected mRNA (Ricciardi, Miller, and Roberts 1979). Clone 103, which has the longest piece of cDNA inserted (630 basepairs), was further characterized by restriction digestion and DNA sequencing. It was used in all of our experiments as the specific molecular probe.

Time course of appearance of α-amylase mRNA. The availability of a molecular probe for α-amylase mRNA sequences allowed us to undertake a detailed study of the kinetics of appearance of α-amylase mRNA in aleurones. Total RNA was extracted from barley aleurones treated with 1 μM GA for 1, 2, 3, and 4 hrs by procedures previously described (Muthukrishnan, Chandra, and Maxwell 1979). Equal amounts (20 μg) of RNA, representing different time points, were electrophoresed in agarose gels after treatment with glyoxal, and the RNAs were transferred to nitrocellulose paper according to the procedure of Thomas (1980). After hybridization with "nick-translated" α-amylase clone DNA as the radioactive probe, the nitrocellulose sheet was subjected to autoradiography (Fig. 10.1). α-Amylase mRNA was detected in aleurones as early as 1 hr after addition of the hormone (lane 1) and continued to increase in abundance at 2, 3, and 4 hrs after addition of GA (lanes 2 through 4). Using in vitro translation assays we have shown previously that α-amylase mRNA becomes detectable around 4 hrs and reaches a maximum level around 12 hrs after GA addition (Muthukrishnan, Chandra, and Maxwell 1979). These exper-

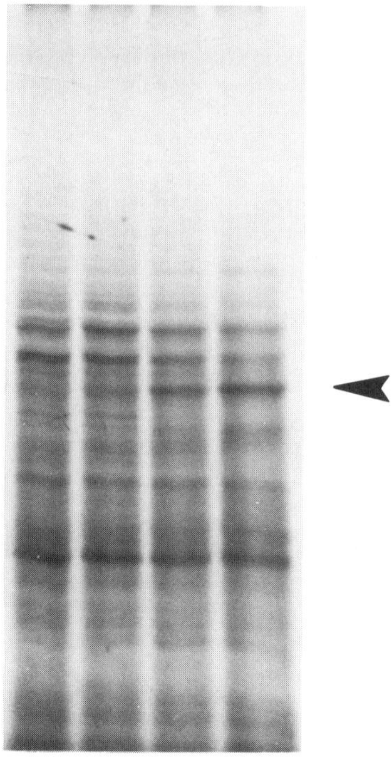

Figure 10.2. Translational products of RNA prepared from aleurones treated with GA and cycloheximide. Aleurones from 180 half-seeds were treated with GA and cycloheximide as indicated, and RNA was prepared from the samples as described previously (Muthukrishnan *et al.* 1979). 0.5 μg of poly (A^+) RNA from each sample was translated in a wheat germ extract using ^{35}S-methionine as the labeled amino acid. The translation products were analyzed by electrophoresis in a 10% polyacrylamide gel followed by autoradiography. Lanes: 1, 2, 3 and 4, cycloheximide (30 μM) added at 0, 2, 4 and 6 hr, respectively, after the initial addition of GA (1 μM). Arrow indicates the migrated position of purified α-amylase.

iments establish that the GA-induced activation of α-amylase gene is an early event, taking place within an hour after addition of the hormone. The action of GA is at the level of RNA synthesis, because there is very little α-amylase mRNA in the aleurones prior to the addition of the hormone (see Fig. 10.5), and GA causes an increase in the level of the α-amylase mRNA in aleurones.

Effect of cycloheximide on the appearance of α-amylase mRNA. Our previous studies had indicated that maximal expression of α-amylase gene requires the synthesis of one or more proteins induced by addition of GA (Muthukrishnan, Chandra, and Maxwell 1979). When an inhibitor of protein synthesis, such as cycloheximide, was added along with GA, the appearance of α-amylase mRNA was severely inhibited as determined by in vitro translation assays. In order to determine whether this requirement for new protein synthesis persisted throughout the induction period, cycloheximide was added to the aleurones at different times after the addition of GA. RNA was extracted from these treated aleurones and translated in a wheat germ extract using ^{35}S-methionine as the labeled amino acid. The translation products were electrophoresed on a polyacrylamide gel and subjected to autoradiography (Fig. 10.2). Addition of cycloheximide along with GA at the beginning of the

Figure 10.3. Influence of cycloheximide on the levels of α-amylase mRNA in aleurones. Total RNA was prepared from 180 aleurones incubated for 12 hr with the indicated additions. 20 μg of total RNA from each sample was treated with glyoxal, electrophoresed on agarose gel and blotted onto nitrocellulose paper as described in the legend to Fig. 1. The blot was probed with 'nick-translated' α-amylase clone 103 DNA (10^7 cpm). Lanes: 1, 1 μM GA only; 2, 1 μM GA and 10 μM cycloheximide; 3, 1 μM GA and 30 μM cycloheximide. The autoradiogram was exposed for 24 hrs.

incubation resulted in nearly complete inhibition of the appearance of the protein band co-migrating with α-amylase. If cycloheximide addition was delayed until 2, 4, and 6 hrs after addition of GA, there was a progressive increase in the intensity of the α-amylase band. This indicates that α-amylase mRNA synthesis continued to some extent during the several hours in which protein synthesis was permitted after GA addition. Even when cycloheximide was added as late as 6 hrs after GA, the amount of functional α-amylase mRNA in the aleurones was less than the level found in aleurones incubated with GA only (data not shown). This suggests that there may be a continuous requirement for a short-lived protein for the accumulation of functional α-amylase mRNA.

Effect of cycloheximide on the accumulation of α-amylase mRNA. From the data described above it is clear that functional α-amylase mRNA fails to accumulate in aleurones in presence of cycloheximide, even though GA is present. Nevertheless, the possibility that cycloheximide treatment inhibits only a processing step in maturation of α-amylase mRNA, resulting in an accumulation of inactive mRNA precursor molecules, could not be eliminated

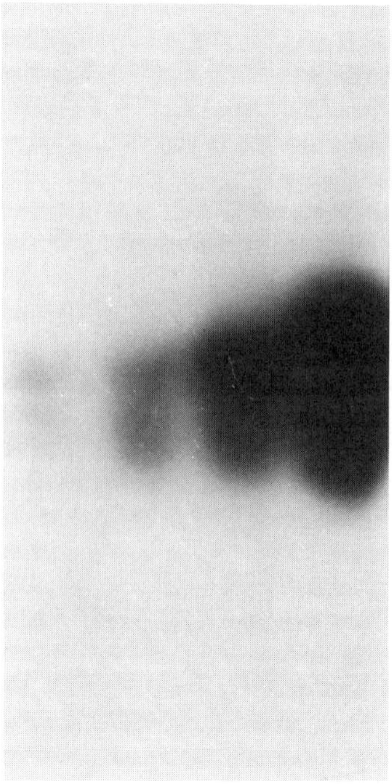

Figure 10.4. α-Amylase mRNA levels in aleurones treated with cycloheximide at different time intervals. GA (1 μM) was added to the aleurones at the beginning of the incubation. Cycloheximide (30 μM) was added at 0 time (lane 1), 2 (lane 2), 4 (lane 3) and 6 hrs (lane 4), and all incubations were continued up to 12 hrs. Total RNA was prepared from each sample (180 aleurones each) and analyzed as described in the legend to Fig. 1. The autoradiogram was exposed for 18 hrs.

by in vitro translation data alone. Both mature and precursor forms of mRNA would be detected by DNA probes containing α-amylase sequences. Therefore, we have analyzed RNA isolated from aleurones treated with increasing concentrations of cycloheximide to determine whether RNA sequences complementary to the probe DNA are present in these preparations. The RNA blots were probed with nick-translated α-amylase clone DNA and subjected to autoradiography. At 10 μM concentration, cycloheximide caused a substantial reduction in the amount of α-amylase mRNA recoverable from aleurones (compare lanes 1 and 2 of Fig. 10.3). When the cycloheximide concentration was increased to 30 μM, there was very little α-amylase mRNA detectable in aleurones (lane 3). The extent of inhibition of the appearance of α-amylase enzyme activity in this medium brought about by 10 μM and 30 μM cycloheximide are 96% and 99%, respectively, of the GA-treated control (data not shown). These results clearly indicate that cycloheximide treatment does not result in an accumulation of inactive or precursor forms of α-amylase mRNA molecules. Two explanations are possible for this failure to detect α-amylase mRNA sequences in cycloheximide-treated aleurones: (a) α-Amylase mRNA synthesis is inhibited in presence of cycloheximide; and (b) α-Amylase mRNA degrades rapidly in presence of cycloheximide, leading to low steady-state levels of α-amylase mRNA. We consider the second possibility to be unlikely,

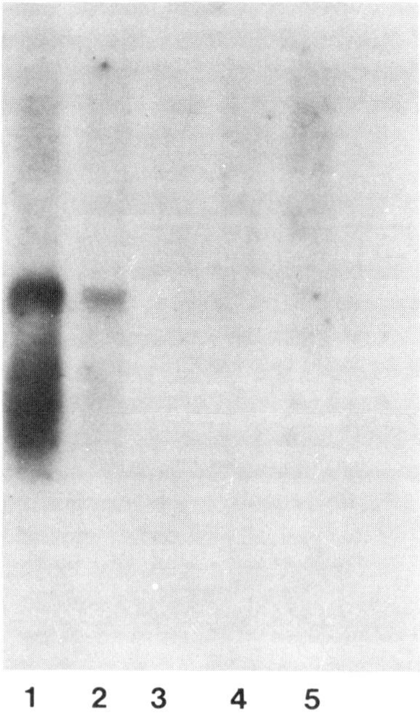

Figure 10.5. Requirement for GA for the appearance of α-amylase mRNA in barley aleurones.

Aleurones were incubated under standard conditions for 12 hrs with the indicated additions. Total RNA was prepared from 180 aleurones, and 20 μg aliquots were analyzed as in Fig. 1 using 'nick-translated' α-amylase clone 103 DNA (20×10^6 cpm) as the labeled probe. The autoradiogram was exposed for 24 hrs.

Lanes: 1, 1μM GA; 2, 1 μM GA and 10 μM cycloheximide; 3, no GA added; 4, no GA, 10 μM cycloheximide; 5, no additions, aleurones processed at 0 time.

because the accumulation of several mRNA's other than α-amylase mRNA is not inhibited by cycloheximide (see Fig. 10.2), ruling out a general degradation of mRNA in the presence of cycloheximide. Degradation of specific mRNAs, including α-amylase, cannot, however, be ruled out. Experiments are in progress to determine whether the effect of cycloheximide is at the level or RNA synthesis or RNA degradation.

From the data shown in Figure 10.2, it is clear that the synthesis and/or accumulation of functional α-amylase mRNA requires the presence of a GA-induced protein. A delay in the addition of cycloheximide permitted the appearance of submaximal amounts of α-amylase mRNA. The effect of delaying the addition of cycloheximide upon the final α-amylase mRNA levels in aleurones at the end of the 12 hr incubation with GA is shown in Figure 10.4. There was only a trace amount of α-amylase mRNA in aleurones when cycloheximide was added at the same time as GA (lane 1). When addition of this inhibitor of protein synthesis was delayed until 2, 4, or 6 hrs after GA addition, the amount of α-amylase mRNA detectable by the probe DNA increased progressively. Thus, there was a substantial increase in the final α-amylase mRNA level when protein synthesis was allowed to proceed for up to 6 hrs instead of up to 2 or 4 hrs after the initial addition of GA (compare lanes 2, 3, and 4). These results imply that even in the presence of GA, inhibition of protein synthesis in the early periods of GA action results in a lower recovery of α-amylase mRNA. This requirement for new protein synthesis apparently continues at least up to 6 hrs. We have not done more detailed experiments to determine whether this requirement is a continuous one or whether it lasts only for a limited period after the initial addition of GA.

GA requirement for α-amylase mRNA induction. In order to establish the role of GA in the induction of α-amylase mRNA, we have compared the levels of this mRNA in aleurone tissue that has been incubated in the presence as well as the absence of GA. As shown in Figure 10.5, there is very little detectable α-amylase mRNA in aleurone tissue before treatment with GA (lane 5) or in tissue that has been incubated for 12 hrs without the addition of GA (lane 3). But in the presence of 1 μM GA there is an enormous increase in the amount of α-amylase mRNA as detected by the α-amylase probe (lane 1). Lanes 2 and 4 show the effect of 20 μM cycloheximide in GA-treated and control aleurone tissues. These experiments clearly establish that GA is needed for the appearance of α-amylase mRNA sequences.

CONCLUSIONS

The induction of α-amylase by GA in barley aleurones has been shown to involve the synthesis of new mRNA for α-amylase. In addition to GA, one or more GA-induced proteins are also needed for the maximal expression of the α-amylase gene. While the nature of this protein is not known at present, it could conceivably be a receptor molecule for GA or a transcription factor specific for α-amylase gene. Alternatively, this protein(s) may be needed to stabilize α-amylase mRNA. This protein appears to be needed at least up to 6 hrs after GA addition and possibly beyond that time for maximal rate of synthesis of α-amylase mRNA. Further experimentation is needed to understand the nature and the role of this protein in α-amylase induction. Our experiments clearly rule out the presence of preexisting α-amylase mRNA or its precursor forms in aleurones prior to GA addition. α-amylase mRNA is shown to be detectable in aleurones as early as 1 hr after GA addition, which is earlier than previously reported. The induction of α-amylase is probably one of the early events following GA addition.

ACKNOWLEDGMENTS

This work was supported by a competitive research grant from USDA. The expert technical assistance of Robert Goeken and Susan Ward is gratefully acknowledged.

LITERATURE CITED

Bernal-Lugo, I., R. Beachy and J. E. Varner. 1981. *The response of aleurone layers to gibberellic acid includes the transcription of new sequences*. Biochem. Biophys. Res. Commun. 102:617–23.

Higgins, T. J. V., J. A. Zwar and J. V. Jacobsen. 1976. *Gibberellic acid enhances the level of translatable mRNA for α-amylase in barley aleurone layers*. Nature (Lond.) 260:166–69.

Ho, D. T., and J. C. Varner. 1976. *Response of barley aleurone layers to abscisic acid*. Plant Physiol. 57:175–78.

Jacobsen, J. V., and J. A. Zwar. 1974. *Gibberellic acid causes increased synthesis of RNA which contains poly(A) in barley aleurone tissue*. Proc. Nat. Acad. Sci. USA 71:3290–93.

Jones, R. L., and J. V. Jacobsen. 1978. *Membrane and RNA metabolism in the response of aleurone cells to gibberellic acid*. Bot. Mag. Tokyo, Special Issue 1:83–89.

Mozer, T. J. 1980. *Partial purification and characterization of the mRNA for α-amylase from barley aleurone layers*. Plant Physiol. 65:834–37.

Muthukrishnan, S., G. R. Chandra, and E. S. Maxwell. 1979. *Hormone induced increase in levels of functional mRNA and α-amylase mRNA in barley aleurones*. Proc. Nat. Acad. Sci. USA 76:6181–85.

———. 1983. *Hormonal control of α-amylase gene expression in barley. Studies using a cloned cDNA probe*. J. Biol. Chem. 258: 2370–75.

Ricciardi, R. P., I. S. Miller, and B. E. Roberts. 1979. *Purification and mapping of specific mRNA's by hybridization-selection and cell-free translation*. Proc. Nat. Acad. Sci. USA 76:4927–31.

Thomas, P. 1980. *Hybridization of denatured RNA and small DNA fragments transferred to nitrocellulose*. Proc, Nat. Acad. Sci. USA 77:5201–205.

Villa-Komaroff, L., A. Efstratiadis, S. Broome, P. Lomedico, R. Tizard, S. P. Naber, W. L. Chick, and W. Gilbert. 1978. *A bacterial clone synthesizing proinsulin*. Proc. Nat. Acad. Sci. USA 75:3727–31.

Yomo, H., and J. E. Varner. 1971. *Hormonal Control of Secretory tissue*, pages 111–14 *in* A. A. Moscona and A. Monroy, eds. *Current Topics in Developmental Biology*. 6. Academic Press, New York.

11] Genetic Analysis of Symbiotic Nitrogen Fixation Genes

by F. M. AUSUBEL,* W. J. BUIKEMA,* S. E. BROWN,*
C. D. EARL,* S. R. LONG,† and G. B. RUVKUN**

ABSTRACT

A 100 kilobase (kbp) region of the *Rhizobium meliloti* genome adjacent to and surrounding the genes for nitrogenase (*nif* genes) was cloned, and the recombinant plasmids were used to construct Tn*5* mutations in the *R. meliloti nif* region. Each of the Tn*5* insertions was assayed for its effect on the ability of *R. meliloti* to nodulate alfalfa (Nod phenotype) and to symbiotically fix nitrogen (Fix phenotype). Tn*5* insertions immediately adjacent to the nitrogenase genes which caused a Fix phenotype (Fix^-::Tn*5*) were found in two clusters, one of 6.3 kbp and a second of at least 5.0 kbp. We found that the *R. meliloti* genes *nifH, nifD,* and *nifK,* which code for the single subunit of the nitrogenase Fe protein and for the two subunits of the nitrogenase MoFe protein respectively, are located in the 6.3-kbp Fix^-::Tn*5* cluster. These genes are transcribed in the order *nifH, nifD, nifK,* which is the same order as in *Klebsiella pneumoniae*. Tn*5* insertions which caused a Nod^- phenotype were found in a cluster only 20 kbp away from the nitrogenase genes. The *nod* and *fix* genes identified in study are located on a large indigenous plasmid which is probably at least 500 kbp.

INTRODUCTION

Bacteria in the genus *Rhizobium* fix nitrogen in specialized nodules that form on the roots of legumes by the symbiotic interaction of bacterium and plant. The formation of an effective nitrogen-fixing nodule involves differentiation of both plant and bacterial cells and is characterized by a series of morphologi-

*Department of Molecular Biology, Massachusetts General Hospital, Boston, Massachusetts 02114; †Department of Biological Sciences, Stanford University, Stanford, California 94305; and **Department of Biology, Massachusetts Institute of Technology, Cambridge, Massachusetts 02139.

GENETIC ENGINEERING: APPLICATIONS TO AGRICULTURE (Beltsville Symposium 7—Lowell D. Owens, ed.) Rowman & Allanheld, Totowa

cally distinct developmental stages. It is known that bacterial genes control steps in the developmental pathway because, in several species of *Rhizobium,* bacterial mutants have been isolated that cause a spectrum of mutant root and nodule phenotypes, ranging from no nodule at all (Nod$^-$ phenotype) to nodules that appear to be morphologically normal but fail to fix nitrogen (Fix$^-$ phenotype).

The identification and manipulation of symbiotic (*sym*) and nitrogen fixation (*nif*) genes in *Rhizobium* has been complicated by the fact that these genes are not normally expressed when the bacteria are in the free-living state. In order to overcome this difficulty and to identify *Rhizobium sym* and *nif* genes, we have developed a variety of molecular genetic techniques that greatly simplify the isolation of *sym* mutations and the subsequent cloning of *sym* genes. In this chapter, we describe how our laboratory has utilized these techniques in *R. meliloti,* the endosymbiont of alfalfa (lucerne) to identify the *R. meliloti* nitrogenase (*nif*) genes, to demonstrate that these genes are located on a large indigenous plasmid, and to discover that the *nif* genes are closely linked to other *sym* genes on the megaplasmid.

LOCALIZATION OF RHIZOBIUM NIF GENES BY INTERSPECIES DNA HOMOLOGY

Because nitrogenase genes in *R. meliloti* are not expressed in the free-living state, it is extremely tedious to identify these genes on the basis of specific *nif* mutations. The reason for this is that individual clones in a mutagenized culture would have to be individually tested on individual alfalfa plants to screen for symbiotic mutants (Sym$^-$ phenotype). Once a collection of Sym$^-$ mutants had been isolated, each mutant would then have to be analyzed biochemically to identify those mutations that were specifically in nitrogenase genes rather than in other symbiotic genes which would result in an overall Sym$^-$ phenotype. To circumvent the necessity of undertaking this cumbersome experimental strategy, we sought a direct method to identify the *R. meliloti* nitrogenase genes based on our previous work on the cloning and characterization of the *nif* genes from *Klebsiella pneumoniae. K. pneumoniae* is closely related to *Escherichia coli* and fixes nitrogen in a free-living state, utilizing N_2 as its sole nitrogen source. For this reason, the *K. pneumoniae* nitrogenase genes could be readily identified by using a combination of classical bacterial genetic techniques and molecular cloning techniques. Because nitrogenases from all species examined share many biochemical and structural features in common, we hypothesized that nitrogenase genes had been conserved in evolution and that we might be able to utilize the cloned *K. pneumoniae* nitrogenase genes to identify the corresponding *R. meliloti* nitrogenase genes.

Fortuitously, we found that the DNA sequences of nitrogenase gene(s) are sufficiently conserved between *R. meliloti* and *K. pneumoniae that a 6.3-kilobase pair (kbp) K. pneumoniae Eco*RI fragment carrying *nifK, nifD,* and *nifH* specifically hybridizes to a 3.9-kbp *Eco*RI fragment from total *R. meliloti*

DNA (Ruvkun and Ausubel 1980). This interspecies homology was used in a colony hybridization procedure (Grunstein and Hogness 1975; Hanahan and Meselson 1980) to clone the *R. meliloti* 3.9-kbp *Eco*RI fragment. A restriction map of a portion of the *R. meliloti* genome containing the 3.9-kbp *Eco*RI fragment is shown in Figure 11.1.

To verify that the cloned 3.9-kbp *Eco*RI fragment contained *R. meliloti* nitrogenase genes, as well as to locate the *nif* genes on the fragment, we used the method of Maxam and Gilbert (1980) to determine the DNA sequence of both strands of the 3.9-kbp *Eco*RI fragment between *Bgl*II site (b) and *XhoI* site (a) (see Fig. 11.1), which, based on the interspecies DNA hybridization data, was expected to contain the N-terminal coding region of *R. meliloti nifH*. The DNA sequencing experiments showed that the presumptive *R. meliloti nifH* gene contained 61% DNA homology and 72% amino acid homology to the *K. pneumoniae nifH* gene over the 360-base pair (bp) region analyzed (Sundaresan and Ausubel 1981; Ruvkun, Sundaresan and Ausubel 1982; Scott, Rolfe, and Shine 1981). These data indicated that we had indeed cloned the *R. meliloti nifH* gene and that transcription of this gene proceeds from right to left as drawn in Figure 11.1. We also showed, by DNA sequence analysis, that the *R. meliloti nifD* gene, as in *K. pneumoniae,* is located directly distal to the *nifH* gene and is transcribed in the same direction as *nifH*. Results similar to ours have also been obtained by Torok and Kondorosi (1981), who determined the entire DNA sequence of the *nifH* gene from a different strain of *R. meliloti*. Their sequence exactly matches our partial sequence in the protein-coding region.

Based on analogy to the *nif* gene organization in *K. pneumoniae,* we expected to find the *R. meliloti* gene equivalent to *nifK* to the left of *R. meliloti nifD* and that it would be approximately 1.5 kbp long. We previously reported no detectable *R. meliloti* DNA sequence homology to *K. pneumoniae nifK* DNA (Ruvkun and Ausubel 1980). Nonetheless, upon hybridization of a ^{32}P-labeled 2.5-kbp *Eco*RI-*Hind*III *K. pneumoniae* restriction fragment containing *K. pneumoniae nifK* to *Eco*RI digested plasmid pRmR8L2 (see Fig. 11.1) DNA fractionated on a 1.8% agarose gel and transferred to nitrocellulose, a hybridization band at 1.7 kbp was detected. Based on the location of this fragment (shown in Fig. 11.1) in the *R. meliloti nif* region, it most likely contains homology to *K. pneumoniae nifK*.

The analysis of the presumptive *R. meliloti nif* genes described above was based on DNA sequence analysis and comparison of *R. meliloti* DNA sequence with the known DNA and amino acid sequences of *K. pneumoniae* nitrogenase genes and proteins. This physical analysis clearly showed that *R. meliloti* contains genes that are very similar to the *K. pneumoniae nif* genes, however, it did not prove that the presumptive *R. meliloti nif* genes are used during and are essential for symbiotic nitrogen fixation. Therefore, in order to provide genetic evidence that these presumptive *R. meliloti nif* genes are essential symbiotic genes, we devised a general method for introducing a mutation into a specified region of the *Rhizobium* genome, provided that that region had been previously cloned in *E. coli* (Ruvkun and Ausubel 1981). Basically, the technique we developed involves the replacement of wild-type

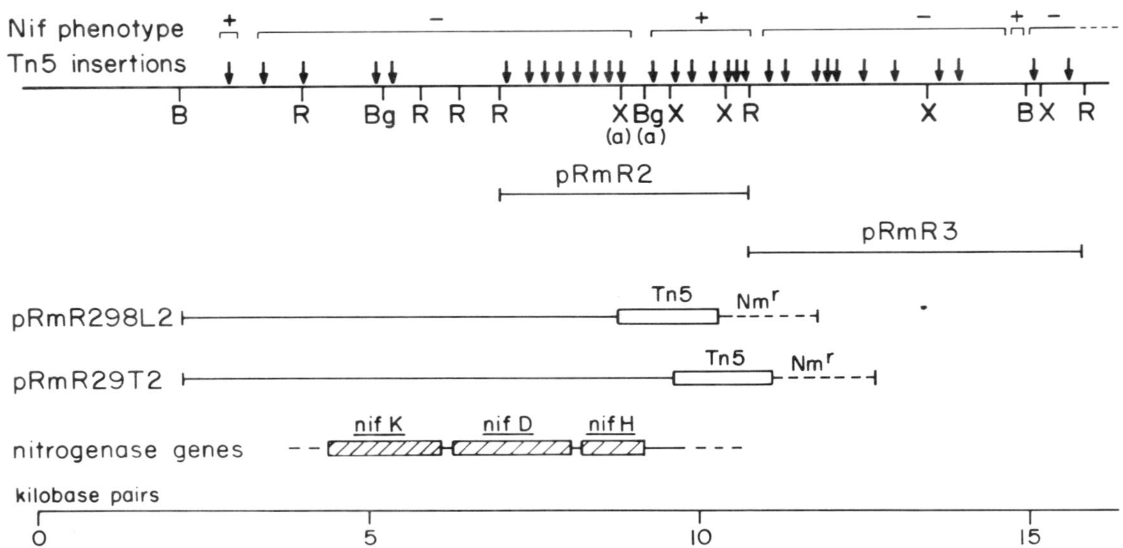

Figure 11.1 The physical genetic map of the *R. meliloti nif* region. (*Top*): Arrows denote Tn*5* insertions, and the symbiotic phenotypes appear above the arrows. Plasmids pRmR298L2 and pRmR29T2 were used in the complementation analysis. (*Bottom*): Orientation of *nif*KDH based on homology to *K. pneumoniae* sequences and complementation analysis (B: *Bam*HI; R: *Eco*RI; Bg: *Bgl*II; X: *Xho*I).

R. meliloti genomic sesquences with the corresponding cloned sequences that have been mutagenized with transposon Tn*5* in *E. coli*.

Specifically, a cloned *R. meliloti* restriction fragment residing on a high copy number plasmid vector, such as pBR322, was mutagenized with Tn*5* in order to generate a collection of plasmids with Tn*5* inserted at different locations. The location of each Tn*5* insertion (±100 bp) was then determined for each Tn*5*-containing plasmid by performing a standard restriction enzyme analysis. Because high copy number *E. coli* cloning vectors do not replicate in *R. meliloti, R. meliloti* fragments containing Tn*5* were recloned into the low copy number broad host-range vector pRK290 (Ditta et al. 1980) that confers tetracycline resistance and contains a single *Eco*RI site for cloning.

Each pRK290 recombinant plasmid containing a Tn*5* mutagenized *R. meliloti* fragment was conjugated into the Fix$^+$ *R. meliloti* strain 1021, and *R. meliloti* transconjugants were selected in which the wild-type *R. meliloti* sequences were replaced by the homologous sequences containing Tn*5* (see Ruvkun and Ausubel 1981 for details). The location of each Tn*5* insertion in the resulting *R. meliloti* strain was mapped using the Southern gel transfer and hybridization technique (Southern 1975), using total DNA isolated from each strain and cloned ^{32}P-labeled *R. meliloti* restriction fragments as hybridization probes. Finally, each *R. meliloti* strain containing a single Tn*5* insertion was inoculated onto several sterile alfalfa seedlings, and after 4 weeks a nitrogenase assay was performed on each plant.

A total of 31 Tn*5* insertions distributed over 14 kbp of the presumptive *R. meliloti nif* region were analyzed by Ruvkun, Sundaresan, and Ausubel (1982) using the above method. The results are summarized in Figure 11.1. The analysis revealed two clusters of Fix$^-$::Tn*5* insertions of 6.3 kbp and at least 5 kbp, separated by a 1.6-kbp cluster of Fix$^+$ insertions. The 6.3-kbp cluster of Fix$^-$ insertions corresponds exactly to the location of *R. meliloti* sequences which are homologous to the *K. pneumoniae* nitrogenase genes (see Fig. 11.1), verifying that the presumptive *R. meliloti nif* genes are indeed required for symbiotic nitrogen fixation.

THE R. MELILOTI NIFH, NIFD, AND NIFK GENES ARE IN AN OPERON

By analogy with *K. pneumoniae* we expected the *R. meliloti nifD* and *nifK* genes to be situated within a single operon. To determine whether this is the case, we performed complementation analysis between selected *nif*::Tn*5* insertions and appropriate plasmids, as described below. Because Tn*5* normally causes polar mutations in an operon, insertion of Tn*5* into any gene in an operon leads to inactivation of all distal genes in that operon. Therefore, complementation analysis using transposon mutations will establish the boundaries of transcription units but not the boundaries of genes within an operon.

The strategy we adopted to perform the complementation analysis was to construct two conjugative plasmids carrying the *nifHDK* genes. One of these plasmids contained the presumptive *nifHDK* promoter, and the other plasmid lacked this promoter due to the insertion of Tn*5* near the terminal end of the *nifH* gene. These plasmids were then conjugated into a variety of *nif*::Tn*5* *R. meliloti* strains, and the resulting merodiploid *R. meliloti* strains were assayed for the ability to fix nitrogen after nodule formation on alfalfa. This analysis showed that the plasmid that contained the presumptive promoter region was able to complement all of the *nifHDK*::Tn*5* mutations, whereas the plasmid that lacked the promoter region failed to complement any of these mutations. These data are consistent with the interpretation that the *nifH, nifD,* and *nifK* genes in *R. meliloti* are present in an operon that is transcribed in the direction *nifH* to *nifK*.

NITROGENASE GENES ARE LOCATED ON A MEGAPLASMID IN R. MELILOTI STRAIN 1021

Because large plasmids in several *Rhizobium* species have been shown to carry genes essential for symbiotic nitrogen fixation (Johnston et al. 1978; Zurkowski and Lorkiewicz 1979; Beynon, Beringer, and Johnston 1980; Banfalvi et al. 1981; Rosenberg et al. 1981), including nitrogenase genes, we sought to determine whether nitrogenase genes were located on plasmids in *R. meliloti* strain 1021. Using the Eckhardt procedure of "in-gel" lysis of the bacteria (Eckhardt 1978), a single extremely large (approx. 500 kbp) plasmid (megaplasmid) was identified in strain 1021 as shown in Figure 11.2A. In order to determine whether the plasmid band observed in Figure 11.2A carries nitrogenase genes, we devised the following internally controlled experimental strategy. The overall idea was to compare strains with Tn*5* insertions in nitrogenase genes (presumptive plasmid location) directly with strains containing Tn*5* insertions in genes necessary for amino acid biosynthesis (presumptive chromosomal location) (see Buikema et al. 1983).

Tn*5* insertions inside the nitrogenase structural genes were constructed as described above, using the site-directed mutagenesis procedure (Ruvkun and Ausubel 1981). Tn*5* insertions within amino acid biosynthetic genes were constructed using a suicide plasmid technique for generating random Tn*5* insertions in the *R. meliloti* genome (Meade et al. 1982). In the case of the two auxotrophic mutants used in the experiment described in Figure 11.2 (Met$^-$ and Pur$^-$), the Tn*5* insertions were correlated with the auxotrophic phenotype by showing that the neomycin-resistant phenotype of Tn*5* in each strain was 100% linked to the auxotrophic phenotype. In addition, the neomycin resistance phenotype was linked to a *pan* marker, which is part of a single circular linkage group that includes at least 20 auxotrophic and drug resistance mutations and that is presumed to represent the *R. meliloti* chromosome (Meade and Signer 1977). In contrast to the auxotrophic strains, the Tn*5* insertions in the nitrogenase genes showed less than 1% linkage to several selected chromosomal markers widely dispersed around the *R. meliloti* chromosome.

The wild-type strain (Fig. 11.2A, lane 1), the strains containing Tn*5* in the nitrogenase genes (lanes 2 and 3), and the strains containing Tn*5* in auxotrophic genes (lanes 5 and 6) all contained a single megaplasmid band. When

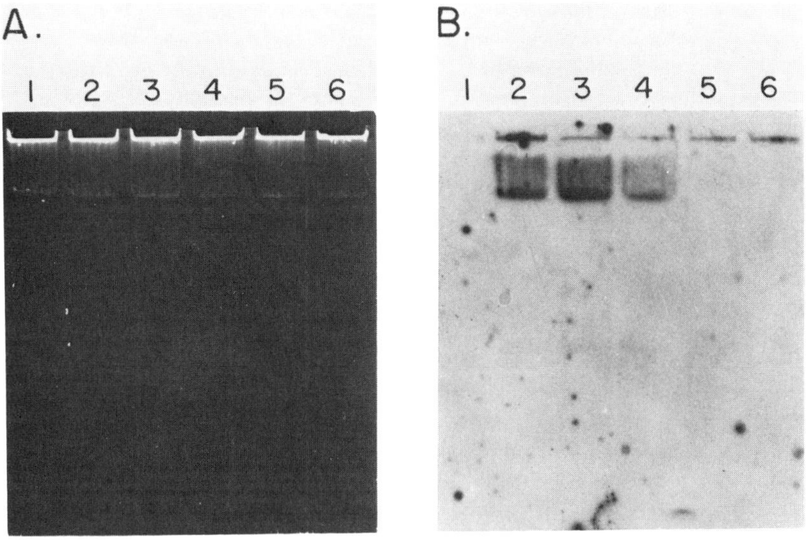

Figure 11.2. Detection of a megaplasmid in *R. meliloti* 1021 and derivative Tn*5* insertion mutants (A), and hybridization of megaplasmids to a Tn5 probe (B). Isolation of the plasmid was according to Ekhardt (1978). A ColEl::Tn*5* plasmid was used as a hybridization probe. Lane 1, Rm1021; lanes 2–3, *nif*− mutants; lane 4, nod− mutant Rm1126; lanes 5–6, auxotrophic mutants.

transferred to nitrocellulose and hybridized with ^{32}P-labelled Tn*5* probe, the megaplasmid band from strains containing Tn*5* insertions in the nitrogenase genes (Fig.11.2B, lanes 2 and 3) hybridized to the probe. In contrast, the Tn*5* probe hybridized only to the top of the gel (where chromosomal DNA is trapped) in the case of the strains containing Tn*5* insertions in the auxotrophic genes (lanes 5 and 6). Figure 11.2 (lane 4) also contains the results obtained with a Tn*5* insertion in a *nod* gene which is located 20 kbp distal of the *nifHDK* operon. As is the case with the *nif*::Tn*5* insertions, the ^{32}P-labelled Tn*5* probe hybridized to the megaplasmid band, and the *nod*::Tn*5* insertion is not genetically linked to chromosomal markers (see next section for more information on *nod* mutations). The reason for the relatively weak hybridization to the top of the gel (chromosomal DNA) is probably due to loss of the DNA during the blotting procedure. Presumably, some megaplasmid DNA is also trapped at the top of the gel and accounts for the hybridization to the chromosomal band in lanes 2 and 3.

To summarize the results of the experiment shown in Figure 11.2, there is a direct correlation between hybridization to the presumptive megaplasmid band and the lack of linkage of Tn*5* to chromosomal markers. The converse is also true: that is, there is a direct correlation between linkage of Tn*5* to chromosomal markers and the lack of hybridization to the megaplasmid. Thus, this experiment demonstrates on the basis of both genetic and physical criteria that the band seen in Figure 11.2A is indeed a plasmid and that this plasmid carries the nitrogenase and linked *nod* genes. These data confirm results published recently by Rosenberg et al. (1981) and by Banfalvi et al. (1981).

IDENTIFICATION OF R. MELILOTI SYMBIOTIC GENES

In contrast to the *nif* genes, many *R. meliloti sym* genes most likely have no functional analogues in *K. pneumoniae* and cannot be identified by interspecies homology with cloned *K. pneumoniae nif* genes. We have therefore adopted two general methods to isolate mutations in and, thereby, identify *R. meliloti sym* genes. First, based on the results obtained recently in several laboratories that a variety of symbiotic genes are located on plasmids (Johnston et al. 1978; Zurkowski and Lorkiewicz 1979; Beynon, Beringer, and Johnston 1980; Banfalvi et al. 1981; Rosenberg et al. 1981), we made the assumption that interesting symbiotic genes would be closely linked to the *nifHDK* genes already identified and cloned on the *R. meliloti* megaplasmid. Working on this assumption, we identified clones spanning approximately 90 kbp of the *R. meliloti* megaplasmid using the cosmid cloning vector pHC79. Overlapping clones, starting at the point of the *nifHDK* locus, were identified by use of a standard stepwise hybridization "walking" protocol. Some of the cosmid clones used to generate a restriction map of the 90-kbp region are shown in Figure 11.3. Selected clones were then mutagenized with Tn*5*. Using the gene replacement technique described above for mutagenesis of the *R. meliloti nif* genes, we have identified several regions in the 90-kbp *nif* region that contain gene(s) required to form nodules (*nod* genes) and genes required for effective fixation in formed nodules (*fix* genes).

At the present time, we have examined a total of 90 Tn*5* insertions in selected areas of the 90-kbp *nif* region, and the results are summarized in Figure 11.3. The most important results from this series of experiments are that both *nod* and *fix* genes are closely linked on the megaplasmid and that the 90-kbp region appears to contain significant regions that do not contain essential *sym* genes. Perhaps the most interesting symbiotic region identified is approximately 20 kbp distal to the *nifHDK* operon, where gene(s) are located that are required for one of the earliest, if not *the* earliest, steps in plant-bacterial interaction. Bacteria that contain insertions in this region fail to induce root hair curling, and inoculated plants show no visible sign of having interacted with the bacteria (also see Long, Buikema, and Ausubel 1982).

The second general method we used for identifying *sym* genes involved the use of a random Tn*5* mutagenesis procedure developed by Van Vliet et al. (1978) and Beringer et al. (1978) to generate *R. meliloti sym* mutants, presumably caused by Tn*5* insertions, with symbiotic defects (Meade et al. 1982). The mutagenesis was accomplished using the "suicide plasmid" pJB4JI, which contains Tn*5* inserted into prophage Mu carried on the broad host range plasmid pPH1JI (Beringer et al. 1978). When pJB4JI is transferred to *Rhizobium* from an *E. coli* host, the plasmid fails to be stably maintained, but Tn*5* transposes at a low frequency from the plasmid before it is lost. Approximately 6000 Nm^r *R. meliloti* transconjugants from this procedure were individually

Table 11.1 Phenotypes and Properties of Symbiotic Mutants Generated by Random Tn*5* Mutagenesis

Strains	Phenotype	Mu sequences present	Tn*5* on plasmid	Tn*5* linkage to chromosome	Tn*5* causes mutation
1111	Fix$^-$	no	yes	no	yes
1128	Fix$^-$	no	yes	no	yes
1126	Nod$^-$	yes	yes	-[a]	yes
1027	Nod$^-$	no	no	yes (pyr)	no
1055	Fix$^-$	no	no	yes (pyr)	no
1058	Fix$^-$	no	no	yes (pyr)	no
1080	Fix$^-$	no	no	yes (trp)	yes
1054	Fix$^-$	no	yes	no	no
1064	Fix$^-$	no	n.t.	yes (pyr)	no
1145	Nod$^-$	yes	no	-[a]	?
1028	Nod$^-$	no	n.t.	yes (pan)	?
1045	Fix$^-$	yes	yes	-[a]	?
1062	Fix$^-$	no	n.t.	-[a]	?

[a]Genetic linkage cannot be determined due to the presence of Mu sequences which kill the transconjugants.

tested for their symbiotic phenotype on alfalfa plants, and 50 mutants deficient in nodulation or fixation were obtained. Each of these mutants was shown to contain a single Tn*5* insertion by use of the Southern gel transfer and hybridization technique, although in 16 of the 50 mutants, phage Mu sequences were found contiguous to Tn*5* in the *R. meliloti* genome. Each mutant was further characterized by a series of physical and genetic tests in order to determine: (a) whether the symbiotic-defective phenotype of each mutant could be correlated with the Tn*5* or the Tn*5* + Mu insertion, and (b) whether the symbiotic mutation was located on the megaplasmid or the chromosome.

One major result from this analysis was that 29 of the 50 mutants were shown to contain a 1.4-kbp endogenous insertion sequence (IsRm1) transposed into the 5.0-kbp *Eco*RI fragment to the right of *nifH* (Ruvkun et al. 1982). This region contains essential *fix* genes, and IsRm1 was shown to be the cause of the symbiotic defect in these strains by cloning the 5.0-kbp *Eco*RI fragment containing IsRml and recombining IsRml back into the genome of wild-type *R. meliloti*. Two additional strains contained IsRm1 inserted into the *nifH* gene (Ruvkun et al. 1982). It appeared that transposition of IsRm1 and Tn*5* in these 31 mutants had occurred independently and that the Tn*5* inserts in these strains were not related to their symbiotic defects. These 31 IsRm1-containing mutants were eliminated from further study. The results obtained from analyzing the remaining 19 mutants are described below and summarized in Table 11.1 (see Buikema et al. 1982).

Based on the results, described above, that symbiotic genes appear to be

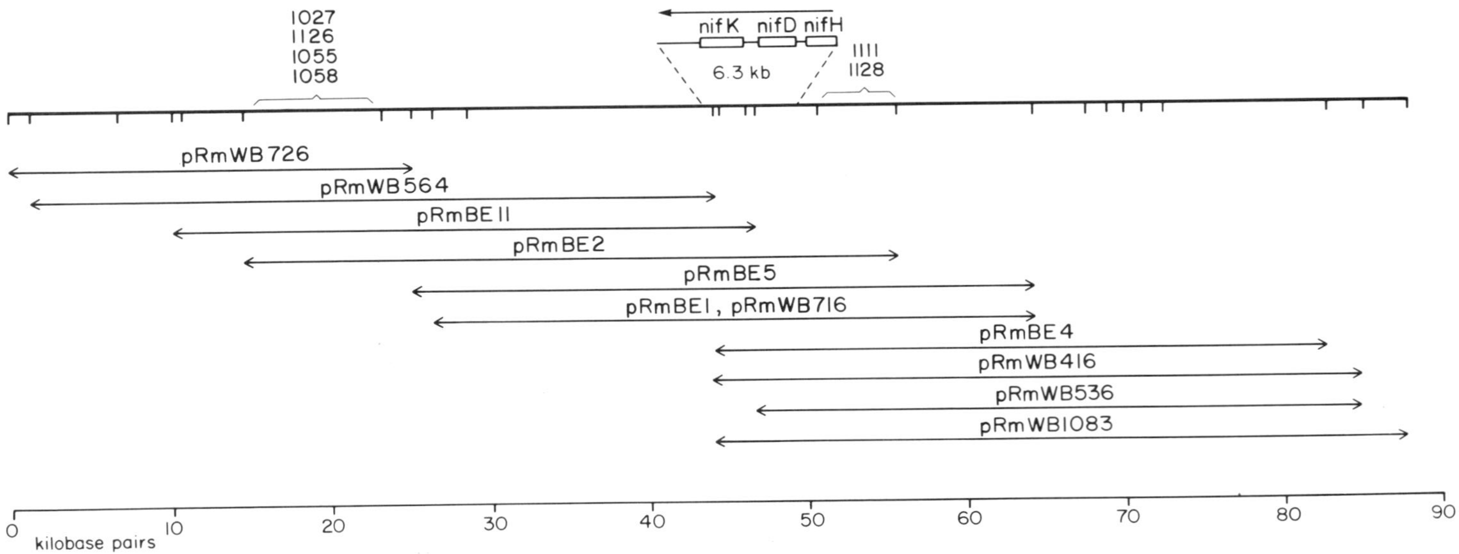

Figure 11.3. Cloned region of the *R. meliloti* megaplasmid. The *Eco*RI restriction map of the 90-kbp region is shown with the position of the *nif*KDH operon and the overlapping cosmids used to generate the map. The positions of six insertion mutations appear above the map.

clustered on the megaplasmid, we screened 14 of the presumptive 19 Tn*5*-induced symbiotic mutants for the presence of Tn*5* on the megaplasmid, using the Eckhardt in gel lysis technique coupled with the Southern blotting and hybridization technique as described in Figure 11.2 (see Table 11.1). In addition, working on the assumption that many symbiotic genes might be linked on the *R. meliloti* megaplasmid, we determined whether the 19 mutants contained an insert within the 90-kbp region surrounding the nitrogenase genes. This was accomplished by probing a Southern blot of *Eco*RI-digested total DNA from the mutants with ^{32}P-labeled pRmBE11 and pRmWB536 DNA (see Fig. 11.3). Six strains contained identifiable inserts in the 90-kbp region (see Fig. 11.3). Strains Rm1111 and Rm1128 contain Tn*5* insertions in the 5.0-kbp *Eco*RI fragment just to the right of the *nifH* gene. These Tn*5* inserts were shown to be the cause of the symbiotic defect by cloning the 5.0-kbp fragments containing Tn*5* and homogenotizing Tn*5* back into wild type strain Rm1021.

The Nod$^-$ mutants Rm1126 and Rm1027 also contain insertions in the 90-kbp region, approximately 20 kbp to the left of *nifK*. Rm1126 contains a 12-kbp insert comprised of Tn*5* + Mu sequences. Due to the presence of Mu DNA in this strain, it was not possible to clone the insertion and characterize it by homogenotization. But it was possible to complement genetically the Nod$^-$ phenotype of Rm1126 in vivo using a specially constructed cosmid clone (pRmSL26) containing the wild-type version of the mutated region in Rm1126 (Long, Buikema, and Ausubel 1982). The other Nod$^-$ mutant, Rm1027, was shown to contain an IsRml insertion (by hybridization with a cloned IsRml probe) and could also be complemented genetically by pRmSL26. To verify that the Tn*5* insert in Rm1027 was not related to its Nod$^-$ phenotype, the *Eco*RI fragment containing Tn*5* was cloned and the Tn*5* insert was homogenotized back into the Rm1021 genome. Rm1021 containing the Tn*5* insert of Rm1027 has a Nod$^+$Fix$^+$ phenotype (Buikema et al. 1982).

The Fix$^-$ mutants Rm1058 and Rm1055 also contain IsRml insertions into the same 8.7-kbp *Eco*RI fragment, which is mutated in Rm1027. As was the case in Rm1027, the Tn*5* insertions in Rm1058 and Rm1055 could not be correlated with the Fix$^-$ phenotype by cloning and homogenotization of the Tn*5* inserts. Attempts to complement the Fix$^-$ phenotype of these strains have not been successful, however, and we have no direct evidence that the IsRml insertion is the cause of the Fix$^-$ phenotype.

Only one mutant, Rm1080, which had a Fix$^-$ phenotype, contains a chromosomal Tn*5* insertion that has been correlated directly to the symbiotic defect. Both Eckhardt gel analysis and genetic linkage data placed the Rm1080 insert on the chromosome. Molecular cloning and homogenotization of the Tn*5* insert showed that the Fix$^-$ phenotype was the result of Tn*5* insertion. This is the only Tn*5*-induced symbiotic mutant that we obtained exhibiting a chromosomal location that was not also associated with an auxotrophic phenotype.

In the case of two Fix$^-$ mutants (Rm1054 and Rm1064), no insertions could be detected within the 90-kbp cloned region, and the symbiotic phenotype could not be correlated to Tn*5* on the basis of cloning and homogenotization of the Tn*5* insert. Rm1064 contains a chromosomal and Rm1054 a megaplasmid Tn*5* insertion. We have not been able to determine the genetic lesions that cause the symbiotic defects in these strains.

The two Nod⁻ mutations Rm1145 and Rm1028 (Hirsch et al. 1982) both contain Tn*5* insertions in the chromosome and no detectable defect in the 90-kbp *nif* region on the megaplasmid. Rm1145 contains Tn*5* + Mu sequences; thus the insert could not be cloned for homogenotization analysis. The Tn*5* insert in Rm1028 has not yet been cloned.

Finally, two Fix⁻ mutants (1045 and 1062) contain Tn*5* + Mu sequences located outside the 90-kbp *nif* region on the megaplasmid. Because of the Mu sequences present in these strains, the insertions can neither be genetically mapped nor readily cloned (Meade et al. 1982). The insert in 1045 is located on the megaplasmid; in 1062 the insert has not yet been located. We have not attempted to correlate either insert with the Fix⁻ phenotype. Six Fix⁻ mutants (not included in Table 11.1) contain Tn*5* (without Mu) inserted in the chromosome, but these inserts have not yet been cloned. Thus we have not yet attempted to correlate the Tn*5* insertions in these strains with the mutant phenotypes by using the homogenotization procedure.

CONCLUDING REMARKS

The results discussed in this chapter argue strongly for an extra-chromosomal location for a number of genes involved in nitrogen fixation in *R. meliloti*. Utilizing the Eckhardt procedure for plasmid visualization, we have demonstrated that transposon Tn*5* insertions within structural genes for the nitrogenase enzyme and within gene(s) involved in nodulation are all located on a very large megaplasmid in strain Rm1021. Recently, Banfalvi et al. (1981) showed that *R. meliloti* strain 41 also contains a megaplasmid that carries nitrogenase and nodulation genes. Moreover, Rosenberg et al. (1981) have demonstrated the presence of similar large megaplasmids that carry nitrogenase genes in a large number of *R. meliloti* isolates from different geographic regions.

The use of cosmid cloning techniques has allowed us to clone approximately 90 kbp of DNA surrounding the *nifHDK* operon. Because of the large size of the inserts, we were able to construct a restriction map of most of the 90-kbp region by taking advantage of overlap between different cosmid clones. Using ^{32}P-labeled cosmids to probe restriction enzyme digests of total mutant DNA, we were able to identify a variety of insertions in the symbiotic mutants. The most abundant type of insertion mutation was caused by the indigenous insertion sequence IsRml. IsRml insertions in 29 mutants were found at a single locus adjacent to the *nifKDH* operon within a symbiotically essential region. IsRml insertions in four other mutants have been detected within two other restriction fragments included in the 90-kbp region that was probed. At this time, however, it is not clear whether the substantial number of unassigned mutants are the result of IsRml transpositions into other essential symbiotic genes elsewhere in the genome.

The physical mapping of insertion mutations has allowed us to show that nodulation genes are closely linked to fixation genes. By physically mapping the insertions in the Nod⁻ mutants Rm1027 and Rm1126 to an 8.7-kbp *Eco*RI fragment, we have shown that nodulation gene(s) are approximately 20 kbp distal of the *nifKDH* operon. We have recently shown that the nodulation deficiency of Rm1126 and Rm1027 can be complemented by a plasmid containing the 8.7-kbp *Eco*RI fragment (Long et al. 1982). We have also recently shown that a Tn*5* insertion 1 kbp from the right end (Fig. 11.3) of the cloned 8.7-kbp fragment causes a Nod$^-$ phenotype when homogenotized into wild-type *R. meliloti* (unpublished data).

We have encountered two major problems in using the suicide plasmid mutagenesis procedure of Van Vliet et al. (1978) and Beringer et al. (1978) with *R. meliloti* strain 1021. One problem is the high frequency of cases (25%) in which Mu sequences are found inserted with the Tn*5*. In these cases, it is not possible to genetically map the Tn*5* insertion, and it is difficult to clone the insertion (Meade et al. 1982). Presumably the Mu sequences are lethal when transferred to a new background. Another problem is the low frequency with which Tn*5* insertions have been shown to be correlated with actual symbiotic lesions. Out of 50 symbiotic mutants presumably caused by Tn*5* mutagenesis, only four have been shown to be caused by a bona fide Tn*5* insertion. In 33 mutants, the Sym$^-$ has been correlated with the insertion of an endogenous insertion sequence (IsRml), and in three others Tn*5* was shown not to be the cause of the Sym$^-$ phenotype, although the actual symbiotic lesion was not identified. The remaining 13 mutants have not yet been fully characterized, and it is possible that some of these mutants could contain Tn*5* in symbiotic genes.

In conclusion, we have found that random Tn*5* mutagenesis using the suicide plasmid pJB4JI has not been as successful as we had hoped in identifying symbiotic genes, and we have had more success using a site-directed mutagenesis procedure on cloned *R. meliloti* DNA fragments. In any case, we have been able to define a region of the *R. meliloti* megaplasmid that contains several genes essential for nodule formation and function. A careful analysis of these genes will, we hope, lead to a better understanding of the symbiotic process.

LITERATURE CITED

Banfalvi, Z., V. Sakanyan, C. Koncz, A. Kiss, I. Dusha, and A. Kondorosi. 1981. *Location of nodulation and nitrogen fixation genes on a high molecular weight plasmid of R. meliloti.* Mol. Gen. Genet. 184:318-25.

Beringer, J. E., J. L. Beynon, A. V. Buchanon-Wollaston, and A. W. B. Johnston. 1978. *Transfer of the drug resistance transposon Tn5 to Rhizobium.* Nature (Lond.) 276:633-34.

Beynon, J. L., J. E. Beringer, and A. W. B. Johnston. 1980. *Plasmids and host-range in Rhizobium leguminosarum and Rhizobium phaseoli.* J. Gen. Microbiol. 120:421-29.

Buikema, W. J., S. R. Long, S. E. Brown, R. C. van den Bos, C. Earl, and F. M. Ausubel. 1983. *Physical and genetic characterization of Rhizobium meliloti symbiotic mutants.* J. Mol. Appl. Genet. (in press).

Ditta, G., S. Stanfield, D. Corbin, and D. Helinski. 1980. *Broad host range DNA cloning system for gram negative bacteria: construction of a gene bank of Rhizobium meliloti.* Proc. Nat. Acad. Sci. USA 77:7347-51.

Ekhardt, T. 1978. *A procedure for the isolation of desoxyribonucleic acid in bacteria.* Plasmid 1:584-88.

Grunstein, M., and D. S. Hogness. 1975. *Colony hybridization: A method for the isolation of cloned DNAs that contain a specific gene.* Proc. Nat Acad. Sci. USA 72:3961.

Hanahan, D., and M. Meselson. 1980. *Plasmid screening at high colony density.* Gene 10:63-67.

Hirsch, A. M., S. R. Long, M. Bang, N. Haskins, and F. M. Ausubel. 1982. *Structural studies of alfalfa roots infected with nodulation mutants of Rhizobium meliloti.* J. Bacteriol. 151:411-19.

Johnston, A. W. B., J. L. Beynon, A. V. Buchanon-Wollaston, S. M. Setchell, P. R. Hirsch, and J. E. Beringer. 1978. *High frequency transfer of nodulating ability between strains and species of Rhizobium.* Nature (Lond.) 276:634-36.

Long, S. R., W. J. Buikema, and F. M. Ausubel. 1982. *Cloning of Rhizobium meliloti nodulation genes by direct complementation of Nod$^-$-mutants.* Nature (Lond.) 298:485-88.

Maxam, A., and W. Gilbert. 1980. *Sequencing end labelled DNA with base specific chemical cleavages.* Meth. Enzymol. 65:499-560.

Meade, H., and E. Signer. 1977. *Genetic mapping of Rhizobium meliloti.* Proc. Nat. Acad. Sci. USA 74:2076-78.

Meade, H. M ., S. R. Long, G. B. Ruvkun, S. E. Brown, and F. M. Ausubel. 1982. *Physical and genetic characterization of symbiotic and auxotrophic mutants of Rhizobium meliloti induced by transposon Tn5 mutagenesis.* J. Bacteriol. 149:114-22.

Rosenberg, C., P. Boistard, J. Denarie, and F. Casse-Delbart. 1981. *Genes controlling early and late functions in symbiosis are located on a megaplasmid in Rhizobium meliloti.* Mol. Gen. Genet. 184:326-33.

Ruvkun, G. B., and F. M. Ausubel. 1980. *Interspecies homology of nitrogenase genes.* Proc. Nat. Acad. Sci. USA 77:191-95.

———. 1981. *A general method for site directed mutagenesis in prokaryotes.* Nature (Lond.); 289:85-86.

Ruvkun, G. B., S. R. Long, H. M. Meade, R. C. van den Bos, and F. M. Ausubel. 1982. *IsRml: A Rhizobium meliloti insertion sequence which preferentially transposes into nitrogen fixation (nif) genes.* J. Mol. Appl. Genet. 1: 405–18.

Ruvkun, G. B., V. Sundaresan, and F. M. Ausubel. 1982. *Directed transposon Tn5 mutagenesis and complementation analysis of Rhizobium meliloti symbiotic nitrogen fixation genes.* Cell 29:551-59.

Scott, K. F., B. G. Rolfe, and J. Shine. 1981. *Biological nitrogen fixation: primary sequence of the Klebsiella pneumoniae nifH and nifD genes.* J. Mol. Appl. Genet. 1:71-81.

Southern, E. M. 1975. *Detection of specific sequences among DNA fragments separated by gel electrophoresis.* J. Mol. Biol. 98:503–17.

Sundaresan, V, and F. M. Ausubel. 1981. *Nucleotide sequence of the gene coding for the nitrogenase iron protein from Klebsiella pneumoniae.* J. Biol. Chem. 256:2808-12.

Torok, I., and A. Kondorosi. 1981. *Sequence of the R. meliloti nitrogenase reductase (nifH) gene.* Nucl. Acids. Res. 9:5711-23.

Van Vliet, F., B. Silva, M. Van Montagu, and J. Schell. 1978. *Transfer of RP4::Mu plasmids to Agrobacterium tumefaciens.* Plasmid 1:446-55.

Zurkowski, W., and Z. Lorkiewicz. 1979. *Plasmid-mediated control of nodulation in Rhizobium trifolii.* Arch. Microbiol. 123:195-201.

12] Host Genes Involved in Symbiosis with Rhizobium

by D. P. S. VERMA,* J. D. BEWLEY,† S. AUGER,**
F. FULLER,* S. PUROHIT,* AND P. KUNSTNER*

ABSTRACT

Development of a symbiotic association between a legume plant and *Rhizobium* leading to the fixation of atmospheric nitrogen is the result of coordinated modulation of expression of several host and bacterial genes. Using immunological and molecular techniques, we have identified a group of plant genes that are induced following infection of the soybean by *Rhizobium japonicum*. These genes are transcribed into a moderately abundant mRNA population, which follows similar kinetics of induction as that of leghemoglobin. We have identified several of these nodule-specific host sequences in a library of cDNA clones prepared from nodule mRNA. In addition, some of the sequences that are common between root and nodule tissue increase significantly in concentration in the nodule. Part of the latter response appears to be due to the effect of the growth hormone, indole acetic acid (IAA), produced by *Rhizobium*. One such hormone-regulated gene product has been detected by two-dimensional gel electrophoresis of proteins from *Rhizobium*-infected and hormone-treated seedlings. The expression of the nodule-specific sequences in moderate abundance suggests that they play an important role in root nodule symbiosis.

INTRODUCTION

The successful infection of leguminous roots by *Rhizobium* spp. culminating in the formation of an intracellular association in a specialized organ called the root nodule, requires the expression of specific genes of both host and the endosymbiont (see Verma 1981 for review). Any perturbation of this process

*Department of Biology, McGill University, Montreal, Quebec, Canada H3A 1B1; †On leave from the Department of Biology, University of Calgary, Calgary, Alberta, Canada T2N 1N4; and **Present address Allied Chemical, Syracuse, New York.

GENETIC ENGINEERING: APPLICATIONS TO AGRICULTURE (Beltsville Symposium 7—Lowell D. Owens, ed.) Rowman & Allanheld, Totowa

caused either by the host or *Rhizobium* renders this association ineffective (unable to fix nitrogen). Classical genetic experiments have indicated that a number of plant genes are involved in the process of infection and development of nodules, and in their effectiveness in fixing nitrogen (Nutman 1969; Caldwell and Vest 1977). The fact that a fully competent strain of *Rhizobium* effective on one host, forms ineffective nodules on the other, suggests that host genes also play an important role in the coordinated modulation of expression of *Rhizobium* genes. Nevertheless, the precise function of these host genes is unknown.

We believe that only when the molecular basis of interactions between a legume plant and *Rhizobium* spp. are fully understood will it be possible to genetically manipulate this association. Furthermore, due to the large number of bacterial genes involved in nitrogen fixation and development of effective symbiosis, and because of the specific requirements of the nitrogenase enzyme (i.e., low oxygen tension), it would not be possible to achieve nitrogen fixation by simply introducing these bacterial genes into a plant genome. Yet it is likely that the transfer of some plant genes that are obligatory for association with nitrogen-fixing bacteria will render the new host capable of developing a nitrogen-fixing association. As a first step toward improving and manipulating the process of symbiotic nitrogen fixation, we are attempting to unfold the complexity of this association.

Although leghemoglobins are the most abundant host gene products in nodules, their synthesis is not sufficient to ensure effectiveness in symbiotic nitrogen fixation (Maier and Brill 1976; Verma et al. 1981). Using an immunological approach, we have been able to identify a group of host gene products (nodulins) that appear to be involved in root nodule symbiosis (Legocki and Verma 1980). Kinetic fractionation of total nodule-cDNA allowed us to isolate a set of nodule-specific sequences that are present in the moderately abundant mRNA population of root nodules (Auger and Verma 1981). We have recently developed a nodule-cDNA library of soybean and have identified a number of host genes that appear to be expressed specifically in nodules. Further characterization of these genes and their products may enable us to elucidate the molecular basis of interactions between the plant and nitrogen-fixing bacteria.

IDENTIFICATION OF THE NODULE-SPECIFIC HOST GENES AND THEIR PRODUCTS

Analysis of the total soluble proteins from nodules effective in nitrogen fixation indicated that 30% of this fraction was leghemoglobin. No other abundant polypeptide was detected; however, subfractionation revealed another peptide of 35,000 MW that was present in all nodules irrespective of their effectiveness in fixing nitrogen (Legocki and Verma 1979). Preparation of a nodule-specific antibody mixture by absorbing total nodule antibodies with control root proteins, followed by reaction of nodule-specific antibodies with the host polysomal translation products, indicated the presence of several small molec-

Table 12.1 Expression of Various Host Sequences in Nodule Tissue

Type of sequence	Number of sequences	mRNA molecules/cell	
		Root	Nodule
Leghemoglobin	1[a]	0	110,500
Nodule-specific sequences	19	0	2,400
Common-abundant sequences	440	8	125

[a]Since all leghemoglobin sequences cross-hybridize, they are treated as one sequence (see Auger, Baulcombe, and Verma 1979).

ular weight proteins (nodulins) that appear to be specific to nodule tissue. Furthermore, a detailed comparison of two-dimensional gels of translation products from nodule and control tissues showed reduction and complete disappearance of some peptides in nodules, indicating that some genes may be repressed following infection of the plant by *Rhizobium*. In addition, the existence of a set of nodule-specific mRNA sequences was detected by kinetic fractionation of nodule-cDNA (Auger and Verma 1981). This fractionation of cDNA also yielded a population of sequences (common sequences) that are moderately abundant in nodules but are present in low abundance in uninfected root. Finally, we have identified three distinct host gene products (mRNA) by molecular cloning of nodule-cDNA (see below).

INDUCTION AND THE REGULATION OF EXPRESSION OF HOST GENES IN NODULES

Leghemoglobins are induced only following infection of the legume plant by *Rhizobium*. The fact that their induction occurs prior to the appearance of nitrogenase in *Rhizobium* and that all ineffective nodules tested contain low levels of leghemoglobin mRNA (Verma et al. 1981) suggests that these host genes are derepressed soon after infection. It is possible that this represents a host response to invasion by a "pathogenic bacteria." Several other nodule-specific genes appear to be induced along with leghemoglobin; however, their sequences in a mature nodule are much less abundant (Table 12.1). The population of moderately abundant common sequences, which differ by an order of magnitude in root and nodule tissues, follows the same induction profile as leghemoglobins (Verma et al. 1981). The nodule-specific sequences appear to be regulated at the transcriptional level, since they are not detected in the nuclear RNA of the control tissue (Auger and Verma 1981). It is not known what controls the expression of these genes but, since they are not expressed in uninfected tissue, it seems that either *Rhizobium* provides a

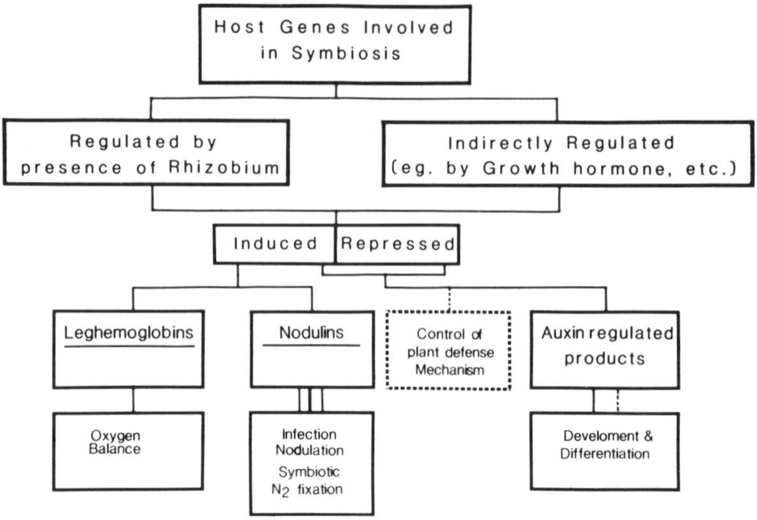

Figure 12.1 Classification of various sets of host genes that appear to be involved in the development of root nodule symbiosis between a legume plant and *Rhizobium* sp.

specific component that is responsible for their induction, or the physical presence of *Rhizobium* influences the host cellular environment in such a way that the genes are induced. Figure 12.1 summarizes our current conceptual understanding of various host genes that may be involved in the process of symbiotic nitrogen fixation and that are directly or indirectly regulated by *Rhizobium*. At present there is no evidence that *Rhizobium* manifests any control over the host defense mechanism. It would appear, however, that *Rhizobium* escapes the normal defense of the host during effective nodulation.

ROLE OF PLANT HORMONES IN REGULATION OF HOST GENES INVOLVED IN SYMBIOSIS

It has long been known that rhizobia growing in liquid culture medium produce a substance that stimulates cell division and elongation in plant tissue (Molliard 1912). It is now clearly established that *Rhizobium* synthesize IAA from tryptophan (Link 1937) and that root nodules have a much greater concentration of IAA than noninfected tissue (Thimann 1936). Nevertheless, the role of IAA in the events of infection, nodulation, and nitrogen fixation is still unknown.

We reasoned that if IAA does modulate the expression of host genes in nodule tissue, then application of the hormones to a localized uninfected root area might have an effect similar to infection upon gene expression. In fact, it has been shown that externally applied IAA induces the synthesis of specific enzymes such as cellulase (Verma et al. 1975), which appears to play a role in the development of nodules (Verma, Zogbi, and Bal 1978). Accordingly, IAA

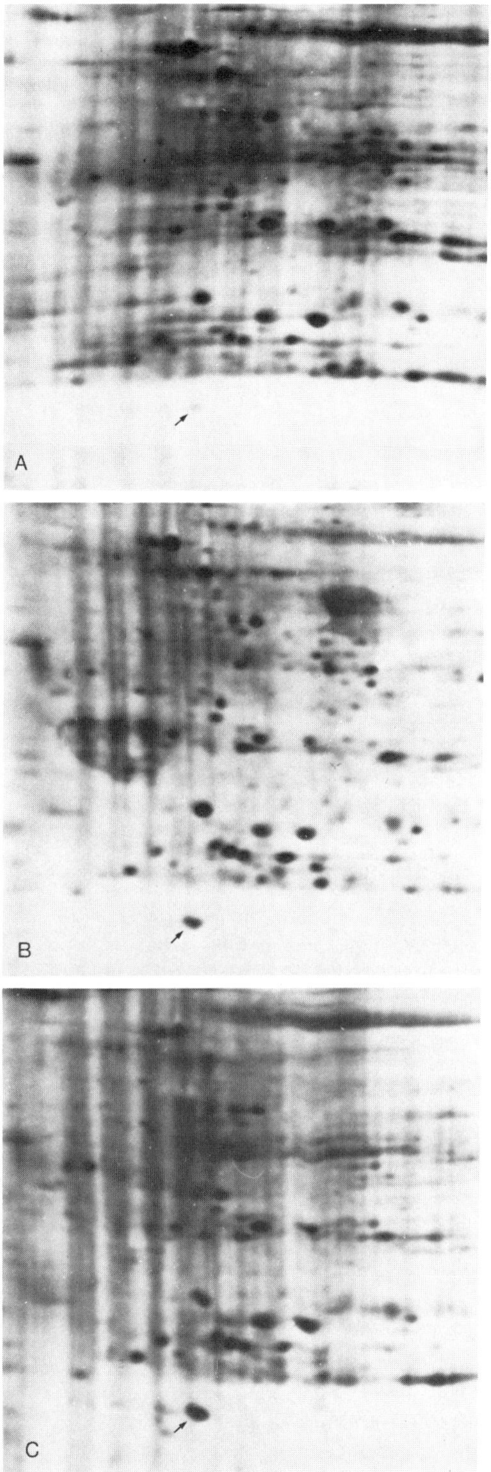

Figure 12.2 Two-dimensional gel electrophoretic analysis of proteins from (a) uninfected roots (control); (b) auxin (IAA)-treated; and (c) *Rhizobium* infected root tissue of soybean. Three-day-old seedlings were either treated with 0.1% IAA in lanolin paste or infected with *R. japonicum* (strain 61A76), and 300 mg of tissue (from the site of swelling, for IAA–treated tissue, or from the site of incipient infection, for *Rhizobium*-infected roots) were isolated and labeled with 200 μCi of ^{3}H-leucine by incubating for 3 hours at room temperature. Soluble proteins were extracted and analyzed on two-dimensional gels (O'Farrel 1976). The arrow points to a spot which increases substantially following either infection by *Rhizobium* or treatment with IAA. This spot appears to consist of at least 3 polypeptides. *Note:* IAA induces swelling about 1.5 cm below the site of application.

in a lanolin paste was applied to a localized area of root tissue of 3-day germinated seedlings. At intervals of 1 to 4 days following application, tissue was harvested from the treated area and frozen in liquid nitrogen. Control seedlings received identical treatment of lanolin paste without IAA. Seedlings were infected in parallel with *Rhizobium* and harvested at similar time intervals. Polyadenylated polysomal RNA was prepared for hybridization to cDNA probes, and ^{3}H-leucine-labeled cytoplasmic protein fractions were prepared for two-dimensional gel electrophoretic analysis.

The majority of polypeptides visualized by two-dimensional gel electrophoresis of proteins from IAA—treated, *Rhizobium*-infected and control root tissues were common to all tissues (Fig. 12.2). There were, however, several quantitative changes—increases and decreases—in the concentration of a few proteins. In particular, the synthesis of one low molecular weight polypeptide (Fig. 12.2. arrow) increased significantly in both IAA—treated and *Rhizobium*-infected tissue.

Hybridization studies using total nodule-cDNA indicated that a population of mRNA sequences that were rare in the uninfected root increased in relative concentration and became moderately abundunt in nodule tissue (Auger, Baukombe, and Verma 1979). A cDNA probe, Common-M-cDNA, which contains moderately abundant sequences common to both root and nodule (Auger and Verma 1981) was used as a probe in studies on the effect of auxin on host genes. The results shown in Figure 12.3 illustrate the effect of IAA and infection upon the concentration of sequences represented in Common-M cDNA. IAA-treatment and infection result in a small but significant increase in the relative concentration of this population of sequences over that in control seedlings. Although these results may imply that auxin treatment and infection produce the same changes in the expression of Common-M sequences, several points should be borne in mind regarding interpretation of these data. First, the expression of these sequences in the control tissue undergoes substantial change during the time period examined; second, they represent a rather complex population of about 400 mRNA sequences, and these sequences may not increase in relative concentration to the same extent; third, it cannot be determined from this result whether the sequences that increase in concentration with auxin treatment are the same ones that increase following infection. Despite these caveats, it is apparent that auxin treatment has an effect upon the expression of some host sequences that may be involved in nodule development. It should be pointed out, however, that IAA treatment does not have an effect on leghemoglobin or any other nodule-specific abundant sequences (data not shown).

MOLECULAR CLONING OF NODULE-SPECIFIC HOST GENES

In order to obtain molecular probes to study the control and kinetics of induction of nodule-specific plant genes, a library of 5700 cDNA-containing clones was prepared from poly (A) mRNA isolated from 3-week nodules (Fuller, et al. 1983). About 750 (13%) of these clones hybridized with cDNA prepared against control root poly (A) mRNA and were removed from the

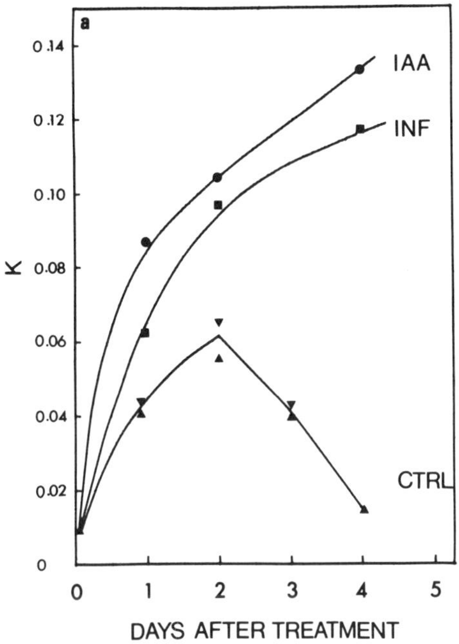

Figure 12.3 Hybridization of cDNA representing common, moderately abundant nodule sequences to poly (A) mRNA from control (▲) auxin-treated (●) and infected (■) root tissues. Changes in the relative concentration of sequences are depicted by the rate constant (K) of the hybridization reaction. See Auger and Verma (1981) for details of hybridizations.

library. About 2100 clones hybridized with homologous, 3-week nodule-cDNA. Clones that did not hybridize with nodule-cDNA presumably contain sequences which are in low abundance in nodule tissue. Nearly 900 of the 2100 nodule-cDNA hybridizing clones contained sequences that hybridized to a previously isolated leghemoglobin clone (Sullivan, et al. 1981). Therefore, leghemoglobin-coding clones represent about 15% of the library, which is in good agreement with hybridization data and indicates that leghemoglobin mRNA represents about 20% by mass of the nodule poly (A) mRNA (Auger, Baulcombe, and Verma 1979).

A subset of the library, containing 24 clones picked at random and 10 clones which were picked as hybridizing to nodule-cDNA, was analyzed further. Dot-blot analysis of these clones (not shown) indicated that 13 of the 24 randomly selected clones hybridized intensely with nodule-cDNA but not with root-cDNA. Five of these were leghemoglobin-coding clones. Of the 11 clones that did not hybridize to nodule-cDNA, one hybridized weakly to root-cDNA. Of the other 10 clones picked initially as hybridizing to nodule-cDNA, three were leghemoglobin-coding clones, three hybridized intensely to nodule-cDNA but

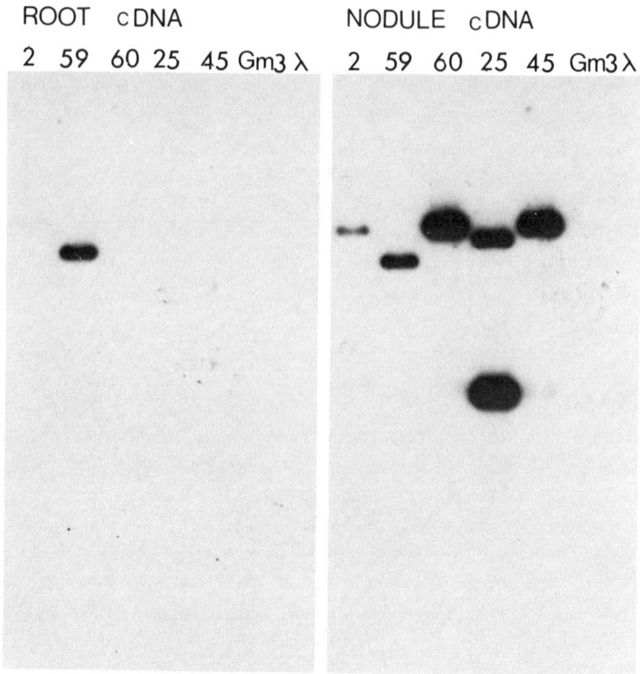

Figure 12.4 Southern-blot analysis of nodule-cDNA clones. DNAs from five nodule-cDNA clones were cleaved with *Hinc*II restriction endonuclease, electrophoresed on 1% agarose gels, blotted to nitrocellulose paper and hybridized to radio-labeled nodule (first panel) or root (second panel) cDNA probes. Clone 2 and 59 sequences hybridize to both root- and nodule-cDNAs while clone 60, 25, and 45 sequences hybridize only to nodule cDNA and thus appear to be nodule-specific. Also included on the blots are GM3 DNA, which is a phage lambda clone containing the soybean rDNA sequences, and phage lambda DNA. No hybridization to these control DNAs is detectable with either probe.

not to root-cDNA, and one clone hybridized well to both root-cDNA and nodule-cDNA.

A synopsis of the above results is depicted in Figure 12.4. Southern blots of five *non*-leghemoglobin clones were probed with nodule-cDNA or root-cDNA. Clones 60, 25, and 45 hybridize intensely to nodule-cDNA, while no hybridization to root-cDNA is detectable. Clone 59 hybridizes equally well to both root- and nodule-cDNAs, while clone 2 hybridizes moderately to nodule-cDNA but only weakly to the root-cDNA probe. Root and nodule cDNA probes do not hybridize to the soybean ribosomal DNA clone GM3, indicating that the cloned sequences that hybridize to these probes are not of ribosomal origin which may occur in cDNA cloning (Sullivan, Brisson, and Verma 1980).

Several cloned DNA fragments have been removed from the vector (plasmid) DNA and cross-hybridized to the 1300 *non*-leghemoglobin clones that hybridize to total nodule-cDNA. Clone 25 cross-hybridizes to about 300 clones, while 45 and 60 each cross-hybridize to about 60 distinct clones. These clones represent sequence species, which we term NodA, NodB and NodC, whose abundancies were estimated to be 6%, 1% and 1%, respectively, in nodule poly (A) mRNA. The three clones do *not* hybridize to rhizobial DNA, but do hybridize to unique restriction fragments of soybean genomic DNA in Southern blots (data not shown), indicating that these nodule-specific sequences, are of host origin.

SUMMARY

To form effective nodules in response to rhizobial infection, legume root cells must undergo developmental changes in the pattern of their gene expression. In particular, the induction of several nodule-specific host genes would appear to be required. We have detected several proteins (besides leghemoglobin) in nodule tissue that are much less abundant or absent in root tissue. In addition, we have observed one polypeptide whose synthesis appears to be induced to a very high level when root tissue is either treated with the plant hormone IAA, or infected with *Rhizobium*. Several quantitative changes in the levels of other proteins have also been observed. Furthermore, a class of poly (A) RNAs, which is moderately abundant in nodules and is much less abundant in root, was found to increase in concentration in root tissues treated with IAA. Therefore, the observed production of relatively high levels of this hormone by bacteroids may play a role in regulating plant gene expression during nodule development.

Analysis of cDNA clones of some nodule-specific sequences indicates that there are at least three major nodulation products (in addition to leghemoglobin) and that these are encoded by the plant genome. Isolation and characterization of more "nodulation" cDNA clones should provide us with the probes needed to test various inducers of the nodulation response, and to determine the extent of pleiotropy of nodulation-defective mutants of both host and endosymbiont. Nucleotide sequence analysis of these clones will provide the amino acid sequence of the nodule-specific gene products which, in turn, should aid in elucidating the biological function of these polypetides. These probes will also allow the isolation of host genes involved in symbiosis, which should facilitate genetic manipulation of other potential host plants.

ACKNOWLEDGMENTS

This work was supported by Research Grants from the NSERC of Canada and FCAC Quebec to Desh Pal S. Verma, who is also a E. W. R. Steacie Fellow (1981–82). Forrest Fuller is a NATO-Post doctoral Fellow. We would like to thank Truyen Nguyen for excellent technical assistance and Yvette Mark for typing the manuscript.

LITERATURE CITED

Auger, S., D. Baulcombe, and D.P.S. Verma. 1979. *Sequence complexities of the poly (A) containing mRNA in uninfected soybean root and the nodule tissue developed due to the infection by Rhizobium*. Biochem. Biophys. Acta. 563: 496–509.

Auger, S., and D.P.S. Verma. 1981. *Induction and expression of nodule-specific host genes in effective and ineffective root nodules of soybean*. Biochem. 20: 1300–1306.

Caldwell, B.E., and H.G. Vest. 1977. *Genetic aspects of nodulation and dinitrogen fixation by legumes: The macrosymbiont*. Pages 557–75 in R.W.F. Hardy and W.S. Silver, eds., *A Treatise on Dinitrogen Fixation III* Wiley-Interscience, London.

Fuller, F., P. W. Künster, T. Nguyen and D. P. S. Verma. 1983. *Soybean nodulin genes: Construction and analysis of cDNA clones reveals several tissue specific sequences expressed in nitrogen fixing root nodules*. Proc. Nat. Acad. Sci. USA (in press).

Legocki, R.P., and D.P.S. Verma. 1979. *A nodule-specific plant protein (Nodulin-35) from soybean*. Science 205: 190–93.

———.1980. *Identification of "nodule-specific" host proteins (nodulins) involved in the development of Rhizobium-legume symbiosis*. Cell 20: 153–63.

Link, G.K.K. 1937. *Role of heteroauxones in legume nodule formation-beneficial effects on nodules, and soil fertility*. Nature (Lond.) 140: 507.

Maier, R.J., and W.J. Brill. 1976. *Ineffective and non-nodulating mutant strains of Rhizobium japonicum*. J. Bacteriol. 127: 763–69.

Molliard, M. 1912. *Action hypertrophiante des produits elaborés par le Rhizobium radicicola Beijer*. Compt. Rend. Acad. Sci. (Paris) 155: 1531–34.

Nutman, P.S. 1969. *Genetics of symbiosis and nitrogen fixation in legumes*. Proc. Roy. Soc. (Lond.) B. 172: 417–37.

O'Farrell, P.H. 1975. *High resolution two-dimensional electrophoresis of proteins*. J. Biol. Chem. 250:4007–4021.

Sullivan, D.E., N. Brisson, B. Goodchild, D.P.S. Verma, and D.Y. Thomas. 1981. *Molecular cloning and organization of two leghemoglobin genomic sequences of soybean*. Nature (Lond.) 289:516–18.

Sullivan, D.E., N. Brisson, and D.P.S. Verma. 1980. *Reverse transcription of 25S soybean ribosomal RNA in the absence of exogenous primer*. Bochem. Biophys. Res. Commun. 94:144–50.

Thimann, K.V. 1936. *On the physiology of formation of nodules in legume roots*. Proc. Nat. Acad. Sci. USA 22:511–14.

Verma, D.P.S., G.A. Maclachlan, H. Bryne, and E. Ewings. 1975. *Regulation and in vitro translation of messenger ribonucleic acid for cellulase from auxin-treated pea epicotyls*. J. Biol. Chem. 250:1019–28.

Verma, D.P.S., V. Zogbi, and A.K. Bal. 1978. *A cooperative action of plant and Rhizobium in dissolving host cell wall during development of symbiosis*. Plant Sci. Lett. 13:137–42.

Verma, D.P.S. 1981. *Plant-Rhizobium interactions in symbiotic nitrogen fixation*. Pages 437–66 in H. Smith and D. Grierson, eds., *Molecular Biology of Plant Development*. Blackwell Scientific Publ., London.

Verma, D.P.S., R. Haugland, N. Brisson, R.P. Legocki, and L. Lacroix. 1981. *Regulation of the expression of leghemoglobin genes in effective and ineffective root nodules of soybean*. Biochim. Biophys. Acta. 653:98–107.

Verma, D.P.S., R.P. Legocki, S. Auger. 1981. *Expression of nodule-specific host genes in soybean*. Pages 205–208 in A.A. Gibson, and W.E. Newton, eds., *Current Perspectives in Nitrogen Fixation*. Australian Academy of Science, Canberra.

13] Viroid cDNA—Uses in Viroid Detection and Molecular Biology

by ROBERT A. OWENS,* DEAN E. CRESS,†
and T. O. DIENER*

ABSTRACT

Viroids are a novel class of subviral pathogens—unencapsidated, small, circular RNAs that can be isolated from plants afflicted with specific diseases. As the emphasis in viroid research begins to shift from physical-chemical studies of viroid structure to studies of mechanisms of viroid-host interaction, cloned viroid cDNA is proving to be a very versatile tool.

Potato spindle tuber disease poses a potentially serious threat to efforts to adapt the potato to growth in subtropical and tropical climates. Exclusion of potato spindle tuber viroid (PSTV) from seed potatoes will be essential, because the presence of PSTV causes severe damage, even total crop loss, in warm climates. Disease diagnosis has often been very difficult, however. Hybridization between highly radioactive recombinant PSTV cDNA and viroid RNA that has been attached to a solid support provides a sensitive and reliable new method for the detection of PSTV.

Viroid cDNA is also expected to be extremely useful in studies of viroid-host interactions. Current efforts to identify regions of the viroid RNA sequence that are involved in viroid-host interaction(s) depend upon comparisons of the sequences of known viroids and suffer important limitations. Directed in vitro mutagenesis of cloned viroid cDNAs offers a promising alternative approach.

INTRODUCTION

Viroids are low molecular weight RNAs that are present in certain species of higher plants afflicted with specific diseases. They are not detectable in healthy

*Plant Virology Laboratory, Plant Protection Institute, and †Tissue Culture and Molecular Genetics Laboratory, Plant Physiology Institute, U.S. Department of Agriculture, Beltsville, Maryland 20705.

GENETIC ENGINEERING: APPLICATIONS TO AGRICULTURE (Beltsville Symposium 7—Lowell D. Owens, ed.) Rowman & Allanheld, Totowa

individuals of the same species but, after introduction into such individuals, they are replicated autonomously and cause the appearance of the characteristic disease syndrome (Diener 1979a). Unlike viral nucleic acids, viroids are not encapsidated.

Viroids constitute a novel class of subviral pathogens; they are the smallest known agents of infectious disease and represent a minimal genetic and biological system. To date, viroids are definitely known to exist only in higher plants, and each consists of a unique, covalently closed, circular RNA molecule. The five viroids whose nucleotide sequence has been determined contain 244 to 371 nucleotides. Although viroids have been discovered because they cause readily recognizable disease symptoms in certain hosts, viroids are often replicated in other hosts without causing obvious damage.

In recent years viroids have received increasing attention from molecular biologists, with the result that our knowledge of their physical-chemical properties has increased dramatically (reviewed by Diener 1979b; Gross and Riesner 1980). Knowledge of viroid-host plant interactions has accumulated at a much slower rate. As the emphasis in viroid research begins to shift from physical-chemical studies of viroid structure to studies of mechanisms of viroid-host interaction, cloned viroid cDNA will find many uses. Only two of the many potential applications of cloned potato spindle tuber viroid (PSTV) cDNA will be considered here—first, the development of a sensitive and reliable new method for detection of PSTV and, second, identification of regions of the viroid RNA sequence involved in viroid-host interactions by directed in vitro mutagenesis of cloned viroid cDNA.

MOLECULAR CLONING OF PSTV cDNA

Determination of the 359-nucleotide sequence of PSTV established viroids as the first naturally occuring examples of covalently closed circular RNA (Gross et al. 1978). The molecular cloning and characterization of double-stranded PSTV cDNAs (Owens and Cress 1980) has been facilitated by knowledge of the RNA sequence. In the latter work, double-stranded (ds) PSTV cDNA was synthesized from a polyadenylated linear PSTV template and inserted in the *Pst*I endonuclease site of plasmid pBR322 using the oligo(dC)·oligo(dG) tailing procedure. Although one recombinant clone (pDC-29) contained a larger than expected 460-base pair (bp) insert, restriction endonuclease mapping and nucleotide sequence determinations demonstrated that all recombinants contained less than a complete copy of PSTV. Figure 13.1 shows the strategy used to construct clones containing full-length dsPSTV cDNA.

Restriction analysis of the PSTV-specific inserts from two of these clones, pDC-29 and pDC-22, suggested that these clones were partially overlapping and could be used to construct a full-length clone. The 285-bp *Ava*II-*Hae*III fragment from pDC-29 was ligated at the *Ava*II site to the contiguous 74-bp *Ava*II-*Hae*III fragment from pDC-22. *Hind*III oligodeoxynucleotide linkers were added to the *Hae*III blunt ends of this linear ligation product and, following digestion with *Hind*III, this fragment was cloned in the *Hind*III site

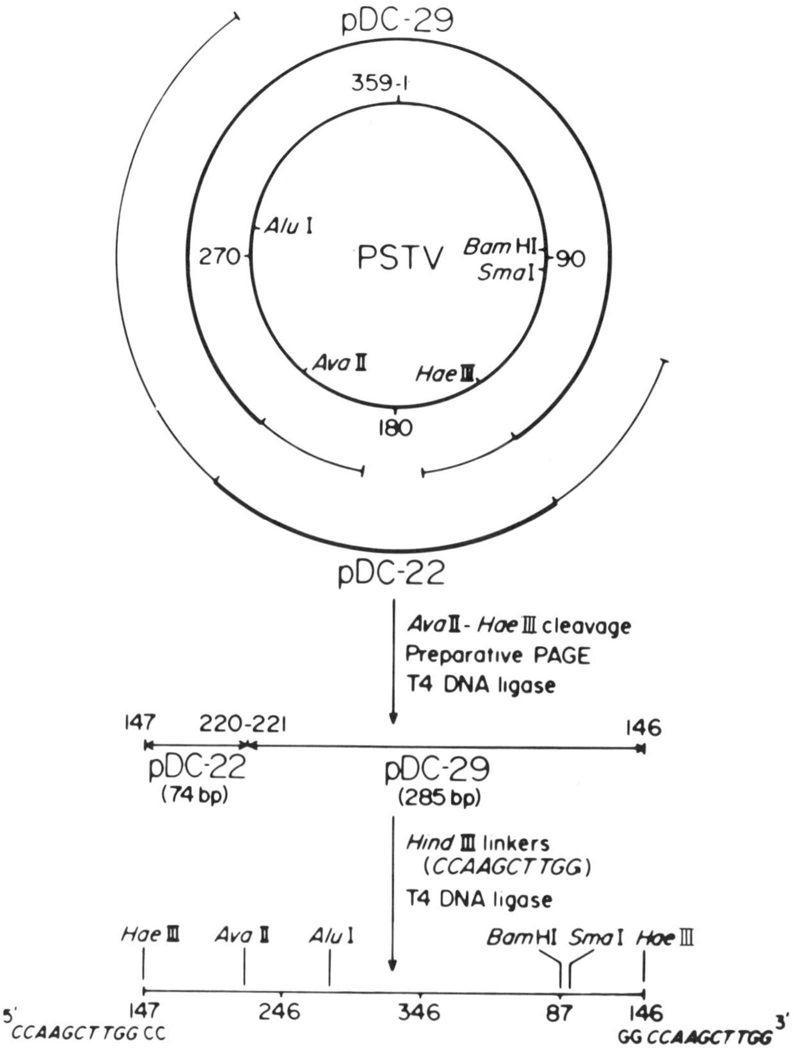

Figure 13.1. Construction and molecular cloning of full-length dsPSTV cDNA.

of pBR322. *Hind*III digestion of the resulting clones released the expected 365-bp PSTV-specific fragment.

Since use of this particular *Hind*III decanucleotide linker reconstructed the terminal *Hae*III sites of the PSTV insert, digestion of the recombinant DNA with *Hae*III releases a 359-bp fragment. This fragment is a full-length dsPSTV cDNA. Determination of the complete nucleotide sequence of this DNA by a combination of the Maxam-Gilbert chemical degradation and the M13 dideoxy chain termination methodologies has shown the cloned DNA sequence to be identical to that predicted by the RNA sequence published by Gross et al. (1978).

SENSITIVE AND RAPID DIAGNOSIS OF PSTV

Practical consequences of PSTV infection. The spindle tuber disease of potato was first described by Martin (1922) in Irish Cobbler potatoes being grown for seed purposes in southern New Jersey. The initial source of PSTV seems to have been potatoes imported from Maine, where the disease had long been recognized by local growers (Schultz and Folsom 1923). The disease is common in the potato-growing regions of northern and northeastern USA and Canada and has been reported to be one component of the potato "gothic" virus complex in the USSR (reviewed by Diener, 1979a).

A number of independently isolated PSTV strains have been described and can be classified as either severe or mild strains on the basis of the symptoms produced in Rutgers tomato. Tomato has been used as a diagnostic host for PSTV because symptoms in some potato cultivars are indistinct. Severe strains cause extreme shortening of the internodes, severe epinasty, shortening of petioles and midribs, and necrosis of stems, petioles, and midribs in tomatoes. Symptoms of mild strains, on the other hand, are slow to develop and often so mild that they are easily overlooked (Fernow 1967). The complete nucleotide sequences of a severe and a mild strain of PSTV differ in only three of the 359 nucleotides (Gross et al, 1981).

Depending on the cultivar, strain of PSTV present, and environmental conditions, symptoms of PSTV infection in potato may vary considerably. Foliage symptoms are often obscure, and severity of the characteristic tuber symptoms—elongation with the appearance of prominent bud scales and growth cracks—depend upon temperature and the length of infection. While the potato spindle tuber disease causes neither total crop destruction nor storage losses, it can cause a serious reduction in total production. As long as control is effective in keeping the incidence of disease low, yield losses are of little consequence in temperate growing areas (1%–2%). Inadequate control, however, can lead to catastrophic losses (~20% for mild strains or ~60% for severe strains) in a relatively short time (reviewed by Diener 1979a).

Although symptoms are often indistinct or lacking at moderate temperature, PSTV may cause severe damage or even total crop loss in plants grown at high temperature. Major efforts are underway to supplement the world's food supply by adapting the potato to growth in subtropical and tropical climates,

Tissue sample

↓ *Homogenize in* 200 mM K_2HPO_4
10 mM DIECA
5 mM DTT
0.1% Triton X-100
Clinical centrifugation

Clarified Sap

↓ *Transfer 3-5 μl aliquot to nitrocellulose membrane*
Bake 2 hrs @80°C in vacuo

PSTV Bound to Nitrocellulose Membrane

↓ *Hybridization with* [^{32}P] *recombinant DNA*
Wash @55°C

[^{32}P] DNA•PSTV Hybrids Bound to Nitrocellulose Membrane

↓ *X-ray film*
Intensifying screen

Detection of Hybrids by Autoradiography

Figure 13.2. PSTV detection by nucleic acid hybridization.

and a lowland tropical potato cultivar has already been developed (Sawyer 1979). PSTV poses a potentially serious threat to these efforts because of its ready transmission through vegetative propagation, foliage contact, and true seed and pollen. The importance of a sensitive, rapid, and reliable method for PSTV diagnosis is apparent.

Because PSTV lacks the antigenic protein coat characteristic of viruses, an assay based on the sensitive and widely used ELISA (enzyme-linked immunosorbent assay) technique has not been reported. Bioassay on suitable tomato cultivars (Raymer and O'Brien 1962) and polyacrylamide gel electrophoresis of extracted nucleic acids (for example, Pfannenstiel, Slack, and Lane 1980) have been used to detect PSTV. Although Fernow, Peterson, and Plaisted (1969) have demonstrated that a double-inoculation technique can be used to detect and eliminate both mild and severe strains of PSTV from potato seed stocks before planting, bioassays on tomato are slow and often unreliable. Polyacrylamide gel electrophoresis methods are laborious and expensive. Neither method is suitable for the rapid screening of thousands of seed potato tubers. A third method, hybridization of highly radioactive PSTV cDNA with PSTV bound to a membrane filter and autoradiographic dection of the resulting DNA-RNA hybrids, is sensitive, rapid, and reliable (Owens and Diener 1981). Automation of the testing procedure appears feasible.

PSTV diagnosis by nucleic acid hybridization. The steps involved in the diagnosis of PSTV infection by nucleic acid hybridization are schematically depicted in Figure 13.2. To simplify sample preparation when large numbers of samples must be processed, clarified plant sap rather than purified nucleic acid serves as the source of PSTV. The relatively high ionic strength and diethyldithiocarbamate concentration of the extraction buffer are required to release PSTV from nuclei and inhibit enzymatic polyphenol oxidation. After any PSTV present in the sap samples is bound to nitrocellulose membranes, the immobilized PSTV is hybridized with [^{32}P] recombinant DNA and the resulting [^{32}P] PSTV cDNA-PSTV hybrids detected by autoradiography.

Even though the presence of sap constituents (presumably proteins) reduces the binding of PSTV to the membrane approximately tenfold compared to purified nucleic acid samples, <100 pg of PSTV can still be detected when the specific activity of the [^{32}P] PSTV cDNA probe is 2×10^8 cpm/mg (Owens and Diener 1981). The entire procedure requires 4 days for completion and is approximately ten times more sensitive than polyacrylamide gel electrophoresis. A significant improvement over the tomato bioassay method (Fernow, Peterson, and Plaisted et al. 1969) is the ability to detect both mild and severe strains *directly*.

The feasibility of this method for the rapid and sensitive detection of PSTV by nucleic acid hybridization was demonstrated with tubers harvested from commercially important U.S. potato varieties intentionally infected with PSTV (Owens and Diener 1981). Because PSTV poses a significant threat to potato germplasm collections and potato-breeding programs aimed at adapting the potato to cultivation in warm climates, we have extended our initial studies in collaboration with Dr. Luis F. Salazar at the International Potato Center (CIP), Lima, Peru. Table 13.1 summarizes some of the results obtained.

Tubers from 24 of the many potato clones currently under investigation at CIP were selected essentially at random for testing at Beltsville. Most of the selections had been tested by polyacrylamide gel electrophoresis at CIP, but the results of these analyses were not revealed to us until the tests at the Beltsville Agricultural Research Center (BARC) were complete. As expected, clones which had tested "PSTV plus" by polyacrylamide gel electrophoresis at CIP also tested "PSTV plus" by nucleic acid hybridization at BARC (results not shown). Comparison of the polyacrylamide gel electrophoresis and initial hybridization results for nine clones classified as "PSTV minus" by polyacrylamide gel electrophoresis at CIP, however, suggested that eight of the nine were actually "PSTV plus" (Table 13–1, columns labeled "original results"). This result was quite gratifying because it seemed to confirm the greater sensitivity of the hybridization technique.

Three different analyses were performed at BARC to confirm these initial results: bioassay on tomato, polyacrylamide gel electrophoresis of low molecular RNA extracted from foliage of the inoculated tomatoes, and a second hybridization analysis of those potato clones where the tuber originally tested was no longer available. Hybridization analysis of tomato foliage was also used to distinguish uninfected plants from those containing mild strains of PSTV in the bioassay. Except for clone MB 15.25II, which had given an ambiguous

Table 13.1 Screening of Potato Tuber Clones for the Presence of PSTV

	Original results		New results		
Clone	PAGE[a] (CIP)	cDNA (BARC)	Bioassay	cDNA	PAGE[a]
BR 63.15.I	–[b]	+[b]	Mild	+	
BR 63.15 II	–	–	T-1[c]: – T-2[c]:Mild	– +[d]	– +
MB 6.II	+	+	Severe	+	
MB 6.42	–	+	Severe	+	
MB 6.1	–	+	Mild	+	
MB 30.2	–	+	Mild	+	+
MB 34.97 I	–	+	Mild	+	+
MB 15.25 II	–	±	–	–	
MB 117.36	–	+	Severe	+	
PSP 5.6	–	–	T-1:Not Done, tuber lost T-2:Severe	+[d]	+

[a]Polyacrylamide gel electrophoresis.
[b]The presence of PSTV is indicated by "+"; the absence of PSTV is indicated by "–".
[c]T-1 and T-2 designate individual tubers from a single clone.
[d]Positive reaction also with extract from tuber.

result in the first hybridization analysis, our initial results were confirmed. One possible explanation for our inability to transmit PSTV from clone MB 15.25II to tomato is a low PSTV titer in that particular tuber. This possibility is supported by the results obtained with clones BR 63.15II and PSP 5.6, where only one of the two tubers received proved to be infected.

The results summarized in Table 13.1 extend our previously published results (Owens and Diener 1981) and illustrate the potential usefulness of cloned PSTV cDNA to current efforts to adapt the potato to cultivation in warm climates.

MECHANISMS OF VIROID-HOST INTERACTION

As mentioned earlier, viroids were the first naturally occurring examples of covalently closed circular RNA to be described (Gross et al. 1978). The origin of viroids is unknown, but with the discovery of split genes and RNA splicing in eucaryotic organisms it has been suggested that viroids may have originated by circularization of excised intervening sequences (Diener 1979b; Crick 1979). If such introns contained the appropriate recognition sequences, they might be transcribed (replicated) by a host enzyme able to function as an RNA-directed RNA polymerase and thus escape host cell control mechanisms. Such a hypothetical model for the origin of viroids has several interesting implications for the mechanisms of viroid-host interaction because disease induction is most readily understood by postulating interference by the viroid with gene regulation (Diener 1979b).

Comparisons of the nucleotide sequences of PSTV and PSTV-complementary RNA (Owens and Cress 1980) with the consensus sequence of eucaryotic intron-exon boundaries and the 5′-sequence of U1 RNA have been independently performed by three separate investigators (Dickson 1981; Diener 1981; Gross et al. 1981). Both Diener (1981) and Gross et al. (1982) have noted the striking complementarity between the 5′-terminus of U1 RNA and nucleotides 257 to 279 of the PSTV complement. Comparison of the corresponding regions from two other viroids, chrysanthemum stunt and citrus exocortis, has shown that this region is highly conserved (Gross et al. 1982).

Dickson (1981) has proposed a different model for the involvement of viroids in RNA splicing—one involving complementarity between plant intron sequences and two separate regions of PSTV itself, nucleotides 113 to 118 and 307 to 311. The minor sequence differences between a mild and a severe strain of PSTV fall within these regions (Gross et al. 1981), and a comparison of the sequences of PSTV and chrysanthemum stunt viroid showed that the degree of complementarity to plant mRNA splice junctions could be correlated with severity of disease induced in chrysanthemum. Although these hypotheses are intriguing and susceptible to experimental verification as described below, the mechanism(s) of viroid pathogenesis must be more complex because neither avocado sunblotch viroid nor its complement contains these sequences (Symons 1981).

Genetic analysis of viroid-host interaction depends on the isolation and characterization (sequencing) of mutants with defined phenotypic differences. Until now such analyses (i.e., those discussed in the preceding paragraph) have depended upon sequence comparisons of known viroids or viroid strains. Lack of reliable local lesion hosts has hampered efforts to select desired viroid mutants following random mutagenesis. The development of methods for specific cleavage and enzymatic manipulation of DNA, cloning of DNA fragments, nucleotide sequence analysis, and rapid chemical synthesis of oligonucleotides of defined sequence allows mutations to be constructed at predetermined sites in a cloned DNA molecule (reviewed by Shortle, DiMaio, and Nathans 1981). Racaniello and Baltimore (1981) have shown that transfection of cultured mammalian cells with pBR322 recombinant plasmids containing a complete cloned cDNA copy of the RNA genome of poliovirus leads to the synthesis of infectious poliovirions. If the same would be true for cloned PSTV cDNA, genetic analysis of viroid-host interaction would be greatly facilitated.

Figure 13.3 shows a specific example of the potential power of the directed mutagenesis technique as applied to viroids. The underlined regions of the sequence of PSTV severe strain are those whose possible involvement in viroid-host interaction was discussed above. Cloned PSTV cDNA contains two recognition-cleavage sites for the restriction endonuclease *MspI,* and one of those sites falls within the region (nucleotides 257–279) identified by two

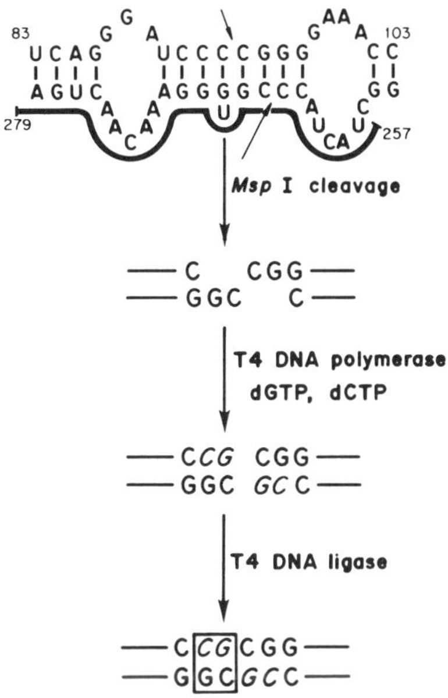

Figure 13.3. Identification of viroid-host interactions by directed mutagenesis of cloned dsPSTV cDNA. Regions of the PSTV or complementary RNA sequence whose involvement in viroid-host interaction has been postulated are underlined. The sites of cleavage of the dsPSTV cDNA by *Msp*I endonuclease are indicated by arrows. Sequential treatment of recombinant DNAs containing dsPSTV cDNA with *Msp*I, T4 DNA polymerase, and T4 DNA ligase would allow isolation of derivatives containing a 361 bp dsPSTV cDNA. Half of these derivatives would contain a 2-bp insertion in the region (nucleotides 257–279) that exhibits a striking complementarity with the 5′-terminus of U1 RNA.

groups (Diener 1981; Gross et al. 1982) as possibly involved in PSTV pathogenesis. If the infectivity of cloned PSTV DNA can be demonstrated, it would be comparatively simple to introduce defined nucleotide insertions (shown in Fig. 13.3) or deletions into this region and determine the biological effect of these changes. Introduction of defined mutations in this and other regions of the PSTV sequence would complement current attempts to correlate viroid structure and function by RNA sequence comparison.

CONCLUSIONS

Although cloned viroid cDNAs have been available only for approximately two years, they have already been applied to problems in both practical and basic agricultural research. Hybridization between highly radioactive recombinant PSTV cDNA and PSTV that has been attached to a solid support provides a sensitive and reliable method for the detection of PSTV (Owens and Diener 1981). This technique will facilitate elimination of PSTV from potato varieties that are currently being developed for growth in tropical and subtropical climates. Using the appropriate homologous viroid cDNA clones, the same procedure could be used to detect other economically important viroids.

Cloned partial-length viroid cDNAs have also been important tools in basic studies of viroid replication and structure. Owens and Cress (1980) used strand-specific hybridization probes to demonstrate that PSTV-infected tomato tissue contains full-length RNA molecules complementary to PSTV, and Symons (1981) has used viroid cDNA fragments as primers for cDNA synthesis during the determination of the complete 247-nucleotide sequence of avocado sunblotch viroid. We may anticipate even more exciting results in the future if the infectivity of newly available, full-length PSTV cDNA clones can be demonstrated. Directed mutagenesis of infectious cloned PSTV cDNA will greatly facilitate genetic analysis of viroid-host interactions.

LITERATURE CITED

Crick, F. 1979. *Split genes and RNA splicing*. Science 204:264–71.

Dickson, E. 1981. *A model for the involvement of viroids in RNA splicing*. Virology 115:216–21.

Diener, T. O. 1979a. *Viroids and viroid diseases*. John Wiley & Sons, New York.

———. 1979b. *Viroids: Structure and function*. Science 205:859–66.

———. 1981. *Are viroids escaped introns?* Proc. Nat. Acad. Sci. USA 78:5014–15.

Fernow, K. H. 1967. *Tomato as a test plant for detecting mild strains of potato spindle tuber virus*. Phytopathol. 57:1347–52.

Fernow, K. H., L. C. Peterson, and R. L. Plaisted. 1969. *The tomato test for eliminating spindle tuber from potato planting stock*. Am. Potato J. 46:424–9.

Gross, H. J., H. Domdey, C. Lossow, P. Jank, M. Raba, H. Alberty, and H. L. Sänger. 1978. *Nucleotide sequence and secondary structure of potato spindle tuber viroid*. Nature (Lond.) 278:203–8.

Gross, H. J., G. Krupp, H. Domdey, M. Raba, P. Jank, C. Lossow, H. Alberty, K. Ramm, and H. L. Sänger. 1982. *Nucleotide sequence and secondary structure of citrus exocortis and chrysanthemum stunt viroid*. Eur. J. Biochem. 121:249–57.

Gross, H. J., U. Liebl, H. Alberty, G. Krupp, H. Domdey, K. Ramm, and H. L. Sänger. 1981. *A severe and a mild potato spindle tuber viroid isolate differ in three nucleotide exchanges only*. Biosci. Reports 1:235–41.

Gross, H. J., and D. Riesner. 1980. *Viroids: A class of subviral pathogens*. Angewandte Chemie 19:231–43.

Martin, W. H. 1922. *"Spindle Tuber," a new potato trouble*. Hints to Potato Growers, N. J. State Potato Assoc. 3(8):4pp.

Owens, R. A., an D. E. Cress. 1980. *Molecular cloning and characterization of potato spindle tuber viroid cDNA sequences*. Proc. Nat. Acad. Sci. USA 77:5302–6.

Owens, R. A., and T. O. Diener. 1981. *Sensitive and rapid diagnosis of potato spindle tuber viroid disease by nucleic hybridization*. Science 213:670–2.

Pfannenstiel, M. A., S. A. Slack, and L. C. Lane. 1980. *Detection of potato spindle tuber viroid in field-grown potatoes by an improved electrophoretic assay*. Phytopathol. 70:1015–18.

Racaniello, V. R., and D. Baltimore. 1981. *Cloned poliovirus complementary DNA is infectious in mammalian cells*. Science 214:916–19.

Raymer, W. B. and M. J. O'Brien. 1962. *Transmission of potato spindle tuber virus to tomato*. Am. Potato J. 39:401–8.

Sawyer, R. L. 1979. *Annual Report, 1978*. International Potato Center, Lima, Peru, p. iii.

Schultz, E. S. and D. Folsom. 1923. *Transmission, variation, and control of certain degeneration diseases of Irish potatoes*. J. Agr. Res. 25:43–117.

Shortle, D., D. DiMaio, and D. Nathans. 1981. *Directed mutagenesis* Ann. Rev. Genet. 15:264–94.

Symons, R. H. 1981. *Avocado sunblotch viroid: Primary sequence and proposed secondary structure*. Nucl. Acids Res. 9:6527–37.

five

GENETIC MODIFICATION

14] Gene Vectors for Higher Plants

by J. SCHELL,*† M. VAN MONTAGU,† **
J. P. HERNALSTEENS,** H. DE GREVE,**
J. LEEMANS,** C. KONCZ,* †† L. WILLMITZER,*
L. OTTEN,* J. and G. SCHRÖDER*

ABSTRACT

Tumor-inducing (Ti) plasmids, carried by *Agrobacterium tumefaciens* have been shown to be responsible for crown gall formation in plants. Ti plasmids are natural gene vectors with which *Agrobacteria* achieve the transfer and stable maintenance of a defined DNA segment (called T-region) into the nucleus of transformed plant cells. Using site-specific mutagenesis, it was possible to introduce mutations in different parts of the T-region. The transcription of the T-DNA in wild-type and mutant crown galls was compared, and it was found that the induction of specific developmental patterns could be correlated with the absence of specific T-DNA transcripts. Double mutants were obtained in which the expression of all the "onc" genes was abolished. Tobacco, potato, and petunia plant cells harboring such inactivated T-DNAs were shown to regenerate normal, fertile plants that transmit the T-DNA segment as a single Mendelian locus. Structural genes, coding for opine synthase enzymes, were shown to be fully active. Several T-DNA genes were sequenced, and transcription promoter and termination signals were identified.

INTRODUCTION

Convincing evidence has accumulated over the past few years indicating that tumor-inducing (Ti) plasmids in *Agrobacterium* are efficient natural gene vectors which transfer a number of genes into the plant nucleus, where they are stably maintained by covalent integration in plant chromosomal DNA and

*Max-Planck-Institut für Züchtungsforschung, 5000 Cologne 30, FRG; †Laboratorium voor Genetika, Rijksuniversiteit Gent, 9000 Gent, Belgium; **Vrije Universiteit Brussels, Laboratorium voor Genetische Virologie, 1040 St. Genesius-Rode, Belgium; and ††Biological Research Center of Hungarian Academy of Sciences, Institute of Genetics, 6701 Szeged, Hungary.

GENETIC ENGINEERING: APPLICATIONS TO AGRICULTURE (Beltsville Symposium 7—Lowell D. Owens, ed.) Rowman & Allanheld, Totowa

expressed by the plant transcriptional and translational system. Today no other gene-vector system has yet been developed to a similar degree. We shall therefore confine this chapter to a review of the recent work with Ti plasmids and to a description of the results we obtained in our attempts to develop Ti plasmids as practical experimental gene vectors for plants.

Ti plasmids are harbored by a group of soil bacteria *(Agrobacteria)* and are responsible for the capacity of these bacteria to induce so-called crown gall tumors on most dicotyledonous plants. (For recent reviews see Gordon 1979; Schell et al. 1979; Schilperoort et al 1979; Van Montagu and Schell 1979; Schell and Van Montagu 1980; Schilperoort et al. 1980; Van Montagu et al. 1980.) The plant tumor cells, unlike nontransformed plant tissues, can be cultured under axenic conditions on synthetic media in the absence of growth hormones, i.e., cytokinins and auxins (Braun 1956). The tumor cells also produce low molecular weight compounds, called opines, not found in untransformed plant tissues. The type of opine produced defines crown galls as octopine-, nopaline-, or agropine-type tumors (Guyon et al. 1980). Ti plasmids have been found to be responsible for most of the typical properties of *Agrobacteria*: (a) crown gall tumor induction; (b) specificity of opine synthesis in transformed plant cells; (c) catabolism of specific opines; (d) agrocin sensitivity; (e) conjugative transfer of Ti plasmids; and (f) catabolism of arginine and ornithine (Van Larebeke et al. 1974; Zaenen et al. 1974; Engler et al. 1975; Schell 1975; Van Larebeke et al. 1975; Watson et al. 1975; Bomhoff et al. 1976; Genetello et al. 1977; Kerr, Manigault, and Tempé 1977; Firmin and Fenwick 1978; Klapwijk, Scheulderman, and Schilperoort 1978; Petit, Dessaux, and Tempé 1978; Petit et al. 1978; Ellis et al. 1979; and Guyon et al. 1980). The crown gall tumors contain a DNA segment (called T-DNA) derived from Ti plasmids that is homologous and colinear with a defined fragment of the corresponding Ti plasmid present in the tumor-inducing bacterium (Chilton et al. 1977; Schell et al. 1979; Lemmers et al. 1980; Thomashow et al. 1980a; De Beuckeleer et al. 1981). The T-DNA is covalently linked to plant DNA (Yadav et al. 1980; Zambryski et al. 1980; Thomashow et al. 1980b) in the nucleus of the plant cell (Chilton et al. 1980; Willmitzer et al. 1980).

The T-DNA is transcribed in the transformed plant cell, and T-DNA-encoded proteins have been found in several octopine crown gall lines, one of them being the octopine-synthesizing enzyme lysopine dehydrogenase (LpDH) (Schröder et al. 1981a). Therefore, Ti plasmids are a natural gene vector for plant cells, evolved by and for the benefit of the bacteria that harbour Ti plasmids (Schell et al. 1979). Crown gall cells, as a direct result of genetic transformation by the Ti plasmid, produce various opines. Free-living *Agrobacteria* utilize these opines as sources of carbon and nitrogen. The genetic information both for the synthesis of opines in transformed plant cells and for their catabolism by free-living *Agrobacteria* is carried by Ti plasmids. Bacteria are known to be able to conquer an ecological niche by acquiring the capacity to catabolize certain organic compounds not readily degradable by most other bacterial species. In several cases, the genes determining this degradative capacity have been found to be part of extrachromosomal plasmids. Several groups of soil bacteria, especially those living in and around the rhizosphere of plants, are able to decompose organic compounds released by plants. Clearly, with the advent of Ti plasmids, *Agrobacteria* have carried this capacity one step further by genetically forcing plant cells—via a gene transfer mechanism—to produce specific compounds (opines) that they are uniquely equipped to catabolize. This novel type of parasitism has therefore been called "genetic colonization" (Schell et al. 1979). The question of whether and how the Ti plasmid can be used and developed as a general, experimental gene vector for plants (Schell and Van Montagu 1977; De Beuckeleer et al. 1978) is approached in this chapter.

THE DEVELOPMENT OF THE TI PLASMID AS AN EXPERIMENTAL GENE VECTOR

The T-DNA region is defined as that segment from the Ti plasmid that is homologous to sequences present in crown gall cells. The sequences which are transferred from the Ti plasmid to the plant and determine tumorous growth have been called T-DNA. The T-region of octopine and nopaline Ti plasmids has been studied in great detail both physically and functionally. The T-regions, roughly 23 kbp in size, are only a portion of the entire plasmids (Lemmers et al. 1980; Thomashow et al. 1980a; De Beuckeleer et al. 1981; Engler et al. 1981). Southern blotting and cross-hybridization of restriction endonuclease digests of the two types of plasmids as well as electron microscope heteroduplex analyses have revealed that 8 to 9 kbp of the T-DNA regions are conserved and common to both octopine and nopaline types of plasmids (Chilton et al. 1978; Depicker, Van Montagu, and Schell 1978; Engler et al. 1981). All attempts to reveal homology between the T-DNA region of either octopine or nopaline Ti plasmids and plant DNA have failed thus far.

Detailed analysis of some nopaline lines (Lemmers et al. 1980; Yadav et al. 1980; Zambryski et al. 1980, 1982) suggests that the mechanism of T-DNA transfer is rather precise, since the same continuous segment of the Ti plasmid is always present. Some lines appear to contain a single T-DNA copy, whereas in others the T-DNA occurs in multiple copies that are organized in a tandem array.

Several octopine tumor lines have been studied, and the data suggest that the octopine T-DNA is more variable (Thomashow et al. 1980a, b; De Beuckeleer et al. 1981; Ooms et al. 1982). A left T-DNA region (TL) containing the conserved sequence (which is also part of the nopaline T-DNA region) is always present; this region is usually 12 kbp in size, but one *Petunia* tumor line is shortened at the right end of TL by about 4 kbp (De Beuckeleer et al. 1981). In addition, there is often a right T-DNA region (TR) which contains sequences which are adjacent but not contiguous in the octopine Ti plasmid. In some tumor lines TR is amplified, whereas TL is not (Merlo et al. 1980; Thomashow et al. 1980a). Recent observations indicate that TL can also be part of a tandem array (Holsters et al. 1983).

T-DNAs of nopaline and octopine tumor lines are not identical. Yet there are some similarities: both T-DNAs contain a common or conserved DNA, and both T-DNAs can be amplified either as a whole, as in the nopaline T-DNA, or in parts, as in octopine T-DNA. It is not known whether integration is the result of plant or Ti plasmid-specific functions, but it is likely that both are involved.

Genes inserted in the T-region of Ti plasmid are co-transferred to the plant nucleus as part of the T-DNA. In view of the observed involvement of the "ends" of the T-region in the integration of T-DNA, it could be expected that any DNA segment inserted between these "ends" would be co-transferred, provided no function essential for T-DNA transfer and stable maintenance was inactivated by the insertion. The genetic analysis of the T-region by transposon-insertion provided Ti plasmid mutants to test this hypothesis. A Tn7 insertion in the nopaline synthase locus produced a Ti plasmid able to initiate T-DNA transfer and tumor formation (De Beuckeleer et al. 1978; Van Montagu and Schell 1979; Hernalsteens et al. 1980). Analysis of the DNA extracted from these tumors showed that the T-region containing the Tn 7 sequence had been transformed as a single, 38-kbp segment without any major rearrangements.

Several different DNA sequences have since been introduced in different parts of the T-region of octopine and nopaline Ti plasmids (see, e.g., Leemans et al. 1981, 1982b). The preliminary observations with these mutant Ti plasmids fully confirm our initial expectations: most DNA sequences inserted between the "ends" of the T-region are co-transferred with and become a stable part of the T-DNA of the plant tumor cells transformed with these mutant Ti plasmids. If the experimental insert were to inactivate a function essential for the transfer of the T-region or for the integration of the T-DNA, such a mutant Ti plasmid would not be able to transform plant cells. With one possible exception, no such inserts have as yet been characterized. This indicates that T-DNA transfer and integration is either coded for by genes outside the T-region, by a very limited number of genes in the T-region, or by a combination of functions with different genetic localization.

These observations have, therefore, firmly established that the Ti plasmids can be used as experimental gene vectors and that large DNA sequences (of at least 28 kbp) can be transferred stably to the nucleus of plant cells as a single DNA segment.

The T-DNA contains transcription promoter signals. The concentration of total T-DNA-specific RNA in the octopine tumor cell line A6-S1 is between 0.0005 and 0.001%. Eight polyadenylated RNAs have been identified that range in size from approximately 900 to 3000 nucleotides and vary considerably in their relative abundance (Willmitzer et al. 1981; Willmitzer, Schmalenbach, and Schell 1981; Willmitzer, Simons, and Schell 1982). Probably all these transcripts are initiated from promoters within the T-DNA. One of the more abundant RNAs was shown to be the mRNA coding for the octopine synthase (or lysopine dehydrogenase) (Schröder et al. 1981a, b). The 5′ end of the mRNA was accurately mapped by sequencing a T-region DNA fragment hybridized to this mRNA and thus protected from degradation by the single-strand–specific S1 nuclease (De Greve et al. 1982b). The complete base sequence of the octopine synthase gene was determined.

The promoter sequence that has thus been identified is more eucaryotic than procaryotic in its recognition signals, and no introns interrupt the open reading frame which starts at the first AUG codon following the 5′ start of the transcript. Similar work was also performed for the nopaline synthase gene (Depicker et al. 1982; Barnes and Chilton 1983). Essentially the same conclusions were reached.

It is important to note that the promoter sequences for the octopine synthase appear to escape possible control mechanisms since they remain active in all tissues of plants regenerated from tobacco cells transformed with mutant Ti plasmids (De Greve et al. 1982a). To further test the notion that the sequence immediately upstream of the sequence coding for the 5′ end of the transcript is the functional promoter for the plant genes and to answer the question of how many nucleotide base pairs (bp) upstream of this 5′ transcription start constitute a functional promoter, the following experiments were performed (C. Koncz and J. Schell, unpublished): The *Bam*H1 fragment 17a from the octopine Ti plasmid Ach5 T-region was cloned into the *Hind*III-23 fragment of the T-region of pTiC58 (a nopaline Ti plasmid). This *Bam*H1-17a fragment contains the whole of the octopine synthase structural gene together with 140 bp upstream of the 5′ transcription start. This cloned fragment was introduced in the pTiC58 plasmid via an intermediate vector as described by Leemans et al. (1982b). The *Agrobacterium* strain carrying this recombinant plasmid induced normal tumors on tobacco, in which, however, no octopine synthase activity could be detected.

In an analogous experiment a fragment containing about 300 bp upstream of the 5′ transcription start was transferred to tobacco cells after introduction in the T-region of a vector Ti plasmid. In this case normal expression of the introduced opine synthase gene was observed.

These experiments demonstrate that these plant gene promoters indeed determine an independent and functional transcription unit and consist of between 140 and 300 bp upstream of the 5′ initiation point of transcription. These observations are crucial to the development of the Ti plasmid as a gene vector, since they provide us with the possibility of inserting coding sequences behind these promoters with a reasonable chance that these sequences will be expressed in plant cells. In this respect we also studied the expression of the Tn7 insertion in the nopaline synthase locus as a model system for other genes inserted in this area. Tn7 codes for a dihydrofolate reductase, which is resistant to methotrexate (Tennhammer-Ekman and Sköld 1979). Suspension cultures of both untransformed and crown gall tobacco tissue were found to be completely inhibited by 2 μg/ml methotrexate. In contrast, a culture established with crown gall cells containing Tn7 grew well on media containing 2 μg/ml methotrexate. In addition, nuclei isolated from these methotrexate-resistant tobacco lines were shown to synthesize Tn7 transcripts homologous to Tn7, and a poly(A) mRNA fraction derived from purified polysomes was also shown to contain Tn7 transcripts. These initial positive results led us to the optimistic view that possibly the plant transcriptional mechanism would accomodate procaryotic promoters. We therefore transferred a number of procaryotic genes into tobacco by inserting them into the nopaline synthase gene of pTiC58. In no other instance, however, could we observe transcription of a procaryotic promotor introduced into tobacco via the Ti plasmid. The exact reason for the exceptional transcription of Tn7 remains to be elucidated.

In a similar way the human gene for fibroblast interferon was transferred to tobacco by insertion into the nopaline synthase locus of pTiC58 (J. Leemans, M. Van Montagu, W. Fiers, and J. Schell, personal communication). No transcription could be observed from this animal promoter. It is therefore very likely that expression of "foreign" genes in plants will require precise constructions linking the coding sequences of these genes to appropriate plant promoters. Such constructions are presently being made in our laboratory by use of octopine and nopaline synthase promoters to initiate transcription. The 3′ polyadenylated terminus of the octopine synthase transcript was also analyzed and a polyadenylation signal 5′-AAUAA-3′ was found about ten nucleotides from the start of the poly(A) sequence. This appears to be a general feature of eucaryotic mRNAs, since it has been observed in a number of animal mRNAs (Fitzgerald and Schenk 1981). It would therefore appear that these opine-synthesizing genes were designated to function in eucaryotic cells rather than in procaryotic cells.

Functions coded for by T-DNA. RNA transcripts homologous to T-DNA have been shown to be present in all crown gall tissues studied thus far (Drummond et al. 1977; Gurley et al. 1979; Gelvin et al. 1981; and Willmitzer et al. 1981a). In order to determine whether these transcripts determine well-defined functions, three complementary approaches have been used.

1. The number, sizes, and location of the transcribed T-DNA segments were studied in both octopine and nopaline tumors (Willmitzer, Simon and Schell 1982; Willmitzer et al. 1983). Tumor-specific RNAs were detected and mapped by hybridization of ^{32}P-labeled Ti plasmid fragments to polyadenylated RNA which had been separated on agarose gels and then transferred to DBM paper. The results show that octopine tumors contain a total of eleven distinct transcripts that differ in their relative abundance and in their sizes. They all bind to oligo(dT)-cellulose, indicating that they are polyadenylated. Thus, the T-DNA transferred from a procaryotic organism provides specific poly(A)-addition sites. The direction of transcription was determined, and the locations of the approximate 5′- and 3′-ends were mapped on the TL- and TR-DNA.

All RNAs mapped within the T-DNA sequence. This, and the observation that transcription is inhibited by low concentrations of α-amanitin (Willmitzer, Schmalenbach and Schell 1981), suggests that each transcript is determined by a specific promoter site on the T-DNA recognized by plant RNA polymerase II. Considering the groupwise orientation of several transcripts, the simplest model would assume one promoter site per group of transcripts. If so, one would expect that inactivation of a 5′-proximal gene of a group would also lead to disappearance of the transcripts from the 5′-distal genes. Nevertheless, analysis of some cell lines containing the T-DNA of Ti plasmid mutants indicated that groupwise inactivation of genes does not occur (Leemans et al.

1982a). The results available so far are consistent with the assumption that each gene on the T-DNA has its own signals for transcription in the eucaryotic plant cells.

2. In order to determine whether these mRNAs are translated into proteins, a hybridization selection procedure was developed that was sufficiently sensitive and specific to detect mRNAs that represent about 0.0001% of the total mRNA activity in the plant cells. This procedure was used to enrich for T-DNA–derived mRNAs by hybridization to Ti plasmid fragments covalently bound to microcrystalline cellulose; the hybridized RNAs were eluted and translated in vitro in a cell-free system prepared from wheat germ.

The results obtained with this approach (Schröder and Schröder 1982) showed that tumor cells contain at least three T-DNA-derived mRNAs that are translated in vitro into distinct proteins, and that the coding regions correlate with those of three of the previously identified transcripts. The protein encoded at the right end of the TL-DNA (Mw 39,000) was of particular interest, since previous genetic analysis indicated that this region is responsible for octopine synthesis (Koekman et al. 1979; De Greve et al. 1981; Garfinkel et al. 1981).

The in vitro-synthesized protein was shown to be identical in size to the octopine-synthesizing enzyme in octopine tumors. Immunological studies showed that this protein was recognized by antiserum against the tumor-specific synthase (Schröder et al. 1981b). These results demonstrate that the structural gene for the octopine-synthesizing enzyme is on the Ti plasmid. So far, this is the only protein product of the T-DNA with known enzymatic properties; the possible functions of two smaller T-DNA–derived proteins are not yet known.

As noted above, the mRNAs for the three in vitro–synthesized proteins each represent about 0.0001% of the total mRNA activity in polyribonsomal RNA, and this appears to be the detection limit at present for translatable RNA. Some of the other transcripts detected by hybridization experiments are present at even lower concentrations. Assuming that they possess mRNA activity, this is likely to be the reason why the corresponding proteins have not so far been identified by in vitro translation.

A different approach was therefore developed to search for coding regions on the T-DNA and their protein products. Fragments from the T-region were cloned into *E.coli* plasmids and analyzed for gene expression in *E.coli* minicells (Schröder et al. 1981a). At least four different coding regions within the TL-DNA can be expressed in minicells into distinct proteins from promoters that are active in the procaryotic cells. The four regions expressed in *E.coli* correlated with four regions transcribed into RNA in plant cells. The plant transcripts are larger than the corresponding proteins in *E.coli,* and the regions expressed in minicells appear to lie within the regions transcribed in plant cells. One can therefore speculate that plant cells and *E.coli* at least partly express the same coding regions.

3. Specific mutations were introduced in the T-DNA regions of octopine and nopaline Ti plasmids to produce transformed plant cells in which one or more T-DNA–derived transcripts would not be expressed. By observing the pheno-

types of the plant cells harboring such partially inactivated T-DNAs, it was possible to assign functions to most of the different transcripts (Leemans et al. 1982a; Joos et al. 1983). Thus, it was found that none of the T-DNA transcripts was essential for the transfer and stable maintenance of T-DNA segments in the plant genome.

Essentially two different functions were found to be determined by T-DNA transcripts: (a) *Transcripts coding for opine synthases.* Octopine tumors contain either one or two such genes. One of them is located on the right end of TL and codes for octopine synthase; the other is located at the right end of TR and codes for agropine and mannopine synthase. Tumors that contain both TL and TR-DNA, therefore, produce both octopine and agropine, whereas tumors containing only TL-DNA produce octopine but not agropine. Nopaline tumors also contain at least two transcripts coding for different opines. One is located at the right end of the T-DNA and codes for nopaline synthase, whereas the other one is located in the left part of the T-DNA and codes for agrocinopine (Ellis and Murphy 1981). Nopaline tumors therefore usually produce both types of opines. (b) *Transcripts (after translation into proteins) directly or indirectly responsible for tumorous growth.* These transcripts were found to be derived from the T-DNA sequences that are common to both octopine and nopaline tumors. In total, six different well-defined transcripts were found to be derived from this common region of octopine and nopaline T-DNAs. Probably all of these transcripts act by suppressing plant organ development. It was observed that shoot and root formation are suppressed independently and by different transcripts. Thus, two transcripts (1 and 2) were identified that specifically prevent shoot formation. The effect of these T-DNA gene products is in many ways analogous to that of auxin-like plant growth hormones. Another transcript (4) was found to prevent root formation specifically, and the effect of this T-DNA gene can therefore be compared to the effects observed when normal plant cells are grown in the presence of high concentrations of cytokinins.

That both the shoot inhibition and the root inhibition resulting from the activity of these genes may be due to the fact that they directly or indirectly determine the formation of auxin- and cytokinin-like growth hormones is further substantiated by our observation that these genes, respectively, inhibit shoot or root formation both in T-DNA–containing and in T-DNA–negative (normal) cells, provided that both types of cells are growing as one mixed tissue. This interpretation of the possible function of transcripts 1 and 2 (auxin-like), on the one hand, and 4 (cytokinin-like), on the other hand, is also consistent with recent measurements of endogenous levels of auxin and cytokinin in teratoma and unorganized tobacco crown gall tumor tissues (Amasino and Miller 1982). Even more convincing are the recent data obtained by R. Morris (personal communication) indicating that the cytokinin-to-auxin ratio in wild-type tumors is 0.22, whereas it is 0.02 in root-forming tumors induced by Ti mutants in gene 4, and 14.4 in shoot-forming tumors induced by Ti mutants in genes 1 or 2.

In addition to this hormone-like activity, the T-DNA codes for at least three other transcripts (5, 6a, and 6b) that suppress development only in the cells in which they are formed. The combination of two of these transcripts (6a and 6b) with the cytokinin-like activity of transcript 4 was shown to be sufficient to suppress development of transformed cells and to allow their hormone-independent growth. Another transcript (5) was found to inhibit the organization of transformed cells into leaf bud structures. Elimination of this transcript, along with the shoot-inhibiting auxin-like genes (1 and 2) resulted in transformed cells organizing themselves as teratomas.

Normal and fertile plants can be regenerated from plant cells containing active T-DNA genes. The ultimate aim of many gene transfer attempts in plants is to produce fertile cultivars harboring and transmitting new genetic properties. It was, therefore, essential to determine whether T-DNA transfer could be dissociated from neoplastic transformation and whether normal, fertile plants could be derived from T-DNA–transformed plant cells. In order to answer this question a mutant plasmid (pGV2100), which was clearly less oncogenic on tobacco and sunflower hypocotyls when compared to the wild-type plasmid, was used. Tumors appeared only after prolonged incubation time. Furthermore, shoots proliferated from the greenish tumors, in contrast to the undifferentiating white tumors induced by strains harboring the wild-type TiB6S3 plasmid. The Tn7 insertion in pGV2100 was mapped and found to be located in the gene coding for transcript 2. Shoot-forming tumors induced with pGV2100 were shown to contain all TL-DNA transcripts except for transcript 2. As a test for transformation, tumor tissue and shoots on tobacco were assayed for the presence of lysopine dehydrogenase. The tumor tissue was found to be positive; most of the shoots were negative, but some of the proliferating shoots were positive. One such shoot was grown further on growth-hormone–free media and found to develop roots and later to grow into a fully normal, flowering plant. Each part of this plant—leaves, stem, and roots—was found to contain lysopine dehydrogenase activity, and polysomal RNA was found to contain T-DNA transcripts homologous to the opine synthesis locus. No transcripts of the conserved segment of the T-region were observed (De Greve et al. 1982a).

These observations, therefore, demonstrate that normal plants can be obtained from plant cells transformed with Ti plasmids genetically altered in specific segments of the T-region. Seeds obtained by self-fertilization of these plants produced new plants with active T-DNA–linked genes, thus demonstrating that genes introduced in plant nuclei, via the Ti plasmid, can be sexually inherited. A series of sexual crosses was therefore designed to study the transmission pattern of the T-DNA–specified genes. The results of these crosses are presented in Table 14.1 and demonstrate very convincingly that the T-DNA–linked genes (LpDH) are transmitted as a single Mendelian factor through both the pollen and the eggs of the originally transformed plant. These crosses also showed that the original transformed plant was a hemizygote containing T-DNA on only one locus of a pair of homologous chromosomes. By these crosses tobacco plants homozygous for the altered T-DNA were obtained (Otten et al. 1981).

Subsequent experiments have demonstrated that pGV2100 and other similar

Table 14.1 Mendelian Transmission of LpDH in rGV-1 Tobacco

Crosses Male	Crosses Female	Number of progeny tested		LpDH positive		LpDH negative
E40.2	X E40.2	145		110 (76%)		35 (24%)
			semi-quant. test	++ 42 (21%)	+ 95 (48%)	– 63 (31%)
E40.2	X wild-type	248		124 (50%)		124 (50%)
Wild-type	X E40.2	187		81 (43%)		106 (57%)
Plantlet derived from anther cultures of E40.2		102		47 (46%)		55 (54%)

Conclusions:
- LpDH is transmitted through meiosis as a single dominant factor.
- LpDH$^+$ eggs and pollen are fertile.
- LpDH gene behaves as a single locus on a single chromosome.
- E40.2 is HEMIZYGOTE LpDH/–.

mutant Ti plasmids reproducibly give rise to normal plants in tobacco, petunia, and potato. In all these cases the plants were shown to contain and express the octopine and/or agropine synthase genes of the mutant T-DNA.

The use of "intermediate vectors" for the in vitro introduction of selected genes in the T-region of Ti plasmids. Because of the size of Ti plasmids (about 130 Md) and because a large number of Ti plasmid genes are involved in the transformation mechanism, it did not appear to be feasible to develop a "mini-Ti" cloning vector with unique cloning sites at appropriate locations within the T-region and with all functions essential for T-DNA transfer and stable maintenance. An alternative method was therefore developed to introduce genes at specified sites in the T-region of a functional Ti plasmid. The principle was to make an "intermediate vector" (Van Montagu et al. 1980; Schell et al. 1981; Leemans et al. 1981, 1982b) consisting of a common *E.coli* cloning vehicle, for example the pBR322 plasmid, into which an appropriate fragment of the T-region of a Ti plasmid is inserted. Single restriction sites in this T-region fragment can then be used to insert the chosen DNA sequence. This "intermediate vector" is subsequently introduced, via mobilization, into an *A.tumefaciens* strain carrying a Ti plasmid that has been made constitutive for transfer (Holsters et al. 1980) and that already carries antibiotic resistance markers (e.g., streptomycin, sulfonamide) cloned into the same T-region restriction site as that chosen to insert the DNA to be transferred. The resistance markers were introduced into this site by a procedure essentially identical to the one described here. Recombination in vivo will transfer the DNA of interest into the appropriate site of the Ti plasmid. The Ti plasmid, thus engineered to contain the desired DNA, will transfer this DNA into plant cells (Leemans et al. 1981).

CONCLUSIONS

Evidence has been presented demonstrating that Ti plasmids can efficiently be used as experimental gene vectors. The main limitations of this type of gene vector are (a) its limited host range, (b) the relative complexity involved in the use of "intermediate vectors," and (c) the absence of selectable markers other than autonomous growth.

Host range. Most dicotyledonous plants that have been studied were found to be sensitive to one or more *Agrobacterium tumefaciens* strains. Some *Agrobacterium* strains have a broad host range, whereas others have been shown to have a limited host range. It was recently shown that these differences in host range were determined by the type of Ti plasmids carried by these strains (Thomashow, Knauf, and Nester 1981).

It is unfortunate that no strains have thus far been found to transform cereals. Lippincott and Lippincott (1978) have suggested that monocotyledonous plants lack *Agrobacterium* adherence sites essential for the initial stages of transformation. If this is the case, one could hope that the introduction of Ti plasmid DNA via liposomes or microinjection would allow transformation of cereals. Recent reports indicate that the use of liposomes (S. Dellaporte et al. unpublished) or direct transformation (Krens et al. 1982) and/or fusion between plant cells and bacterial protoplasts (Hasezawa, Nagata, and Syono 1981) will provide alternative ways to introduce T-DNA sequences experimentally into plant cells and possibly also into cereals.

A simplified use of T-region vectors. Intermediate vectors, with which one can introduce a specific DNA sequence at a predetermined site in the T-region of a complete Ti plasmid, have been applied successfully. Their use, however, is somewhat cumbersome, because it is difficult to work with DNA sequences that are not linked to a genetic marker with which one can monitor the different recombinational events. For these reasons, further research is essential in order to simplify the use of these intermediate vectors. The following general approach is currently under study: transfer of a cloned and modified T-region by complementation with a Ti plasmid from which the T-region has been deleted. The cloned and modified T-region should have the following properties: (a) it should contain the necessary recognition sequences to initiate its transfer to the plant cells (by complementation); (b) it should contain the necessary recognition sequences to promote integration in the plant nuclear DNA; (c) it should contain genes for the enzymes necessary for integration, provided that such enzymes are indeed coded for by Ti plasmid genes; (d) it should contain genetic markers allowing selection of transformed cells; (e) it should contain one or several unique restriction sites for convenient cloning and be located so as to allow or promote expression of the cloned DNA; and (f) it should not contain any function that prevents regeneration of whole plants from the transformed plant cells.

Such a vector might function either directly after uptake in plant protoplasts by liposome or protoplast fusion, or by PEG treatment, or after being reintroduced into *Agrobacterium* strains containing a resident Ti plasmid providing, in trans, the missing oncogenic functions. Introduction of plasmid DNA into *Agrobacteria* can be achieved by transformation (Holsters et al. 1978) at a fairly low frequency or by conjugation from an *E.coli* donor. The T-region vector plasmid can be constructed such that it will replicate both in *E.coli* and in *Agrobacterium*.

The development of selectable markers in the T-region. Our observation that the Tn7 gene for a methotrexate-resistant dihydrofolate reductase is transcribed at a low level in transformed tobacco cells, and might be responsible for the capacity of this tumor line to grow in the presence of low concentrations of methotrexate (2 μg/ml), leads to the prediction that this and other bacterial genes could be expressed in plant cells after insertion at an appropriate site in the T-region. Plants such as tobacco are very sensitive to a number of antibiotics, such as G418, kanamycin, neomycin, and gentamicin. Bacterial genes that code for functions that inactivate these toxic substances have been described and have been isolated. Some of these genes have been transferred to animal cells (P. Berg, personal communication) and to yeast cells (Jimenez and Davies 1980), and found to be expressed. By inserting the coding sequence of some of these genes behind the T-DNA promoters for opine synthases, one might expect to achieve good expression and possibly sufficient resistance to allow for a direct selection of the transformed plant cell.

ACKNOWLEDGMENTS

The investigations reported here were supported by grants from the "Onderling Overlegde Akties" (O.O.A. 12052179) and by the Max-Planck-Gesellschaft, Munich. JPH is Research Associate at the National Fund for Scientific Research, Belgium, and LO was supported by a long-term EMBO fellowship.

LITERATURE CITED

Amasino, R. M., and C. Miller. 1982. *Hormonal control of tobacco crown gall tumor morphology*. Plant Physiol. 69:389–92.

Bevan, M., W. M. Barnes, and M.-D. Chilton. 1983. *Structure and transcription of the nopaline synthase gene region of T-DNA*. Nucl. Acids Res. 11: 369–85.

Bomhoff, G., P. M. Klapwijk, H. C. M. Kester, R. A. Schilperoort, J. P. Hernalsteens and J. Schell. 1976. *Octopine and nopaline synthesis and breakdown genetically controlled by a plasmid of Agrobacterium tumefaciens*. Mol. Gen. Genet. 145:177–81.

Braun, A. C. 1956. *The activation of two growth substance systems accompanying the conversion of normal to tumor cells in crown gall*. Cancer Res. 16:53–56.

Chilton, M.-D., H. J. Drummond, D. J. Merlo, D. Sciaky, A. L. Montoya, M. P. Gordon, and E. W. Nester. 1977. *Stable incorporation of plasmid DNA into higher plant cells: The molecular basis of crown gall tumorigenesis*. Cell 11:263–71.

Chilton, M.-D., M. H. Drummond, D. J. Merlo, and D. Sciaky. 1978. *Highly conserved DNA of Ti-plasmids overlaps T-DNA, maintained in plant tumors*. Nature (Lond.) 275:147–49.

Chilton, M.-D., R. K. Saiki, N. Yadav, M. P. Gordon, and F. Quetier. 1980. *T-DNA from Agrobacterium Ti-plasmid is in the nuclear DNA fraction of crown gall tumor cells*. Proc. Nat. Acad. Sci. USA 77:4060–64.

De Beuckeleer, M., M. De Block, H. De Greve, A. Depicker, R. De Vos, G. De Vos, M. De Wilde, P. Dhaese, M. R. Dobbelaere, G. Engler, C. Genetello, J. P. Hernalsteens, M. Holsters, A. Jacobs, J. Schell, J. Seurinck, B. Silva, E. Van Haute, M. Van Montagu, F. Van Vliet, R. Villarroel, and I. Zaenen. 1978. *The use of the Ti-plasmid as a vector for the introduction of foreign DNA into plants*. Pages 115–26 in M. Ridé ed. *Proceedings IVth International Conference on Plant Pathogenic Bacteria*. Angers, I.N.R.A.

De Beuckeleer, M., M. Lemmers, G. De Vos, L. Willmitzer, M. Van Montagu, and J. Schell. 1981. *Further insight on the transferred-DNA of octopine crown gall*. Mol. Gen. Genet. 183:283–88.

De Greve, H., H. Decraemer, J. Seurinck, M. Van Montagu, and J. Schell. 1981. *The functional organization of the octopine Agrobacterium tumefaciens plasmid pTiB6S3*. Plasmid 6:235–48.

De Greve, H., J. Leemans, J. P. Hernalsteens, L. Thia-Toong, M. De Beuckeleer, L. Willmitzer, L. Otten, M. Van Montagu, and J. Schell. 1982a. *Regeneration of normal and fertile plants that express octopine synthase, from tobacco crown galls after deletion of tumour-controlling functions*. Nature 300, 752–55.

De Greve, H., P. Dhaese, J. Seurinck, M. Lemmers, M. Van Montagu, and J. Schell. 1982b. *Nucleotide sequence and transcript map of the Agrobacterium tumefaciens Ti plasmid–encoded octopine synthase gene*. J. Mol. Appl. Genet. 1: 499–512.

Depicker, A., M. Van Montagu, and J. Schell. 1978. *Homologous DNA sequences in different Ti plasmids are essential for oncogenicity*. Nature (Lond.) 275:150–53.

Depicker, A., S. Stachel, P. Dhaese, P. Zambryski, and H. M. Goodman. 1982. *Nopaline synthase: Transcript mapping and DNA sequence*. J. Mol. Appl. Genet. 1:561–74.

Drummond, M. H., M. P. Gordon, E. W. Nester, and M.-D. Chilton. 1977. *Foreign DNA of bacterial plasmid origin is transcribed in crown gall tumors*. Nature (Lond.) 269:535–36.

Ellis, J., A. Kerr, J. Tempé, and A. Petit. 1979. *Arginine catabolism: A new function of both octopine and nopaline Ti plasmids of Agrobacterium*. Mol. Gen. Genet. 173:263–69.

Ellis, J. D., and P. J. Murphy. 1981. *Four new opines from crown gall tumours—their detection and properties*. Mol. Gen. Genet. 181:36–43.

Engler, G., M. Holsters, M. Van Montagu, J. Schell, J. P. Hernalsteens, and R. A. Schilperoort. 1975. *Agrocin 84 sensitivity: A plasmid determined property in Agrobacterium tumefaciens*. Mol. Gen. Genet. 138:345–49.

Engler, G., A. Depicker, R. Maenhaut, R. Villarroel, M. Van Montagu, and J. Schell. 1981. *Physical mapping of DNA base sequence homologies between an octopine and a nopaline Ti plasmid of Agrobacterium tumefaciens*. J. Mol. Biol. 152:183–208.

Firmin, J. L., and G. R. Fenwick. 1978. *Agropine—a major new plasmid-determined metabolite in crown gall tumors*. Nature (Lond.) 276:842–44.

Fitzgerald, M., and T. Schenk. 1981. *The sequence 5'-AAUAAA-3' forms part of the recognition site for polyadenylation of late SV40 mRNAs*. Cell 24:251–60.

Garfinkel, D. J., R. B. Simpson, L. W. Ream, F. F. White, M. P. Gordon, and E. W. Nester. 1981. *Genetic analysis of crown gall: Fine structure map of the T-DNA by site-directed mutagenesis*. Cell 27:143–53.

Gelvin, S. B., M. P. Gordon, E. W. Nester, and A. I. Aronson. 1981. *Transcription of the Agrobacterium Ti-plasmid in the bacterium and in crown gall tumors*. Plasmid 6:17–29.

Genetello, C., N. Van Larebeke, M. Holsters, A. Depicker, M. Van Montagu, and J. Schell. 1977. *Ti plasmids of Agrobacterium as conjugative plasmids*. Nature (Lond.) 265:561–63.

Gordon, M. P. 1979. *Tumor formation in plants*. Pages 531–70 in A. Marcus, ed., *Proteins and Nucleic Acids*. The Biochemistry of Plants, vol. 6. Academic Press, New York.

Gurley, W. B., J. D. Kemp, M. J. Alber, D. W. Sutton, and J. Gallis. 1979. *Transcription of Ti-plasmid derived sequences in three octopine-type crown gall tumor lines*. Proc. Nat. Acad. Sci. USA 76:2828–32.

Guyon, P., M.-D Chilton, A. Petit, and J. Tempé. 1980. *Agropine in "null-type" crown gall tumors: evidence for the generality of the opine concept*. Proc. Nat. Acad. Sci. USA 77: 2693–97.

Hasezawa, S., T. Nagata, and K. Syono. 1981. *Transformation of Vinca protoplasts mediated by Agrobacterium spheroplasts*. Mol. Gen. Genet. 182:206–10.

Hernalsteens, J. P., M. Holsters, M. Van Montagu, and J. Schell. 1980. *The Agrobacterium tumefaciens Ti plasmid as a host vector system introducing foreign DNA in plant cells*. Nature (Lond.) 287:654–56.

Holsters, M., D. De Waele, A. Depicker, E. Messens, M. Van Montagu, and J. Schell. 1978. *Transfection and transformation of Agrobacterium tumefaciens*. Mol. Gen. Genet. 163:181–87.

Holsters, M., B. Silva, F. Van Vliet, C. Genetello, M. De Block, P. Dhaese, A. Depicker, D. Inzé, G. Engler, R. Villarroel, M. Van Montagu, and J. Schell. 1980. *The functional organization of the nopaline A.tumefaciens plasmid pTiC58*. Plasmid 3:212–30.

Holsters, M., R. Villarroel, J. Gielen, J. Seurinck, H. De Greve, M. Van Montagu, and J. Schell. 1983. *An analysis of the boundaries of the octopine TL-DNA in tumors induced by Agrobacterium tumefaciens*. Mol. Gen. Genet., submitted.

Jimenez, A., and J. Davies. 1980. *Expression of a transposable antibiotic resistance element in Saccharomyces*. Nature (Lond.) 287:869–71.

Kerr, A., P. Manigault, and J. Tempé. 1977. *Transfer of virulence in vivo and in vitro in Agrobacterium*. Nature (Lond.) 265:560–61.

Klapwijk, P. M., T. Scheulderman, and R. A. Schilperoort. 1978. *Coordinated regulation of octopine degradation and conjugative transfer of Ti-plasmids in Agrobacterium tumefaciens: Evidence for a common regulatory gene and separate operons*. J. Bacteriol. 136:775–85.

Koekman, B. P., G. Ooms, P. M. Klapwijk, and R. A. Schilperoort. 1979. *Genetic map of an octopine Ti-plasmid*. Plasmid 2:347–57.

Krens, F. A., L. Molendijk, G. J. Wullems, R. A. Schilperoort. 1982. *In vitro transformation of plant protoplasts with Ti-plasmid DNA*. Nature (Lond.) 296:72–74.

Leemans, J., R. Deblaere, L. Willmitzer, H. De Greve, J. P. Hernalsteens, M. Van Montagu, and J. Schell. 1982a. *Genetic identification of functions of TL-DNA transcripts in octopine crown-galls*. EMBO J. 1:147–52.

Leemans, J., H. De Greve, J. P. Hernalsteens, C. Shaw, L. Willmitzer, L. Otten, M. Van Montagu, and J. Schell. 1982b. *Ti-plasmids and directed genetic Engineering*. Pages 537–45 in G. Kahl and J. Schell, ed. *Molecular Biology of Plant Tumors*. Academic Press, New York.

Leemans, J., C. Shaw, R. Deblaere, H. De Greve, J. P. Hernalsteens, M. Maes, M. Van Montagu, and J. Schell. 1981. *Site-specific mutagenesis of Agrobacterium Ti plasmids and transfer of genes to plant cells*. J. Mol. Appl. Genet. 1:149–64.

Lemmers, M., M. De Beuckeleer, M. Holsters, P. Zambryski, A. Depicker, J. P. Hernalsteens, M. Van Montagu, and J. Schell. 1980. *Internal organisation, boundaries and integration of Ti-plasmid DNA in nopaline crown gall tumors*. J. Mol. Biol. 144:355–78.

Lippincott, J. A. and B. B. Lippincott. 1978. *Cell walls of crown gall tumors and embryonic plant tissues lack Agrobacterium adhesive sites*. Science 199:1075–78.

Merlo, D. J., R. C. Nutter, A. L. Montoya, D. J. Garfinkel, M. H. Drummond, M.-D. Chilton, M. P. Gordon, and E. W. Nester. 1980. *The boundaries and copy numbers of Ti plasmid T-DNA vary in crown gall tumors*. Mol. Gen. Genet. 177:637–43.

Ooms, G., A. Bakker, L. Molendijk, R. A. Schilperoort, M. P. Gordon, and E. W. Nester. 1982. *Organization of T-DNA and cellular composition of octopine crown gall tissues induced in vivo and in vitro by A. tumefaciens: Impact of T-DNA on tumor phenotype*. Cell 30:589–97.

Otten, L., H. De Greve, J. P. Hernalsteens, M. Van Montagu, O. Schieder, J. Straub, and J. Schell. 1981. *Mendelian transmission of genes introduced in plants by the Ti plasmids of Agrobacterium tumefaciens*. Mol. Gen. Genet. 183:209–13.

Petit, A., Y. Dessaux, and J. Tempé. 1978. *The biological significance of opines: I—A study of opine catabolism by A. tumefaciens*. Pages 143–52 in M. Ridé, ed. *Proceedings IVth International Conference Plant Pathogenic Bacteria*. Angers, I.N.R.A.

Petit, A., J. Tempé, A. Kerr, M. Holsters, M. Van Montagu, and J. Schell. 1978. *Substrate induction of conjugative activity of Agrobacterium tumefaciens Ti plasmids*. Nature (Lond.) 271:570–72.

Schell, J. 1975. *The role of plasmids in crown gall formation by A. tumefaciens*. Pages 163–81 in L. Ledoux, ed. *Genetic Manipulations with Plant Materials*. Plenum Press, New York.

Schell, J. and M. Van Montagu. 1977. *The Ti plasmid of Agrobacterium tumefaciens, a natural vector for the introduction of NIF genes in plants?* Pages 159–79 in A. Hollaender, ed. *Genetic Engineering for Nitrogen Fixation*. Plenum Press, New York.

———. 1980. *The Ti-plasmids of Agrobacterium tumefaciens and their role in crown gall formation*. Pages 453–70 in Ch. Leaver, ed., *Genome Organisation and Expression in Plants*. Plenum Press, New York.

Schell, J., M. Van Montagu, M. De Beuckeleer, M. De Block, A. Depicker, M. De Wilde, G. Engler, C. Genetello, J. P. Hernalsteens, M. Holsters, J. Seurinck, B. Silva, F. Van Vliet, and R. Villarroel 1979. *Interactions and DNA transfer between Agrobacterium tumefaciens, the Ti-plasmid and the plant host*. Proc. R. Soc. Lond. B 204:251–66.

Schell, J., M. Van Montagu, M. Holsters, J. P. Hernalsteens, J. Leemans, H. De Greve, L. Willmitzer, L. Otten, J. Schröder, and C. Shaw 1981. *The development of host vectors for directed gene-transfer in plants*. Pages 557–75 in D. Brown and C. F. Fox, eds., *Developmental Biology using Purified Genes*. ICN-UCLA Symposia on Molecular and Cellular Biology, vol. 33. Academic Press, New York.

Schilperoort, R. A., P. M. Klapwijk, G. Ooms, and G. J. Wullems. 1980. *Plant tumors caused by bacterial plasmids: Crown gall*. Pages 87–108 in F. J. Cleton and J. W. Simons, eds., *Genetic Origins of Tumor Cells*. Martinus Nijhoff, The Hague.
eds., *Plasmids of Medical, Environmental and Commercial Importance*. Elsevier, Amsterdam.

Schilperoort, R. A., P. M. Klapwijk, G. Ooms, and G. J. Wullems. 1980. *Plant tumors caused by bacterial plasmids: Crown gall*. Pages 87–108 in F. J. Cleton and J. w. Simons, eds., *Genetic Origins of Tumor Cells*. Martinus Nijhoff, The Hague.

Schröder, J., A. Hillebrand, W. Klipp, and A. Pühler. 1981a. *Expression of plant tumor-*

specific proteins in minicells of Escherichia coli: A fusion protein of lysopine dehydrogenase with chloramphenicol acetyltransferase. Nucl. Acids Res. 9:5187–5202.

Schröder, G. and J. Schröder. 1982. *Hybridization selection and translation of T-DNA encoded mRNAs from octopine tumors*. Mol. Gen. Genet. 185:51–55.

Schröder, J., G. Schröder, H. Huisman, R. A. Schilperoort, and J. Schell. 1981b. *The mRNA for lysopine dehydrogenase in plant tumor is complementary to a Ti-plasmid fragment*. FEBS Lett. 129:166–68.

Tennhammer-Ekman, B. and O. Sköld. 1979. *Trimethoprim resistance plasmids of different origin encode different drug-resistant dihydrofolate reductase*. Plasmid 2:334–46.

Thomashow, M. F., V. C. Knauf, and E. W. Nester. 1981. *Relationship between the limited and wide host range octopine-type Ti plasmids of Agrobacterium tumefaciens*. J. Bacteriol. 146:484–93.

Thomashow, M. F., R. Nutter, A. L. Montoya, M. P. Gordon, and E. W. Nester. 1980a. *Integration and organization of Ti-plasmid sequences in crown gall tumors*. Cell 19:729–39.

Thomashow, M. F., R. Nutter, K. Postle, M.-D. Chilton, F. R. Blattner, A. Powell, M. P. Gordon, and E. W. Nester. 1980b. *Recombination between higher plant DNA and the Ti-plasmid of Agrobacterium tumefaciens*. Proc. Nat. Acad. Sci. USA 77:6448–52.

Van Larebeke, N., G. Engler, M. Holsters, S. Van den Elsacker, I. Zaenen, R. A. Schilperoort, and J. Schell, 1974. *Large plasmids in Agrobacterium tumefaciens essential for crown gall-inducing ability*. Nature (Lond.) 252:169–70.

Van Larebeke, N., C. Genetello, J. Schell, R. A. Schilperoort, A. K. Hermans, J. P. Hernalsteens, and M. Van Montagu (1975). *Acquisition of tumor-inducing ability by non-oncogenic Agrobacteria as a result of plasmid transfer*. Nature (Lond.) 255:742–43.

Van Montagu, M., M. Holsters, P. Zambryski, J. P. Hernalsteens, A. Depicker, M. De Beuckeleer, G. Engler, M. Lemmers, L. Willmitzer, and J. Schell. 1980. *The interactions of Agrobacterium Ti-plasmid and plant cells*. Proc. R. Soc. Lond. B 210:351–65.

Van Montagu, M. and J. Schell. 1979. *The plasmids of Agrobacterium tumefaciens*. Pages 71–96 in K. Timmis and A. Pühler, eds., *Plasmids of Medical, Environmental and Commercial Importance*. Elsevier, Amsterdam.

Watson, B., T. C. Currier, M. P. Gordon, M.-D. Chilton, E. W. Nester. 1975. *Plasmid required for virulence of Agrobacterium tumefaciens*. J. Bacteriol. 123:255–64.

Willmitzer, L., M. De Beuckeleer, M. Lemmers, M. Van Montagu, and J. Schell. 1980. *DNA from Ti plasmid present in nucleus and absent from plastids of crown gall plant cells*. Nature (Lond.) 287:359–361.

Willmitzer, L., P. Dhaese, W. Schmalenbach, M. Van Montagu, and J. Schell. 1983. *Size, location and polarity of T-DNA encoded transcripts present in nopaline crown gall tumors; Evidence for common transcripts present in both octopine and nopaline plant tumors*. Cell 32 (in press).

Willmitzer, L., L. Otten, G. Simons, W. Schmalenbach, J. Schröder, G. Schröder, M. Van Montagu, G. De Vos, and J. Schell. 1981. *Nuclear and polysomal transcripts of T-DNA in octopine crown gall suspension and callus cultures*. Mol. Gen. Genet. 182:255–62.

Willmitzer, L., W. Schmalenbach, and J. Schell. 1981. *Transcription of T-DNA in octopine and nopaline crown gall tumors is inhibited by low concentrations of* α-amanitin. Nucl. Acids Res. 9:4801–12.

Willmitzer, L., G. Simons, and J. Schell. 1982. *The TL-DNA in octopine crown gall tumors codes for seven well-defined polyadenylated transcripts*. EMBO J. 1:139–46.

Yadav, N. S., K. Postle, R. K. Saiki, M. F. Thomashow, and M.-D. Chilton. 1980. *T-DNA of crown gall teratoma is covalently joined to host plant DNA*. Nature (Lond.) 287:458–61.

Zaenen, I., N. Van Larebeke, H. Teuchy, M. Van Montagu, and J. Schell. 1974. *Supercoiled circular DNA in crown gall-inducing agrobacterium strains*. J. Mol. Biol. 86:109–27.

Zambryski, P., A. Depicker, K. Kruger, and H. M. Goodman. 1982. *Tumor induction by Agrobacterium tumefaciens: Analysis of the boundaries of T-DNA*. J. Mol. Appl. Gen. 1:361–70.

Zambryski, P., M. Holsters, K. Kruger, A. Depicker, J. Schell, M. Van Montagu, and H. Goodman. 1980. *Tumor inducing DNA of Agrobacterium tumefaciens: Structure and organization in transformed plant cells*. Science 209:1385–91.

15] Agrobacterium-Mediated Transfer of Foreign Genes into Plants

by J. D. KEMP,*† D. W. SUTTON,*† C. FINK,*
R. F. BARKER,† and T. C. HALL†

ABSTRACT

Techniques are now available for the isolation and integration of foreign genes into nuclear DNA of dicotyledonous plants. These techniques include the cloning and tailoring of DNA carrying a desired gene and the site-specific insertion of that DNA into the tumor-inducing (Ti) plasmid of *Agrobacterium tumefaciens*. When a foreign gene is inserted into the transferred DNA (T-DNA) of the Ti plasmid, *A. tumefaciens* will transfer it to a plant cell during infection. Three foreign genes have been successfully transferred by us to sunflower crown gall cells. These are the neomycin phosphotransferase II gene of Tn5, the phaseolin gene of *Phaseolus vulgaris,* and the nopaline synthase gene of pTi-C58. All three genes were stably integrated into plant nuclear DNA, but only the nopaline synthase gene was fully expressed.

INTRODUCTION

Members of the genus *Agrobacterium* may be unique among procaryotes for their ability to transfer DNA to plants during tumorigenesis (Chilton et al. 1977). The transferred DNA (T-DNA) is a 10–20-kilobase pair (kbp) piece of the large tumor-inducing (Ti) plasmid carried by all virulent strains of the bacterium (Zaenen et al. 1974; Chilton et al. 1978; Thomashow et al. 1980). T-DNA is integrated into plant nuclear DNA (Chilton et al. 1980; Thomashow et al. 1980; Yadav et al. 1980) where it is transcribed into mRNAs (Drummond et al. 1977; Gurley et al. 1979), two of which are known to be translated into functional enzymes (Murai and Kemp 1982a,b). These enzymes are the octopine and nopaline synthases that catalyze the synthesis of the octopine and

*Department of Plant Pathology, University of Wisconsin, Madison, Wisconsin, 53706; †Agrigenetics Advanced Research Laboratory, 5649 E. Buckeye Road, Madison, Wisconsin, 53716.

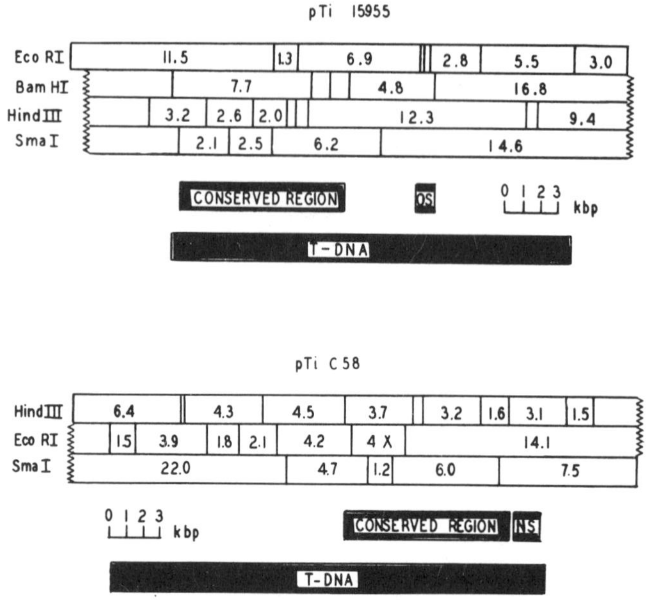

Figure 15.1. Restriction endonuclease map of the T-DNA region of pTi-15955 and pTi-C58. Each fragment is designated by its size in kbp. OS and NS indicate the locations of the octopine synthase and the nopaline synthase genes, respectively.

nopaline families of unusual amino acids, respectively (Kemp et al. 1979; Hack and Kemp 1980). These unusual amino acids, or opines, are specific for crown gall cells and probably serve as a unique source of nutrients for *A. tumefaciens*. The Ti plasmid specifies the particular opine family that the tumor cells will synthesize, as well as the opine that can be catabolized by the inciting bacterium (Petit et al. 1970; Bomhoff et al. 1976; Montoya et al. 1977). For example, if the bacterium contains an octopine Ti plasmid (pTi-15955) it can utilize octopine, and the crown gall it incites synthesizes octopine. The same specificity holds for the nopaline Ti plasmid (pTi-C58). Octopine and nopaline are never synthesized by the same crown gall cell because their respective Ti plasmids are incompatible (Montoya et al. 1978).

Until recently, it was believed that T-DNA was not retained through meiosis (Yang et al. 1980). Yet tumorous tobacco shoots derived from the fusion of normal and crown gall protoplasts contain T-DNA and retain the ability to produce flowers and set seed (Wullems et al. 1981). The seed gave F1 progeny that retained the ability to synthesize nopaline. These results suggest that indeed T-DNA is retained through meiosis. Together, these facts make T-DNA of the Ti plasmid of *A. tumefaciens* an appealing vehicle for the transfer and stable integration of foreign DNA into plants.

A number of functions associated with T-DNA must be considered before one can choose a site for inserting foreign DNA. For instance, virulence is associated with the portion of the T-DNA that is common to most Ti plasmids (Depicker, van Montagu, and Schell 1978; Drummond and Chilton 1978). The conserved area lies on the left side of the physical map of the T-DNA of pTi-15955 and the right side of the T-DNA of pTi-C58 (Figure 15.1). Another function, mapping in the center of the T-DNA of pTi-15955, is the gene for octopine synthase (Hack and Kemp 1980; Murai and Kemp 1982b), a convenient marker of tumorigenesis. A similarly useful gene, nopaline synthase, maps on the far right side of the T-DNA of pTi-C58. The right side of pTi-15955 T-DNA is considered replaceable even though it does show active transcription in crown gall cells (Thomashow et al. 1980; Gurley et al. 1979). The same probably holds for the left side of pTi-C58 T-DNA. When considering where to engineer foreign DNA into T-DNA, the right side of pTi-15955 T-DNA seemed most logical. Not only would one be introducing DNA into an area that should not interfere with the infection process, but one might be able to take advantage of active internal regulatory sites.

Several methods for site-specific insertion into Ti plasmids have been attempted, but without success. These methods require purification of Ti plasmid, followed by linearization at a specific site by forming D-loops or R-loops. The principal reason these approaches failed was because of the inherent instability of pTi due to its size (180–230 kbp). The method outlined in this chapter circumvents this problem by engineering foreign DNA into small, cloned pieces of the T-DNA and relying on in vivo, homologous recombination to insert the engineered fragment back into the Ti plasmid.

TRANSFER OF THE NEOMYCIN PHOSPHOTRANSFERASE II GENE INTO SUNFLOWER CROWN GALL CELLS

Our strategy for engineering the Ti plasmid was similar to that used by Ruvkun and Ausubel (Ruvkun and Ausubel 1981) for site-directed mutagenesis of cloned *Rhizobium meliloti* genes. The strategy includes: (a) cloning selected fragments of T-DNA in *E. coli* using pRK290, a plasmid that can replicate in either *E. coli* or *Agrobacterium;* (b) ligating the neomycin phosphotransferase II (NPT II) gene from Tn5 (Rothstein et al. 1980) into the cloned T-DNA fragment; (c) introducing the recombinant plasmid into *A. tumefaciens* strain 15955 by transformation (Holsters et al. 1978); (d) introducing a second P-1 group plasmid, pPHIJI (Hirch 1980), into the *A. tumefaciens* strains containing pRK290; and (e) selecting for homologous recombination between the T-DNA on pRK290 and its equivalent sequences on the Ti plasmid by simultaneously isolating neomycin (kanamycin Km)-resistant (retention of the NPT II gene) and gentamycin (Gm)-resistant (conferred by pPHIJI) colonies. This strategy is outlined in more detail in Figures 15.2 and 15.3.

Our initial experiments were directed toward gene insertion into the Hind III sites on the right side of the T-DNA of pTi-15955 (Fig. 15.1), since this area is not involved in virulence but is actively transcribed in the crown gall cell (Gurley et al. 1979). These Hind III sites are centered within the 5.5-kbp *Eco*RI fragment of the T-DNA (heavy lines in Fig. 15.3). *Eco*RI fragments are a logical

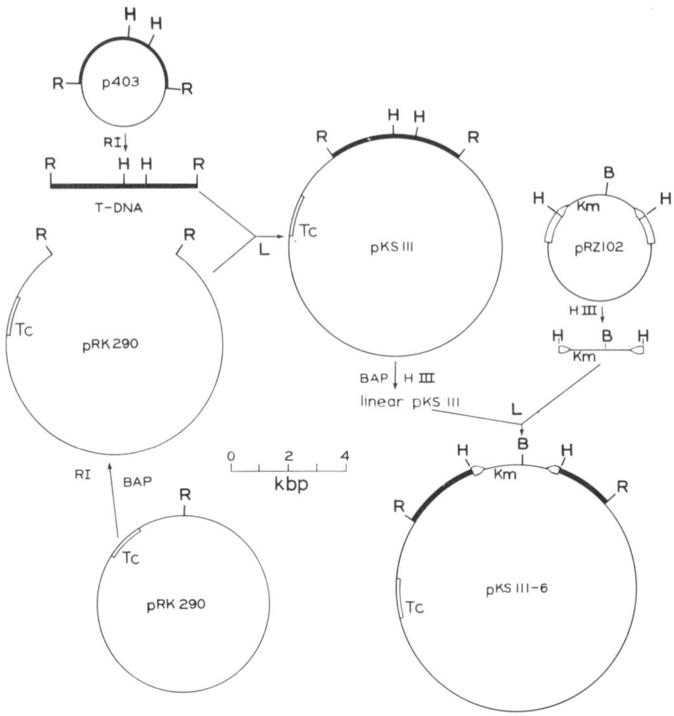

Figure 15.2. Construction of the "Shuttle" Vector pKSIII-K. Preparative amounts of the 5.5-kbp *Eco*RI T-DNA fragment were purified from p403, a pBR322 clone of the fragment, rather than directly from pTi-15955. Equimolar amounts of this fragment and *Eco*RI-treated pRK290 (pKSIII) were ligated. The ligation mixture was used to transform *E. coli* C-600 cells, and Tc-resistant colonies were selected. Restriction endonuclease digests of plasmid purified from one of the transformants confirmed the structure of pKSIII. Equimolar amounts of the 3.3-kbp Hind III fragment from pRZ102 (Rothstein et al. 1980) and Hind III digested pKSIII were ligated. The ligation mixture was used to transform *E. coli* C-600 cells. Transformants were selected on plates containing Km and Tc. Restriction patterns from plasmid digested with *Bam*HI, *Eco*RI, and Hind III yielded results consistant with the above construction (pKSIII-6 = pKSIII – K).

Single letters represent cleavage sites for restriction endonucleases. B = *Bam*HI; H = Hind III; R = *Eco*RI. BAP, RI, HIII, and L are incubations with bacterial alkaline phosphatase, *Eco*RI, Hind III and T4 ligase, respectively. Tc and Km are tetracycline and kanamycin (NPT II) resistance genes, respectively. Each molecule is drawn to scale.

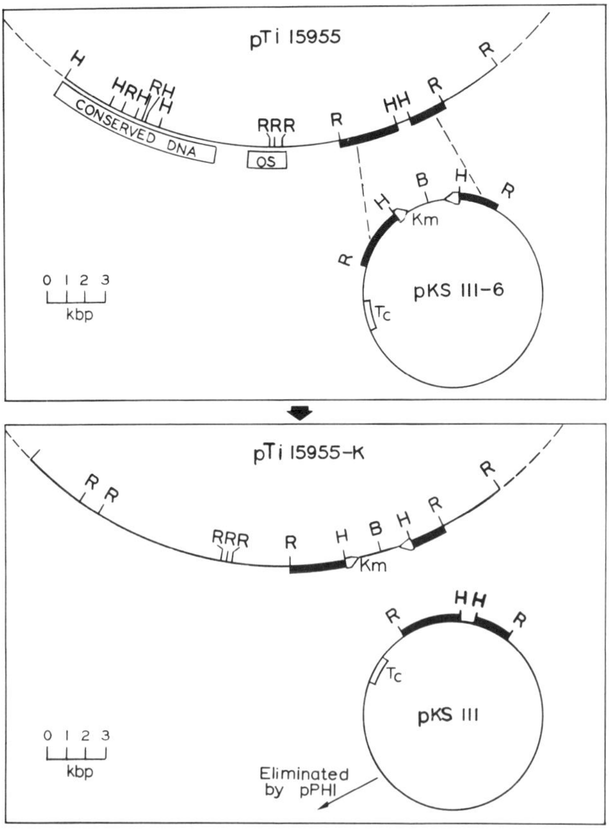

Figure 15.3. Selection of *A. tumefaciens* in which the Km DNA had recombined into pTI. The T-DNA of pTi-15955 is shown as a thin, solid, curved line. Conserved DNA is that portion of the T-DNA found in most Ti plasmids. OS represents the position of the octopine synthase gene. The heavy line represents that portion of the 5.3-kbp *Eco*RI fragment that is common between pTi and pKSIII-K. Km represents that portion of Tn5 DNA carrying the NPT-II gene. See Figure 15.2 for abbreviations.

choice for cloning into pRK290, since this plasmid has a single *Eco*RI site outside the tetracycline (Tc) resistance gene (Ditta et al. 1980). The plasmid resulting from the cloning of the *Eco*RI fragment into pRK290 was designated pKSIII (Fig. 15.2).

We chose the NPT II gene from Tn5 for insertion into pKSIII for several reasons: first, a detailed physical map of the gene is available (Rothstein et al. 1980); second, the gene lies conveniently between the Hind III sites of the inverted repeats of Tn5 (see Fig. 15.2); and third, the gene is expressed in *A. tumefaciens*. The 3.3-kbp Hind III fragment isolated from Tn5 was ligated into the Hind III site of pKSIII. The structure and orientation of pKSIII-K (pKSIII-6) is shown in Figure 15.2. pKSIII-K was transformed into *A. tumefaciens* by the technique of Holsters et al. (1978). One of the transformants was selected for conjugation with *E. coli* containing pPHIJI.

The strategy for isolating cells, where the T-DNA carried by pKSIII-K had recombined with its homologous counterpart on the Ti plasmid, is based on the assumption that recombination occurs naturally in a small subpopulation of cells. The isolation of cells from that subpopulation was effected by conjugation with pPHIJI and simultaneous selection for Gm^r and Km^r. Plasmid DNA was isolated from one of the exconjugants (*A. tumefaciens* 15955-K) to demonstrate that the NPT II gene had recombined into pTi-15955 by flanking T-DNA sequence homology.

Similarly, DNA was purified from a crown gall line (PSCG-15955-K) established from a sunflower plant inoculated with *A. tumefaciens* 15955-K. The tumor DNA contained the 3.3-kbp Hind III fragment of Tn5 DNA integrated into plant DNA and was unmodified. Nevertheless, the NPT II gene was not expressed in the tumor cells at either the mRNA or protein level. This result was not surprising since NPT II is a procaryotic gene and is probably not recognized by the plant cell.

TRANSFER OF THE PHASEOLIN GENE INTO SUNFLOWER CROWN GALL CELLS

Our second attempt to engineer a plant cell using *Agrobacterium* centered on the transfer of a phaseolin gene from *Phaseolus vulgaris* to *Helianthus annuus*. Phaseolin represents the major storage protein of the bean, which makes it extremely important in human nutrition. Engineering with this gene may prove to be beneficial by improving the quality of seed storage proteins. The quality of a protein is judged by whether or not it is efficiently utilized in the diet. Most sources of animal protein are high in quality for human nutrition because they contain a balance of essential amino acids. Plant proteins do not. For example, bean seeds are deficient in methionine and cysteine, whereas sunflower seeds are deficient in lysine. Generally, a deficiency holds for all members of a particular species, which rules out the possibility of effectively breeding for high-quality protein. The result of these deficiencies is that no single plant species is a source of high-quality protein for human nutrition. Man has circumvented this problem by mixing food sources, e.g., corn and bean. A novel approach to creating a high-quality plant protein might occur through interspecies gene transfers at the molecular level. If, for instance, a transferred phaseolin gene were to be expressed in the proper tissue of sunflower and at the proper time, then the new recombinant plant would be a source of high-quality protein for humans.

The best characterized of the phaseolin genomic clones is one that is completely homologous with the sequence derived from one of the full-length cDNA clones of phaseolin mRNA. Sequences characteristic of eucaryotic promoters are present on the clone at positions −74 base pairs (bp) and −28 bp from the mRNA cap site. The cap is followed by 77 bp of noncoding DNA

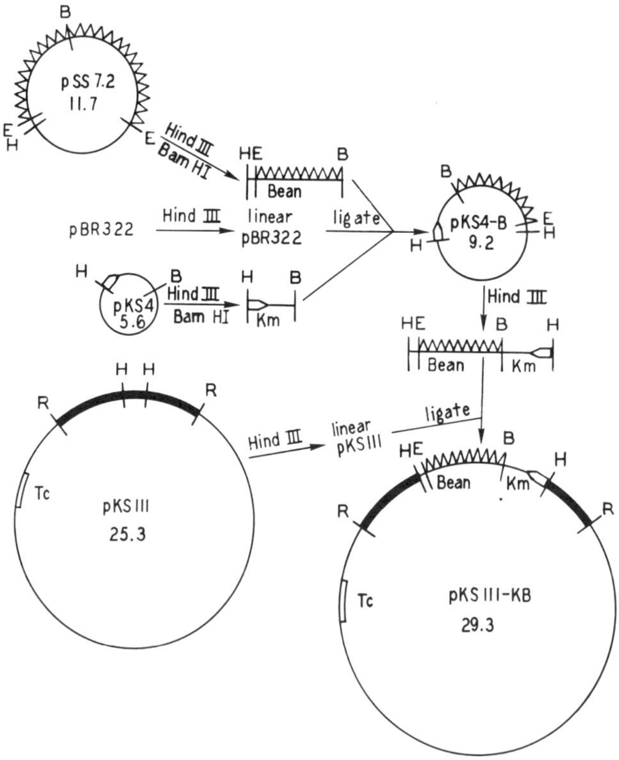

Figure 15.4. Construction of the "Shuttle" Vector pKSIII-KB. A 3.1-kbp Hind III *Bam*HI DNA fragment which contains the *Eco*RI/*Bam*HI fragment of the phaseolin gene was isolated from a pBR322 clone containing the large 7.2-kbp *Eco*RI fragment of bean DNA. The 3.1-kbp fragment contains 31 bp of pBR322 DNA from the Hind III to *Eco*RI site of pBR322. This 3.1-kbp fragment and a 2.1-kbp Hind III/*Bam*HI fragment containing the NPT II gene (from pRZ102, Fig. 15.2) were cloned into the Hind III site of pBR322. The resulting plasmid (pKS4-B) served as a source of the Km/bean DNA. The 5.3-kbp Hind III fragment from pKS4-B was ligated into Hind III site of pKSIII constructing the "shuttle" vector pKSIII-KB. See Figure 15.2 for abbreviations.

and 1263 bp of protein-coding sequence, broken by five intervening sequences. There is a 3′-untranslated region and a poly(A) additional signal. The clone contains an *Eco*RI site 33 bp into the protein-coding sequence and a *Bam*HI site about 1 kbp beyond the 3′-end of the gene.

A 7.2 kbp *Eco*RI fragment of the genomic clone was subcloned into pBR322 (pSS7.2) as part of tailoring the phaseolin gene for insertion into pTi. pSS7.2 served as a source of the *Eco*RI/*Bam*HI fragment for construction of the "shuttle" vector pKSIII-KB (Fig. 15.4). This plasmid, pKSIII-KB, is similar to pKSIII-K except that one of the *Bam*HI/Hind III fragments from Tn5 has been replaced by the phaseolin gene (minus its promoter and the first 33 bp of the coding sequence). Transfer of the Km/phaseolin DNA from the "shuttle" vector to pTi-15955 was accomplished as outlined in the previous section. *A. tumefaciens* harboring the engineered plasmid, pTi-15955-KB, was used to infect sunflower plants, and the resulting crown galls were established in tissue culture (PSCG-15955-KB). The *Eco*RI/*Bam*HI fragment of the phaseolin gene is stably integrated into PSCG-15955-KB DNA, and no reorganization of the phaseolin DNA has been noted since its insertion one year ago.

PSCG-15955-KB tissue appears to be synthesizing an RNA that has sequence homology to the *Eco*RI/*Bam*HI fragment of pSS7.2, but antibodies to phaseolin did not detect this protein in the tissue. A possible reason that the phaseolin gene is only partly expressed may rest with the way the gene was tailored and the place it was inserted into T-DNA. The gene was tailored to exclude its promoter region and sequences coding for the first few amino acids of the protein. It does, however, contain its introns, stop signal, and the poly(A) recognition site. Furthermore, it was inserted into an active T-DNA gene. Without its own promoter, the phaseolin gene may have been transcribed as part of the active T-DNA gene, but it was not translated because the reading frame of the phaseolin DNA did not match that of the T-DNA gene. Two other constructions have been made but not tested: first, the entire phaseolin gene with its own promoter has been transferred to T-DNA; and second, the original *Eco*RI/*Bam*HI fragment has been ligated into another active T-DNA gene in the correct reading frame. These and similar experiments should elucidate the mechanism of foreign gene expression in the crown gall system.

TRANSFER AND EXPRESSION OF THE NOPALINE SYNTHASE GENE

The results of the previous section still leave some doubt as to whether foreign genes transferred by the Ti plasmid can be fully active in their new environment. The following experiments demonstrate that indeed the Ti plasmid can transfer a gene, nopaline synthase, that is expressed. Nopaline synthase (NS) is an enzyme composed of four identical subunits each of 45,000 daltons. The gene that codes for the subunit is located at the right-hand side of the T-DNA of pTi-C58 (Fig. 15.1). The base sequence for this area of the T-DNA shows one continuous open reading frame of 1240 bp starting approximately 350 bp from the right-hand border of the T-DNA. Sequences characteristic of eucaryotic promoters are present upstream from the start of the open reading frame. There is also a poly(A) addition signal approximately 120 bp from the 3′-end of the coding region. An amino acid composition and molecular weight for nopaline synthase were calculated from the sequence data and compared to measured values determined from the purified enzyme (Table 15.1). The match is remarkably similar, strongly suggesting that the structural gene for nopaline synthase lies within the 3.1 kbp Hind III fragment of the T-DNA. The amino acid match also suggests that the enzyme is composed of four *identical* subunits. It is worth noting that this gene is carried by a procaryote and has no

Table 15.1 Amino Acid Composition of Nopaline Synthase

Amino acid	Number of residues/molecule	
	Protein	DNA
Lysine	20	19
Histidine	16	17
Arginine	19	22
Aspartic Acid	44	45
Glutamic Acid	43	33
Threonine	19	22
Serine	29	30
Proline	22	23
Glycine	41	21
Alanine	39	42
Cystine	10	11
Valine	22	26
Methionine	5	6
Isoleucine	25	32
Leucine	31	32
Tyrosine	9	8
Phenylalanine	19	21
Tryptophan	–	3
Total	413	413
Estimated molecular weight	45,000	45,600

intervening sequences, but it has eucaryotic-like promoter sequences and carries a poly(A) addition signal.

The nopaline synthase gene normally functions in a T-DNA environment, but it has no sequence homology with, nor is it ever found associated with an octopine Ti plasmid. For these reasons, NS seemed a logical choice to demonstrate whether an engineered gene could be fully expressed when transferred by pTi. Our cloning strategy was to transfer the 7.5-kbp Sma I fragment from the T-DNA of pTi-C58 to the Hind III sites in the 5.5-kbp *Eco*RI fragment of the T-DNA of pTi-15955 (Fig. 15.1). We would then inoculate sunflower plants with the engineered *A. tumefaciens* and determine the amount of octopine and nopaline in the engineered crown galls.

The 14.1-kbp *Eco*RI fragment that contains the nopaline synthase gene (Fig. 15.1) was cloned into the *Eco*RI site of pPR322 (designated p45). A 7.5-kbp Sma I fragment that lies within the *Eco*RI fragment was isolated, and the Sma I sites were converted to *Bgl* II sites by blunt-end ligation of *Bgl* II linkers (Fig. 15.5). The new *Bgl* II fragment was then cloned into the *Bam*HI site of pKSIII-K. This plasmid (pKSIII-N) is composed of the NS gene, the NPT II gene, and a portion of the T-DNA from pTi-15955 ligated into pRK290. pKSIII-N was

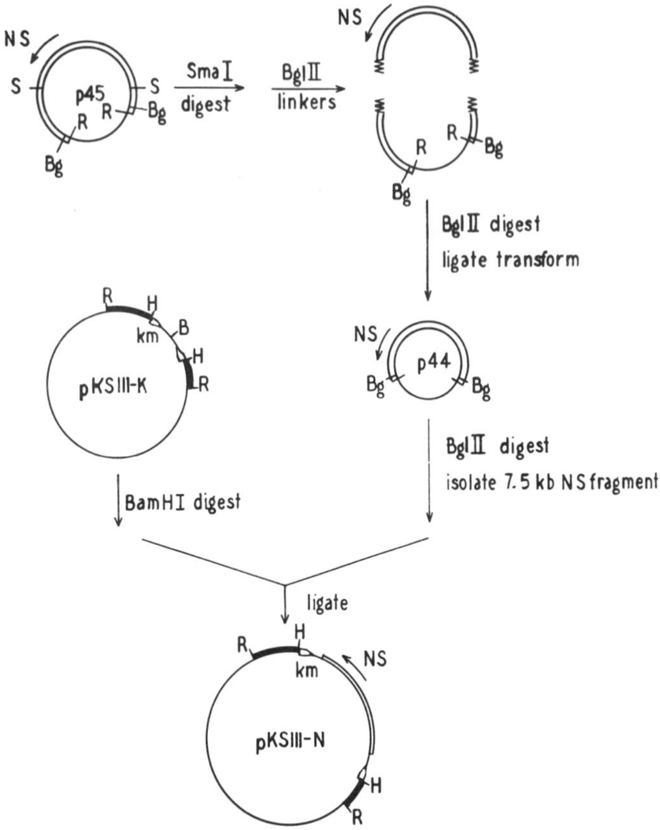

Figure 15.5. Construction of the "shuttle" vector of pKSIII-N. The recombinant plasmid p45 contains the 14.1-kbp *Eco*RI fragment of pTi-C58 inserted into the *Eco*RI site of pBR322. p45 was digested with Sma I, and the digestion products were blunt-end ligated onto kinased *Bgl* II linkers. The ligation products, containing excess attached *Bgl* II linkers, were digested with *Bgl* II. *Bgl* II was then heat inactivated, the mixture was again ligated using T4 DNA ligase, and used to transform calcium shocked *E. coli* K802. The transformants were screened by colony filter hybridization. One colony, p44, that hybridized to the nopaline T-DNA probe was selected for further study.

p44 was digested with *Bgl* II, and the 7.5-kbp fragment containing the nopaline synthase gene was purified from an agarose gel. The *Bgl* II fragment was ligated into the *Bam*HI site of pKSIII-K. Calcium-shocked *E. coli* was transformed with the ligation mix and screened by colony filter hybridization.

Table 15.2 Agrobacterium Strains and Their Opine Characteristics

		Produced by tumor (μmole/gm)		Enzymatic activity (nkat/ml)	
Strain	Bacterial utilization	Octopine	Nopaline	Octopine synthase	Nopaline synthase
15955	octopine	0.22	0.01	0.15	0.01
C58	nopaline	0.01	2.2	0.03	3.45
15955-N	octopine	0.17	1.4	0.71	1.15

Opines in crown gall cells were determined by homogenizing tissue in 10% TCA. The homogenate was centrifuged and the clear supernatant was absorbed onto Aminex A4 ion exchange resin (Kemp 1976). Fractions were assayed for the presence of octopine and nopaline (Johnson et al. 1974). The enzyme assays for the octopine and nopaline synthases were performed as described (Petit et al. 1970; Bomhoff et al. 1976). To test for bacterial utilization, the *A. tumefaciens* strains 15955, C58 and 15955-N were streaked onto minimal media containing 0.1% nopaline or octopine as the sole carbon source. The enzyme activity is expressed in nanokatals. One katal is defined as the amount of enzyme that catalyzes the oxidation of 1 mole of NADPH per second.

moved into *A. tumefaciens* 15955 by transformation. Then bacterial colonies were selected that had transferred the NS gene to the Ti plasmid. The engineered *A. tumefaciens* cells containing pTi-15955-N were used to induce crown galls on sunflower plants, and the resulting crown galls were established in tissue culture (PSCG-15955-N).

PSCG-15955-N tissue cultures synthesized both octopine and nopaline, while the control tissues PSCG-15955 synthesized only octopine, and PSCG-C58 synthesized only nopaline (Table 15.2). The amounts of these products in PSCG-15955-N was approximately the same as in the control tissue. Both octopine and nopaline synthases were also detected in PSCG-15955-N. This evidence indicates that not only is the engineered nopaline synthase gene functional in the plant cell, but it is functional at levels similar to that found in its native environment.

The possibility that the presence of nopaline synthase in PSCG-15955-N tissue was due to the presence of a mixed culture when the plant was inoculated can be ruled out, because *A. tumefaciens* 15955-N could only utilize octopine as a sole source of carbon. Had strain C58 been present, nopaline would also have been utilized. The possibility that *A. tumefaciens* 15955-N contains both pTi-C58 and pTi-15955 can also be ruled out, since these two plasmids are incompatible with each other.

CONCLUSIONS

There are a number of novel approaches to creating genetic diversity in plants. Some of the more interesting include cell and protoplast regeneration, protoplast fusions, and recombinant DNA transformation. When plant cells are grown in tissue culture, one often finds cells that have lost large parts of their chromosomes or have duplicated other parts. Plants regenerated from such cells are abnormal and generally not useful. But plants regenerated from protoplasts often have useful traits not observed in the parental tissue. Furthermore, those traits appear to be stably maintained. For example, potato plants regenerated from protoplasts often have improved disease resistance. Similar results have been found for other plants that are vegetatively propagated.

Another promising approach has been protoplast fusions. This technique involves the mixing of protoplasts from different species and regenerating plants from the fusion products. The technique has the potential of creating interspecies recombinants but, to date, stable fusion products are limited to closely related species.

The prospect for a successful molecular approach to creating new genetic recombinants was greatly increased with the birth of recombinant DNA techniques. These techniques opened the possibility of isolating genes as discrete pieces of DNA, ligating them into appropriate vectors, and transforming plant cell with the recombinants. Potentially, any gene can be moved to any species if the appropriate transfer vehicle is available.

The work reported here clearly demonstrates the feasibility of introducing a foreign gene into a preselected location in pTi and of having it integrated and expressed in the plant cell. Nonetheless, a number of other problems still must be resolved before this transfer system will be generally useful. Transformations by pTi results in a tumor cell that does not easily regenerate, and when it does the plants are often abnormal. A number of T-DNA genes involved in oncogenesis have been mapped and their functions mutated by transposon insertion (as shown in other chapters in this volume). The result of individually mutating these genes are crown gall cells that retain some tumorigenic properties but are usually more easily regenerated into normal plants. The ideal vehicle would have all of the tumorigenic properties removed. If the disadvantages of the pTi can be overcome, pTi or derivatives thereof may become truly useful vehicles for plant genetic engineering.

LITERATURE CITED

Bomhoff, G. H., P. M. Klapwijk, H. C. M. Kester, R. A. Schilperoort, J. P. Hernalsteens, and J. Schell. 1976. *Octopine and nopaline synthesis and breakdown genetically controlled by a plasmid of Agrobacterium tumefaciens*. J. Mol. Gen. Genet. 145:177–81.

Chilton, M-D., M. H. Drummond, D. J. Merlo, and D. Sciaky. 1978. *Highly conserved DNA of Ti plasmids overlaps T-DNA maintained in plant tumours*. Nature (Lond.) 275:147–49.

Chilton, M-D., M. H. Drummond, D. J. Merlo, D. Sciaky, A. L. Montoya, M. P. Gordon, and E. W. Nester. 1977. *Stable incorporation of plasmid DNA into higher plant cells: The molecular basis of crown gall tumorigenesis*. Cell 11:263–71.

Chilton, M-D., R. K. Saiki, N. Yadav, M. P. Gordon, and F. Quetier. 1980. *T-DNA from*

Agrobacterium Ti plasmid is in the nuclear DNA fraction of crown gall tumor cells. Proc. Nat. Acad. Sci. USA 77:4060–64.

Depicker, A., M. van Montagu, and J. Schell. 1978. *Homologous DNA sequences in different Ti plasmids are essential for oncogenicity.* Nature (Lond.) 275:150–58.

Ditta, G., S. Stanfield, D. Corbin, and D. R. Helinski. 1980. *Broad host range DNA cloning system for gram-negative bacteria: Construction of a gene bank of Rhizobium meliloti.* Proc. Nat. Acad. Sci. USA 77:7347–51.

Drummond, M. H., and M-D Chilton. 1978. *Tumor-inducing (Ti) plasmids of Agrobacterium share extensive regions of DNA homology.* J. Bacteriol. 136:1178–83.

Drummond, M. H., M. P. Gordon, E. W. Nester, and M-D. Chilton. 1977. *Foreign DNA of bacterial plasmid origin is transcribed in crown gall tumors.* Nature (Lond.) 269:535–36.

Gurley, W. B., J. D. Kemp, M. J. Albert, D. W. Sutton, and J. Callis. 1979. *Transcription of Ti plasmid-derived sequences in three octopine-type crown gall tumor lines.* Proc. Nat. Acad. Sci. USA 76:2828–32.

Hack, E., and J. D. Kemp. 1980. *Purification and characterization of the crown gall-specific enzyme octopine synthase.* Plant Physiol. 65:949–55.

Hirch, P. 1980. Ph.D. dissertation (n.t.), John Innes Institute, Norwich, England. Pp. 87–91.

Holsters, M., D. de Waele, A. Depicker, E. Messens, M. van Montagu, and J. Schell. 1978. *Transfection and transformation of Agrobacterium tumefaciens.* Mol. Gen. Genet. 163:181–87.

Johnson, R., R. H. Guderian, F. Eden, M-D. Chilton, M. P. Gordon, and E. W. Nester. 1974. *Detection and quantitation of octopine in normal plant tissue and in crown gall tumors.* Proc. Nat. Acad. Sci. USA 71:536–39.

Kemp, J. D. 1976. *Octopine as a marker for the induction of tumorous growth by Agrobacterium tumefaciens strain B_6.* Biochem. Biophys. Res. Commun. 69:816–22.

Kemp, J. D., D. W. Sutton, and E. Hack. 1979. *Purification and characterization of the crown gall specific enzyme nopaline synthase.* Biochemistry 18:3755–60.

Montoya, A. L., M-D. Chilton, M. P. Gordon, D. Sciaky, and E. W. Nester. 1977. *Octopine and nopaline metabolism in Agrobacterium tumefaciens and crown gall tumor cells: Role of plasmid genes.* J. Bacteriol. 129:101–107.

Montoya, A. L., L. W. Moore, M. P. Gordon, and E. W. Nester. 1978. *Multiple genes coding for octopine-degrading enzymes in Agrobacterium.* J. Bacteriol. 136:909–15.

Murai, N., and J. D. Kemp. 1982a. *T-DNA of pTi-15955 from Agrobacterium tumefaciens is transcribed into a minimum of seven polyadenylated RNAs in a sunflower crown gall tumor.* Nucl. Acids. Res. 10:1679-89.

———. 1982b. *Octopine synthase mRNA isolated from sunflower crown gall callus is homologous to the Ti plasmid of Agrobacterium tumefaciens.* Proc. Nat. Acad. Sci. USA 79:86–90.

Petit, A., S. Delhaye, J. Tempé, and G. Morel. 1970. *Recherches sur les guanidines des tissus de crown gall. Mise en evidence d'une relation biochimique specifique entre les souches d'Agrobacterium et les tumeurs qu'elles induisent.* Physiol. Veg. 8:205–13.

Rothstein, S. J., R. A. Jorgensen, K. Postle, and W. S. Reznikoff. 1980. *The inverted repeats of Tn5 are functionally different.* Cell 19:795–805.

Ruvkun, G. B., and F. M. Ausubel. 1981. *A general method for site-directed mutagenesis in prokaryotes.* Nature (Lond.) 289:85–88.

Thomashow, M. F., R. Nutter, A. L. Montoya, M. P. Gordon, and E. W. Nester. 1980. *Integration and organization of Ti plasmid sequences in crown gall tumors.* Cell 19:729–39.

Wullems, G. J., M. Lucy, G. Ooms, and R. A. Schilperoort. 1981. *Retention of tumor markers in F1 progeny plants from in vitro induced octopine and nopaline tumor tissues.* Cell 24: 719–27.

Yadav, N. S., K. Postle, R. K. Saiki, M. F. Thomashow, and M-D. Chilton. 1980. *T-DNA of a crown gall teratoma is covalently joined to host plant DNA*. Nature (Lond.) 287:458–61.

Yang, F., A. L. Montoya, D. J. Merlo, M. H. Drummond, M-D. Chilton, E. W. Nester, and M. P. Gordon. 1980. *Foreign DNA sequences in crown gall teratomas and their fate during the loss of the tumorous traits*. Mol. Gen. Genet. 177:707–14.

Zaenen, I., N. van Larebeke, H. Teuchy, M. van Montagu, and J. Schell. 1974. *Supercoiled circular DNA in crown gall inducing Agrobacterium strains*. J. Mol. Biol. 86:109–27.

16] Morphogenesis of Transformed Cells from Octopine-type Crown Gall

by LOWELL OWENS* and ESRA GALUN†

ABSTRACT

The development of the Ti plasmid as a gene vector for transforming certain higher plants is hampered by the lack of a good method for regenerating whole plants from tumor callus. Recently we have reported the spontaneous, infrequent occurrence of teratoma formation from cultured octopine-type crown gall tissue of tobacco based on the criteria of octopine production and hormome-independent growth. We have extended this investigation to determine whether the teratomas, which originated from noncloned tumor calli, actually arose from transformed tumor cells or were merely chimeras of tumor and normal or revertant cells. More than 60 single-cell clones were obtained by culturing teratoma leaf protoplasts. Most of these clones regenerated teratomas when cultured on hormone-free medium, and each produced octopine. In a separate experiment, leaf explants of the original noncloned teratoma line were forced to grow as calli by incorporation of hormones into the medium. After 6 months of subculturing, the calli sublines were returned to hormone-free medium. All of the calli sublines proceeded to regenerate teratomas whose leaves produced octopine. From these two lines of evidence, we conclude that the initial octopine-type teratoma originated from a variant transformed cell. Structural analyses of the T-DNA in these teratomas may aid in designing a modified Ti plasmid for use as a gene vector.

INTRODUCTION

Because of its natural ability to genetically transform plant cells during infection, the tumor-inducing (Ti) plasmid of *Agrobacterium tumefaciens* is a potential candidate as a gene vector for higher plants (Drummond 1979; Schell

*Tissue Culture and Molecular Genetics Laboratory, Agricultural Research Service, U.S. Department of Agriculture, Beltsville, Maryland 20705; and †Department of Plant Genetics, Weizmann Institute of Science, Rehovot, Israel.

GENETIC ENGINEERING: APPLICATIONS TO AGRICULTURE (Beltsville Symposium 7—Lowell D. Owens, ed.) Rowman & Allanheld, Totowa

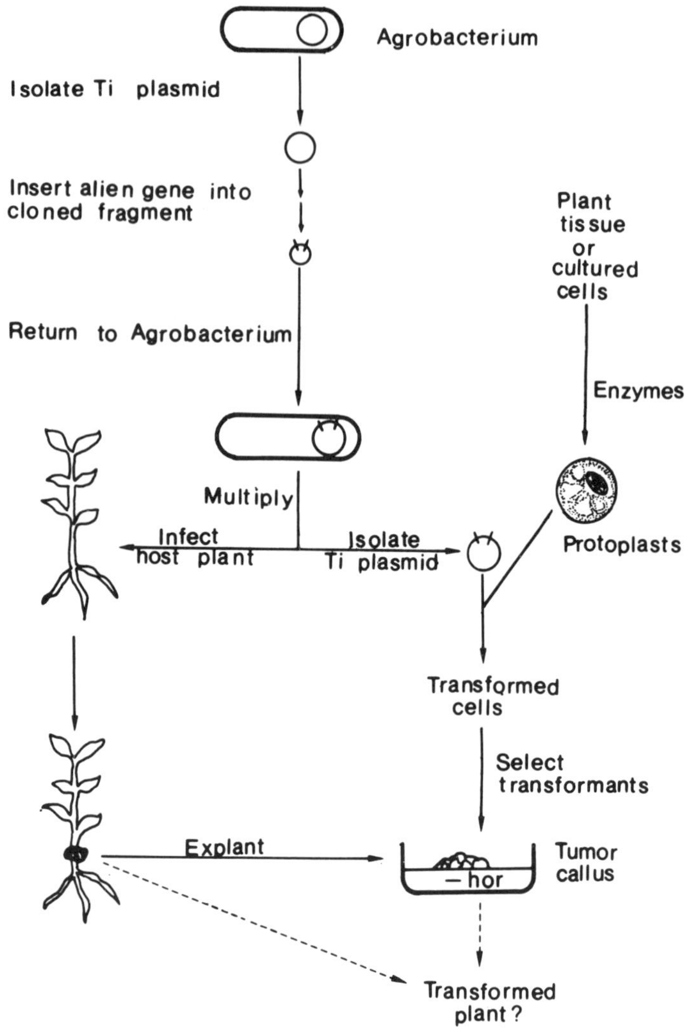

Figure 16.1. Strategies for the genetic transformation of higher plants using the Ti plasmid of *Agrobacterium tumefaciens*.

et al., Chapter 14 and Kemp et al., Chapter 15, this volume). During its tumorigenic transformation of plant cells, a segment of the Ti plasmid is transferred from the bacterium into the plant cell and becomes integrated into nuclear DNA (Chilton et al. 1980; Willmitzer et al. 1980). The transferred segment of plasmid DNA is called T-DNA. The T-DNA codes for several polyadenylated RNA transcripts in crown gall (Gelvin et al. 1982; Murai and Kemp 1982; Willmitzer, Simons, and Schell 1982) and confers upon the transformed cell the ability to grow on medium free of phytohormones and also

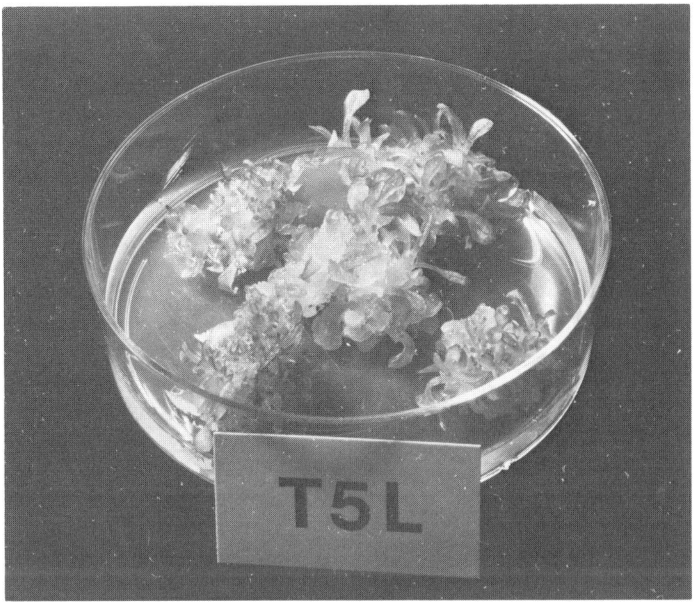

Figure 16.2. Teratoma shoots spontaneously derived from octopine-type tumor callus cultured on phytohormone-free medium.

the ability to produce a novel class of amino acid derivatives, called opines.

Techniques now exist for inserting alien genes into specific sites of the T-DNA region such that it will be transferred into plant cells during transformation (Leemans et al. 1981; Matzke and Chilton 1981). These techniques are depicted schematically in the upper portion of Figure 16.1. The resulting agrobacteria carrying the recombinant Ti plasmid may be used in either of two approaches to transforming plant cells; (a) transformation via infection (Leemans et al. 1981) or (b) transformation via uptake of Ti plasmid by plant protoplasts (Davey et al. 1980; Draper et al. 1982; Krens et al. 1982). With certain important exceptions, the end result of both approaches is cultured tumor callus derived, in the first instance, from an excised crown gall freed of the inciting bacteria or, in the second instance, from transformed cells selected by their ability to grow on medium lacking phytohormones. The exceptions noted refer to the whole transformed plants infrequently obtained from crown galls incited by agrobacteria carrying certain transposon-mutated Ti plasmids (Otten et al. 1981; Schell et al., Chapter 14, this volume) or from revertant tumor cells that have part of their T-DNA deleted (Yang and Simpson 1981). Generally, however, the problem of deriving transformed fertile plants from tumor callus is a serious impediment to the practical use of both approaches to transformation.

This chapter describes our research relating to the problem of regenerating differentiated tissue from undifferentiated tumor callus.

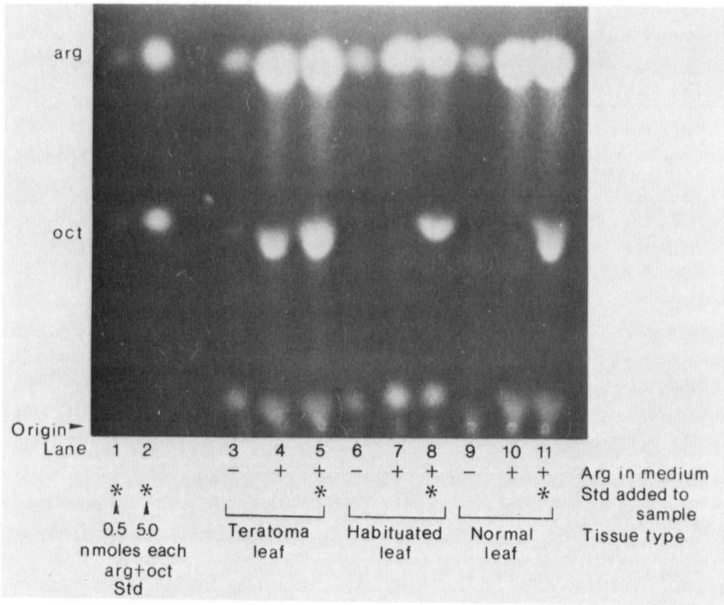

Figure 16.3. Fluorograph of an electrophoretic analysis of excised teratoma (lanes 305), habituated (lanes 6-8) and normal (lanes 9-11) leaves incubated for 3 days with their basal ends inserted into agar medium with or without arginine.

OCTOPINE-TYPE TERATOMAS

Several years ago we established four octopine-type tumor lines from crown galls incited by Ti plasmid B6S3 on four different tobacco (*Nicotiana tabacum* L. cv Bright Yellow) plants. After two or three years of culture on phytohormone-free medium, two of the lines spontaneously formed shoots. These shoots were abnormal in appearance and had the diminutive stature, narrow leaves, and profuse axillary branching characteristic of crown-gall teratomas (Fig. 16.2) commonly associated with nopaline-type tumors (Owens 1982). Since the shoots originated in both cases from tumor lines that had not been cloned from single cells, it was necessary to establish that these shoots were in fact teratomas, i.e., that they had originated from transformed rather than normal plant cells usually present in noncloned tumor lines.

We first asked whether excised leaves of these shoots could express the T-DNA–encoded trait of octopine synthesis. Figure 16.3 shows electrophoretic analyses of extracts from leaves cultured for 3 days in medium containing the octopine-precursor arginine. Octopine was clearly present in extracts of the suspected teratoma leaves (lane 4) but absent in habituated (lane 7) or normal (lane 10) leaves. This was a strong indication of their teratomatous nature.

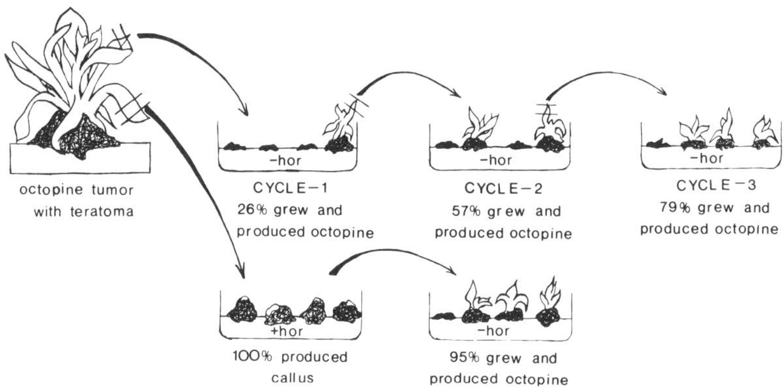

Figure 16.4. Diagrammatic summary of successive tests of teratoma leaf explants for their ability to grow on hormone-free medium and, subsequently, to produce octopine when transferred to a arginine-containing medium.

We next sought to determine whether the leaf tissue could grow on medium lacking phytohormones, another trait of transformed cells. To answer this question small leaf pieces (2×2 mm) were excised and placed on the agar surface of hormone-free medium (Fig. 16.4). About one-fourth of the leaf explants from this planting grew, produced octopine, and regenerated new teratomas. None of the explants from normal leaves displayed these capabilities (Owens 1982). Repetitions of this experiment with newly regenerated teratomas resulted in a progressive increase in the proportion of leaf explants which grew, produced octopine, and regenerated new teratomas (Fig. 16.4). This led us to question whether the initial teratoma leaves were actually chimeras of normal and tumor cells. Perhaps vegetative propagation in the cyclical manner depicted in Figure 16.4 inadvertently selected for tissues having increased proportions of tumor cells and hence increased explant viability when cultured on hormone-free medium.

To investigate this possibility we first cultured leaf explants on medium containing phytohormones, a condition that induced all of the explants to form callus and that also constituted a selection pressure for nontransformed cells. When the calli were returned to hormone-free medium virtually all of the calli continued to grow in an unlimited fashion, produced octopine and regenerated new teratomas (Fig. 16.4). Further, these same characteristics were displayed by 10 calli sublines subcultured on hormone-containing medium for as long as six months before being returned to hormone-free medium (L. Owens, unpublished data). These results allowed us to conclude that each initial teratoma leaf explant contained at least one tumor cell, and probably many more.

SINGLE-CELL CLONING

Still unanswered was the question of whether the teratomas were chimeras of normal and transformed cells and, more important, whether morphogenesis

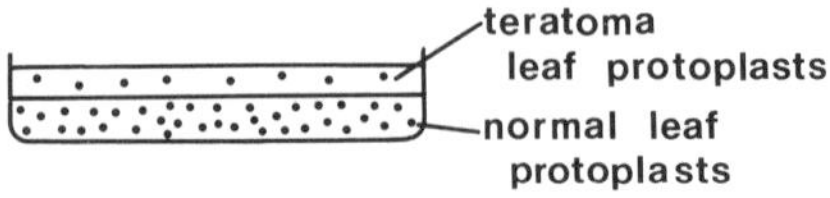

Figure 16.5. Nurse culture for deriving clones from single cells (protoplasts) of teratoma leaves.

was, in fact, dependent upon the presence of normal cells in the shoot tissue. Some bit of circumstantial evidence seemed to point to the latter possibility; the two proven teratomas that arose spontaneously originated from two noncloned tumor lines, while no teratomas have arisen from a number of cloned tumor lines that we have maintained, even though reversion to a nontumorous phenotype might be expected to occur at low frequency in these clonal lines.

To determine the phenotype of cells from the original teratoma line, we decided to derive clonal lines from single leaf cells and observe their behavior. This we accomplished by preparing protoplasts from teratoma leaves and plating them at low cell density in an agar nurse culture (Fig. 16.5). The nurse culture consisted of a bottom layer of protoplasts from normal tobacco leaves plated in agar at double the cell density normally used in such platings. Following solidification of the agar an overlay was made with an equal volume of molten agar containing about 1000 teratoma-leaf protoplasts. When developing clones in the upper layer grew to a diameter of about 1 mm they were carefully removed and placed on a second nurse plate. The latter consisted of a petri dish containing hormone-free agar medium with a moderate-size (1–2 cm dia.) tumor callus situated in the center. The small clones were placed around the periphery of the dish 1 to 2 cm away from the nurse callus in the center. By this technique we obtained more than 60 single-cell clones from teratoma protoplasts.

CHARACTERISTICS OF TERATOMA CLONES

Most of the clones we derived quickly regenerated new teratomas when cultured on medium free of hormones, and also exhibited the additional T-DNA trait of octopine production (L. Owens and E. Galun, unpublished). These results clearly establish that the teratoma leaves are in the main composed of transformed cells and, more important, that teratomas can arise from a single transformed cell. Normal cells evidently are not required for differentiation to occur, although we cannot rule out the possibility that transformed cells reverted at low frequency to a nontumorous phenotype during growth of a clone. Such reversions, possibly resulting from deletion of part or all of the T-DNA, could account for the several clones we obtained that were hormone-requiring and unable to produce octopine. Further investigations are needed to clear up this ambiguity.

CONCLUSIONS

The clonal lines that we derived from transformed teratoma leaf cells exhibit a strong propensity to differentiate new teratomas. In this respect they differ dramatically from the transformed cells of the parent tumor lines from which the teratomas originated. The parent lines, as is generally true for octopine-type tumors, grow as undifferentiated callus.

The teratoma clonal lines resemble in their morphogenetic capabilities the cultured tumors derived by Leemans et al. (1982) from crown galls induced by certain Ti mutants having deletions in their T-region. These tumors also display a high propensity for producing teratomas that lack roots and whose leaves synthesize octopine (also see Schell et al., chapter 14, this volume). We are currently examining the molecular structure of the T-DNA detected in the teratomas described here (D. Cress, unpublished) to determine whether similar deletions exist.

The immediate goal of investigations of this kind is to relate T-DNA structure in transformed cells to their morphogenetic capability. The larger question is whether T-DNA ultimately can be modified in such a way that cells transformed by modified T-DNA can still form crown galls and exhibit hormone-autonomous growth but which, unlike tumors incited by wild-type T-DNA, can be induced by exogenous phytohormones to regenerate fertile plants. The hormone-autonomous growth trait cannot be eliminated altogether, because it is needed either to form crown galls if one is using the infection approach to transformation (Fig. 16.1) or, alternatively, to enable the recognition and isolation of transformed cells occurring at low frequencies, if one is using the protoplast transformation approach. The construction of a Ti plasmid having a hormone-autonomous growth trait that could be overridden by exogenous hormones would constitute a big step toward development of the Ti plasmid as a useful gene vector for higher plants.

LITERATURE CITED

Chilton, M. D., R. K. Saiki, N. Yadav, M. P. Gordon, and F. Quetier. 1980. *T-DNA from Agrobacterium Ti plasmid is in the nuclear DNA fraction of crown gall tumor cells*. Proc. Nat. Acad. Sci. USA 77:4060–64.

Davey, M. R., E. C. Cocking, J. Freeman, N. Pearce, and I. Tudor.s341980. *Transformation of Petunia protoplasts by isolated Agrobacterium plasmids*. Plant Sci. Lett. 18:307–13.

Draper, J., M. R. Davey, J. P. Freeman, E. C. Cocking, and B. J. Cox. 1982. *Ti plasmid homologous sequences present in tissues from Agrobacterium plasmid-transformed Petunia protoplasts*. Plant Cell Physiol. 23:255–62.

Drummond, M. 1979. *Crown gall disease*. Nature 281:343–47.

Garfinkle, D. J., R. B. Simpson, L. W. Ream, F. F. White, M. P. Gordon, and E. W. Nester. 1981. *Genetic analysis of crown gall: Fine structure map of the T-DNA by site-directed mutagenesis*. Cell 27:143–53

Gelvin, S. B., M. F. Thomashow, J. C. McPherson, M. P. Gordon, and E. W. Nester. 1982. *Sizes and map positions of several plasmid-DNA-encoded transcripts in octopine-type crown gall tumors*. Proc. Nat. Acad. Sci. USA 79:76–80.

Hasezawa, S., T. Nagata, and K. Syono. 1981. *Transformation of Vinca protoplasts mediated by Agrobacterium spheroplasts*. Mol. Gen. Genet. 182:206–10.

Krens, F. A., L. Molendijk, G. J. Wullems, and R. A. Schilperoort. 1982. *In vitro transformation of plant protoplasts with Ti-plasmid DNA*. Nature 296:72–74.

Leemans, J., R. Deblaere, L. Willmitzer, H. DeGreve, J. P. Hernalsteens, M. Van Montagu, and J. Schell. 1982. *Genetic identification of functions of TL-DNA transcripts in octopine crown galls*. EMBO J. 1:147–52.

Leemans, J., Ch. Shaw, R. Deblaere, H. DeGreve, J. P. Hernalsteens, M. Maes, M. Van Montagu, and J. Schell. 1981. *Site-specific mutagenesis of Agrobacterium Ti-plasmids and transfer of genes to plant cells*. J. Mol. Appl. Genet. 1:149–64.

Matzke, A. J. M., and M. D. Chilton. 1981. *Site-specific insertion of genes into T-DNA of the Agrobacterium tumor-inducing plasmid: An approach to genetic engineering of higher plant cells*. J. Mol. App. Genet. 1:39–44.

Murai, N., and J. D. Kemp. 1982. *T-DNA of pTi-15955 from Agrobacterium tumefaciens is transcribed into a minimum of seven polyadenylated RNAs in a sunflower crown-gall tumor*. Nucl. Acids Res. 10:1679–89.

Owens, L. D. 1982. *Characteristics of teratomas regenerated in vitro from octopine-type crown gall*. Plant Physiol. 69:37–40.

Otten, L., H. DeGreve, J. P. Hernalsteens, M. Van Montagu, O. Schieder, J. Straub, and J. Schell. 1981. *Mendelian transmission of genes introduced into plants by the Ti-plasmids of Agrobacterium tumefaciens*. Mol. Gen. Genet. 183:209–13.

Willmitzer, L., M. De Buckeleer, M. Lemmers, M. Van Montagu, and J. Schell. 1980. *DNA from Ti plasmid present in nucleus and absent from plastids of crown gall plant cells*. Nature 287:359–61.

Willmitzer, L., G. Simons, and J. Schell. 1982. *The TL-DNA in octopine crown-gall tumors codes for seven well-defined polyadenylated transcripts*. EMBO J. 1:139–46.

Yang, F., and R. B. Simpson. 1981. *Revertant seedlings from crown gall tumors retain a portion of the bacterial Ti plasmid DNA sequences*. Proc. Nat. Acad. Sci. USA 78:4151–55.

17] Mutations and Cell Selections: Genetic Variation for Improved Protein in Rice

by GIDEON W. SCHAEFFER and F. T. SHARPE, Jr.*

ABSTRACT

The rapid progress in microbial genetics in past decades was due in part to short cell cycles and the availability of mutants which prescribed the biochemistry. Hypothetically, large numbers of separated plant cells in liquid suspension provide some of the same benefits to plant biologists. In actual practice, however, plant systems are much more complex: cells rarely exist as single cells, and gene expression is complex and controlled by developmental processes. Plant biochemists urgently need mutants of biochemical events for systematic and predictable progress in plant modification through somatic cell fusion, gene isolation, and gene transfer. This report reviews briefly the current status of plant mutant selections in cell culture, and discusses the potentials and pitfalls of in vitro techniques. The work illustrates useful and deleterious variability recovered by the anther and tissue culture of rice cells. This unexpected variability in the progeny of tissue-cultured material occurred in seed set, tillering, quantity of seed and protein. Advantages and disadvantages of selecting mutants from cells in culture are evaluated.

INTRODUCTION

In the early 1960s scientists began to visualize the specific role of genetics in the control of plant growth and development. Rapid progress in microbial genetics is attributed to the availability of many mutants that could be generated for specific biochemical events. Nonetheless, progress in the biochemical genetics of eucaryotic systems and higher plants in particular has been much slower than with procaryotes, because mutants are not available and not easily generated. In addition, complex interactions, both physical and

*Tissue Culture and Molecular Genetics Laboratory, Agricultural Research Service, U.S. Department of Agriculture, Beltsville, Maryland, 20705.

GENETIC ENGINEERING: APPLICATIONS TO AGRICULTURE (Beltsville Symposium 7—Lowell D. Owens, ed.) Rowman & Allanheld, Totowa

biological, make simple conclusions about the nature of mutants more difficult in higher organisms. Additional levels of cellular and chromosomal organization in plants modulate and control the genetic message coded in complex chromosomes. Mutants of plant cells, particularly crop plants, are required for a precise biochemical understanding of the elements of form and function at all levels of organization.

It appears, at least superficially, that researchers should be able to reduce some of the complexity of plant function by utilizing tissue cultures or protoplasts of plant cells. Numerous schemes have been devised to obtain large numbers of single cells or small aggregates of cells for mutant selection. The protocol is to have large numbers of cells in which specific physical or biochemical selection pressures can be uniformly applied, as has been done with bacterial systems. The hypothetical advantage of treating plant cells as bacterial cells is seldom realized because plant cells, even protoplasts, aggregate rapidly, establish gradients across cell layers, and feed each other. Thus plant cell selections are most often the selection of cells in aggregates. Mutant frequencies are often lower in plant systems, a fact caused, in part, by inefficient recovery techniques, as well as by the presence of multiple gene copies located on different chromosones, and cell-to-cell interactions in even the most ideal suspension culture systems. Even so, tissue cultures, particularly as cells in small aggregates or cells in liquid suspension, have been utilized to recover hundreds of variants and mutants. Examples are numerous and notable: Kao et al. (1970), Widholm (1972, 1974b, 1977a, 1978a); Maliga, Breznovits, and Marton (1973); Maliga, Marton, and Breznovits (1973); Gengenback and Green (1975); Maliga et al. (1975); Liu et al. (1977); Brettell and Ingram (1979); Schaeffer (1979); Shepherd, Bidney, and Shahin (1980); and Hibberd and Green (1982).

There are also numerous reviews on the methods and results of in vitro variant selections in the literature: Chaleff and Carlson (1974); Widholm (1947a, 1977a and b, 1978b); Bottino (1975); Nabors (1976); Day (1977); Green (1977) Kleinhofs and Behki (1977); Nelson (1977); Scowcroft (1977); Maliga (1978); Thomas, King, and Potrykus (1979); Siegemund (1981); and Larkin and Scowcroft (1981). A complete review is not intended here, but emphasis is placed upon the regeneration of plants from intense selection pressures with a brief discussion of recovered variability. Only a few mutants selected in vitro with biochemical selection pressures have been regenerated so that complex whole plant genetic analyses could be accomplished. Achievements in genetic analyses were first accomplished with the *Solanaceae* (Carlson 1973; Maliga, Breznovits, and Marton 1973; Maliga, Marton, and Breznovits 1973; Maliga et al. 1975; Aviv and Galun 1977; Polacco and Polacco 1977; Bourgin 1978; Carlson and Widholm 1978; Chaleff and Parsons 1978b; and Nabors et al. 1980). Notable exceptions are several agriculturally important *Gramineae,* including *Zea* (Gengenbach, Green, and Donovan 1977; Hibberd and Green 1982), *Oryza* (Schaeffer and Sharpe 1981), and *Saccharum* sp. (Heinz et al. 1977; Nickell and Heinz 1977). Examples of whole plant "genetic" analyses of variants generated by tissue culturing are presented in Table 17.1

The tabulation illustrates the effective use of the tobacco model system, the use of protoplasts and suspension culture, and the selection of cells resistant to inhibitors in the aggregate or callus form. The use of biochemical selection pressure for the recovery of variants from cultures of important food crops serves to illustrate the future application of in vitro techniques to crop production.

In some species, variants generated in vitro are ubiquitous. The in vitro culture process may itself be mutagenic. The high rate of instability may be the reason for the relative ineffectiveness of exogenous mutagens. Some of the heterogeneity recovered may be the result of drastic genetic rearrangements during culture and are probably responsible for considerable genetic variation among regenerated plants. The concepts of tissue culture–induced variation and its importance was summarized by Larkin and Scowcroft (1981).

It is clear from the literature that tissue culture produces both useful and deleterious variability. Unselected variability has often been described in the past literature in a detrimental or apologetic manner. It is becoming very clear, however, that such variability can also be valuable and serve as an important source of germplasm for plant improvement and ultimately be a valuable adjunct to plant breeding.

SOMACLONAL VARIATION

The generalized term "somaclones" has been suggested for plants derived from any form of cell culture (Larkin and Scowcroft 1981). At first reading the prefix "soma" seemed inappropriate for cells derived from gametes—for example, the microspore in anther culture. Yet the reprogrammed microspores give rise to sporophytic tissue from which plants are regenerated, and in the sporophytic sense the somaclonal term might be appropriate for anther-derived tissues as well. Tissue and anther culture-derived variability, even though genotype dependent, is pervasive, and is adequately documented in many crops such as sugarcane (Heinz & Mee 1971; Heinz 1973; Nickell & Heinz 1973; Liu et al. 1977); potato (Matern, Strobel, and Shepard, 1978; Wenzel et al. 1979; and Shepard, Bidney, and Shahin, 1980); tobacco (Devreux and Laneri 1974; Burk and Matzinger 1976; Burk et al. 1979); oats (Cummings, Green, and Stuthman 1976); maize (Gengenbach, Green, and Donovan 1977; Green 1977; Brettell and Ingram 1979); barley (Deambrogio and Dale 1980); *Brassica* (Grout and Crisp 1980); *Pelargonium* (Skirvin and Janick 1976 a, b); carrot (Ibrahim 1969); *Chrysanthemum* (Ben-Jaacov and Langhans 1972); lily (Stimart, Ascher, and Zagorski 1980); carnation (Hackett and Anderson 1967); clover (Beach and Smith 1979); sorghum (Gamborg et al. 1977), and many more. Particularly relevant to this report is the work of Nishi, Yamada, and Takahashi (1968) and Henke, Mansur, and Constantin (1978), who reported variations among rice plants regenerated from callus. Detailed analyses of variability recovered from haploid and diploid rice was presented by Oono (1975, 1978a,b). He noted wide variations in fertility, plant height, heading date, and chlorophyll content from anther culture progeny. Only 28% of the progeny from a selfed double haploid appeared normal. Some variation was

Table 17.1 Plant Regeneration and Genetic Characterizations of Tissue Culture Variants Resistant to Selective Inhibitors

Species	Tissue	Mutagenesis	Selection pressure	Type of variant[a]	Genetics[b]	Reference	Year
D. carota (Daucus)	suspensions	–	cycloheximide	1	SRC	Gresshoff, P. M.	1979
D. carota	suspensions	–	5-methyltryptophan	1	SRC	Widholm, J. M.	1974[b]
D. carota	suspensions	+	5-fluorouracil	1	SRC	Sung, Z. R., & S. Jacques	1981
D. innoxia (Datura)	suspensions	–	aminopterin	1	SRC	Mastrangelo, I. A., & H. H. Smith	1977
N. tabacum (Nicotiana)	protoplasts	+	valine	1	CM	Bourgin, J. P.	1978
N. tabacum	suspensions	–	NaCl	1	S	Nabors, M. W., et al.	1980
N. tabacum	protoplasts	+	methionine sulfoximine	1	CM	Carlson, P. S.	1973
N. tabacum	protoplasts	+	N-phenylcarbamate	1	S	Aviv, D. & E. Galun	1977
N. tabacum	suspensions	–	picloram	1	CM	Chaleff, R. S., & M. F. Parsons	1978[a]
N. tabacum	callus	–	streptomycin	1	S	Maliga, P., et al.	1973
N. tabacum	callus	–	carboxin	1	SRC	Polacco, J. C., & M. L. Polacco	1977
N. tabacum	suspensions	+	chlorate	6	SRC	Müller, A., & R. Grafe	1978

N. tabacum	suspensions	–	glycerol	2	CM	Chaleff, R. S., & M. F. Parsons	1978[b]
N. tabacum	suspensions	–	glycine hydroxamate	7	—	Lawyer, A. L., et al.	1980
N. tabacum	callus	–	bentazone phenimedipharm	1	CM	Radin, D. N., & P. S. Carlson	1978
N. sylvestris	callus	+	streptomycin	1	CM	Maliga, P.	1981
O. sativa (Oryza)	callus	–	S-(2-aminoethyl-L-cysteine)	3	S	Schaeffer, G. W., & F. T. Sharpe Jr.	1981
O. sativa	callus	–	5-methyltryptophan	1	SRC	Chen, C. M., & C. C. Chen	1979
S. officinarum (Saccharum)	suspensions	+	H. sacchari toxin	1	SRC	Heinz, D. J., et al.	1977
S. tuberosum (Solanum)	callus	+, –	P. infestans toxin	1	SRC	Behnke, M.	1979
Z. mays (Zea)	embryo callus	–	lysine + threonine	4	SRC	Hibberd, K. A., et al.	1980
						Hibberd, K. A., & C. E. Green	1982
Z. Mays	embryo callus	–	H. maydis toxin	1	CM	Gengenbach, B. G., et al.	1977

[a] Variant type code

1. Resistant to metabolic inhibitor
2. Glycerol utilization
3. Lysine & protein overproducer
4. Threonine overproducer
5. Tryptophan overproducer
6. Nitrate reductaseless
7. Glycine & alanine overproducer

[b] Genetic response code

SRC = sporophytic tissue rechallenged with selective agent.
S = progeny from selfed plants rechallenged.
C = progeny from crosses rechallenged.
CM = progeny from crosses rechallenged — Genetics Mendelian

fixed during the second selfing; other variability segregated in later generations.

ANALOG SELECTIONS FOR NUTRITIONAL VALUE

The in vitro bacterial and plant cell methods for the recovery of cell types with altered metabolic pathways, including variants along the aspartate-lysine pathway, are reasonably well established. Brock, Friederick, and Langridge (1973) proposed a scheme for the selection of mutations for amino acid overproduction by using the analog of lysine, S-(2-aminoethyl)-L-cysteine (S-AEC) as a selective agent. Analogs of amino acids may function as false feedback inhibitors or as a metabolic inhibitor if they are stereo-chemically similar enough to the natural amino acids. Even though S-AEC is incorporated into protein (Bright, Featherstone, and Miflin 1979; Green and Donovan 1980), the analog fits other criteria, at least hypothetically, as a selective agent for the recovery of mutants that overproduce lysine. The selection of plant cells for resistance to S-AEC has been accomplished for a number of cell types, including carrot (Widholm 1978a,b). Halsall, Brock, and Langridge (1972) reported that aspartokinase from rice is sensitive to feedback inhibition by S-AEC. Schaeffer and Sharpe (1981) have shown improvement of rice seed protein and lysine from plants regenerated by S-AEC resistant callus. Since callus in cereals is often difficult to regenerate, researchers have avoided the use of callus tissue and applied pressures directly to isolated embryos or young seedlings. Progress has been reported for the recovery of plants carrying resistance to inhibitory levels of lysine plus threonine in barley. Lysine and threonine synergistically inhibit growth of barley seedlings, and a fertile mutant with altered feedback regulatory properties of aspartate kinase has been recovered (Bright, Miflin, and Rognes 1982). The selection increased total threonine and methionine levels in seeds and vegetative tissue by 6%. Even though the lack of regenerability from culture creates difficulty, the problem is not insurmountable, as the work of Bright, Miflin, and Rognes indicates. The use of mutagenized immature embryos for biochemical selections has also been reported with maize.

Hibberd and Green (1982) used immature embryos for the recovery of maize lines insensitive to feedback inhibition and which overproduced free threonine. These authors included an intermediate tissue culture stage for the increase of the embryogenic tissues. Mutagenesis of the callus with sodium azide produced variants resistant to lysine plus threonine. A single dominant gene in progenies was responsible for the overproduction of threonine in subsequent maize kernels. The choice of embryonic material increased the probability of plant regeneration following cell selection. Thus, tissue culture techniques are being developed for the selection of biochemical variants important in the control of the nutritional quality of seeds of major crop plants.

The remainder of this chapter provides documentation for the recovery of useful variability in the form of improved seed protein and lysine as well as deleterious variability in the form of decreased fertility in rice plants regenera-

Figure 17.1. In vitro sequence leading to regeneration of rice (*Oryza sativa* L.) callus from anther culture and tissue culture of PI 353705.

ted from cells resistant to the analog of lysine, S-AEC. The work also describes the recovery of a phenotype from the tissue/anther culture of a commercial cultivar of rice with enhanced dwarf characteristics. The work provides evidence for unique phenotypic variability of plants regenerated from callus grown in vitro.

THE SELECTION PROCESS

The selection of 34 rice cell lines resistant to S-AEC produced several types of variants (Schaeffer and Sharpe 1981). Nine plants were regenerated from callus, and two produced seed. Progeny from one plant (MA-13) was examined for phenotypic variability in seed lysine, protein, and seed number and other morphological characteristics. The scheme for the cell selections, i.e., resist-

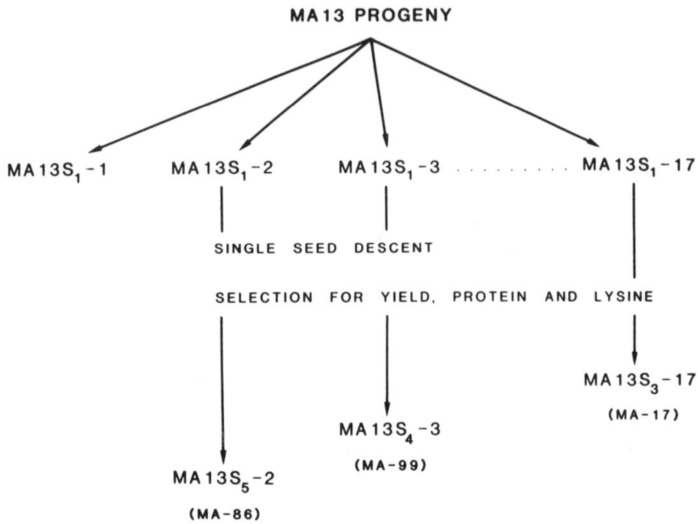

Figure 17.2. Diagram of rice progeny from a plant (MA-13) regenerated from callus resistant to S-aminoethyl-cysteine and selfed for protein and lysine characterization. S_1 is 2nd-generation material, and subsequent generations (S_2 to S_5) were single-plant selections (single-seed descent) from 10 or 15 plants used for analyses. The single plants were selected for optimum seed number, and protein and lysine levels.

ance to S-AEC, is illustrated in Figure 17.1. The procedure included the following sequence, as described earlier by Schaeffer and Sharpe (1981): anther culture of rice, PI 353705, similar to Assam 5 with developing callus; tissue increase and test for morphogenetic potential; selection of cells (small aggregates of cells) resistant to 1 mM S-AEC; recovery of cells resistant to S-AEC; three additional passages on 2 mM S-AEC; regeneration of plants; seed production and progeny analyses, including the 5th selfing of one line derived from a single seed.

Illustrated in Figure 17.2 is the advanced progeny testing scheme. S_1 through S_5 refer to generations of selfings. Seventeen S_1 plants were grown from seed of the original MA-13 plant which had produced over 60 seeds. Three lines were selfed repeatedly for progeny analyses.

INHERITANCE OF S-AEC RESISTANCE

The inheritance of S-AEC resistance was tested by isolating mature embryos from seed and reestablishing tissue cultures of several lines derived from single seeds. All callus from one line was totally resistant to S-AEC and grew better in the presence than the absence of S-AEC. Another line was totally sensitive to S-AEC. Thus the original plant was heterozygous for resistance to S-AEC

Table 17.2 Mean Protein and Lysine Values for Mutant (MA-86 and MA-99, 3rd through 5th generations) and Control (IC, FC) Seed

Plant		Seed Protein	Seed Lysine[a]
type	#	Percent	Percent
MA-86	27	12.7 a	3.89 a
MA-99	47	12.0 a	3.89 a
IC	23	9.5 b	3.90 a
FC	48	8.8 b	3.73 b

[a] Percent of total amino acids—acid hydrolyzate.

Note: Mutant seeds are from plants derived from S-(2-aminoethyl)-L-cysteine resistant callus. Values with different letters are significantly different at $\alpha = 0.05$ based on Duncan's multiple range test.

and/or genetically unstable. We prefer the hypothesis that the original plant was heterozygous for resistance to 2 mM S-AEC and that resistance is dominant. Population analysis of advanced lines is required to test instability as an alternative hypothesis.

PROGENY ANALYSIS FOR PROTEIN, LYSINE, AND RELATED CHARACTERISTICS

Protein and lysine analyses of seed from the original regenerated plant as well as selfed progeny revealed the recovery of an important phenotype: higher protein levels, as well as higher lysine in percent of total amino acids in the protein, than in control materials. This contrasts with the breeding expectations that percent lysine is reduced in high-protein lines (Singh and Axtell 1973). The higher protein and lysine characteristic increases the potential value of this genotype. In the phenotype recovered there was no decrease but an increase in lysine with increased protein. Many other morphological and biochemical variations were recovered in selfed progeny. They included differences in plant height, seed number and size, plant color, extension of inflorescence, stress tolerance (particularly low temperatures), decreased fertility or increased male sterility particularly during the winter months, and a propensity toward the discoloration of floral structures. These observations are consistent with those of Oono (1978a) on the recovery of variability from tissue-cultured rice.

Perhaps the most valuable and consistent changes recovered from S-AEC selected cells were increased protein and lysine (Table 17.2) in seed hydrolyzates. There were statistically significant differences between the mutant

Table 17.3 Variation in Seed Characteristics at Different Levels of Selfing

	Ratio of mutant (MA-99) to field control (FC)			
	Plant	Protein	Lysine	Weight
Generations	(#)	mg/seed	% of total amino acids	mg/seed
S-original	1	1.03	1.10	0.84
S-second	4	0.83	1.09	0.62
S-third-99	7	1.06	1.03	0.76
S-fourth-99	27	1.16	1.09	0.85
S-fifth-99	15	1.13	1.02	0.98
S-fifth-99[a]	8	1.08	1.04	0.98

[a]Seed numbers equal to controls.

progeny and the controls in percent seed protein and lysine after extensive selfing. MA-99 and MA-86 seed had more protein than the internal control (IC) or the non-tissue cultured field control (FC). The IC line had a tissue culture history similar to MA-99 and MA-86 line but was not subjected to the cell selection pressure with S-AEC. The mutants were higher in protein and lysine and maintained the elevated lysine level through advanced generations.

Even though only one plant from those regenerated produced enough seed for progeny testing, there is interesting variability with respect to protein and lysine levels in the progeny from that plant. Two lines designated MA-99 and MA-86 illustrate some of the variability. MA-99 is distinctly different from MA-86 in several ways. MA-86 had significantly higher seed weight and percent protein in the seed. On the other hand, the histogram for seed number is more normal with MA-99 than MA-86. One of the early concerns was the carryover effect from tissue culture. The effects of selfing upon the seed protein, lysine and seed weight in subsequent generations of MA-86 illustrated in Table 17.3. It is clear that the protein level is maintained in advanced generations, and lysine as percent of total amino acids was highest in early tissue culture types and was maintained at levels of 2 to 5% increase in advanced generations over the control. Seed weight appears to have stabilized after the third generation.

Inasmuch as increased seed protein is frequently associated with decreased lysine and yield (Singh and Axtell 1973; Coffman and Juliano 1979), the percent protein and mg lysine/seed were compared in plants with seed numbers equal to and also less than controls. The results of that comparison are shown in Table 17.4. MA-99 plants with seed number and weights equal to the controls maintained the high protein values with no loss in lysine. In fact, the lysine content of the variant is 10% greater than the control, which translates into a gain of 6.0 μg lysine per seed. Similarly, the amino acid profile from seed

Table 17.4 Variations in Characteristics of Advanced Progeny Selected for Normal Seed Production

Plant Type	#	Protein %	Lysine[a] %	Protein (mg/seed)	Lysine (mg/seed)	Seed #	Seed Weight mg	Tiller #
MA-99	8	9.37 b	3.93 a	1.76 a	.068 a	966 a	18.8 a	13.3 b
MA-17	6	10.68 a	3.66 b	1.90 a	.069 a	992 a	17.7 b	16.3 a
FC	13	8.53 c	3.79 b	1.63 b	.062 b	1048 a	19.1 a	13.7 b

[a] Percent of total amino acids—acid hydrolyzate.
Note: MA-99 and MA-17 represent 5th and 4th generation seed, respectively. Values with different letters are significantly different at $\alpha = 0.05$ based on Duncan's multiple range test.

Table 17.5 Total Seed Amino Acids (acid hydrolyzates) of Variant (MA-99) and Field Control (FC)

Amino acids	Ratio: MA-99/FC a	Ratio: MA-99/FC b
ASP	1.05	1.05
THR	1.01	0.97
SER	0.96	0.92
GLU	0.99	1.00
PRO	1.00	1.00
GLY	1.01	1.00
ALA	1.00	1.00
VAL	1.00	1.04
MET	0.99	0.94
ILE	1.05	1.01
LEU	0.98	1.01
TYR	0.98	0.98
PHE	1.01	1.02
HIS	1.01	0.96
LYS	1.04[c]	1.01
ARG	0.99	0.98

[a] Seed/plant of MA-99 equal to FC.
[b] Seed/plant of MA-99 statistically less than FC.
[c] Significantly different at $\alpha = 0.05$.

hydrolyzates (Table 17.5) show increased levels of aspartic acid, isoleucine, and lysine in the mutant lines with seed yields equal to the control.

DELETERIOUS VARIATION FROM CELL CULTURE

The elevated protein and lysine levels recovered as a result of in vitro techniques is an example of useful variation. Deleterious variation was also recovered. One example is the reduction in fertility and seed set in plants regenerated from callus. Fertility may be recovered by selfing, but it does prolong the recovery process and potentially detracts from the full benefits of the techniques. There were significant decreases during early generations in seed weights of the mutants compared with the controls. Seed weights were restored with subsequent selfings. Significant changes occurred in days to flower, but not in growth rates in the genotypes compared.

The one major difference among the S-AEC resistant variants and the IC and FC types was fertility and consequently seed number. The tissue culture–derived lines, both S-AEC selected and internal controls, produced less seed than with non-tissue cultured field controls.

One problem with infertility is that it may produce misleading protein and lysine values. Thus, later experiments were designed to minimize the effect of seed number by selecting types with both high protein and high seed set. The results after the fifth selfing were a 10% increase in protein and a 3 to 4% increase in percent lysine in the protein of plants with full seed set over the field control (Tables 17.3 & 17.4). Thus the analog selection technique has produced rice with improved seed protein. Some components of variability in rice continue to be expressed after four or five selfings. Morphological markers, delayed maturity, and senescence are obvious characteristics that segregated from selfed material for three or more generations. Apparently long-term residual variation occurs in some tissue culture–derived lines.

RECOVERY OF ENHANCED DWARF RICE FROM THE CULTIVAR, CALROSE 76

Calrose 76, a commercial cultivar of rice, was developed for short stature and stiff straw (Rutger, Peterson, and Hu 1977). Doubled haploids, derived by anther culture of Calrose 76, were 25–30% shorter and had 30% more tillering than the parent cultivar (Schaeffer 1982). The variant was stable through three selfings. The original plants regenerated from anther/tissue culture showed variations in leaf size and shape, extension of inflorescence, flowering date, plant height, panicle length, and seeds/panicle (Schaeffer 1982). Equally important as the variability from anther culture is the uniformity of the homozygous doubled haploid progeny from single-anther–derived plants. This is illustrated in Figure 17.3. Compared to the original cultivar (top), the doubled haploid (bottom) had shorter stature, equal yield, increased tillering, and 14% higher

Figure 17.3. Comparison of height and tillering of the original cultivar, Calrose 76 *(top)* with 3rd-generation, spontaneously doubled dihaploid carrying genes for enhanced dwarf characteristics *(bottom)* regenerated from the anther culture of Calrose 76.

seed protein. Seed size was reduced by 5%. The process illustrates useful heterogeneity recovered from the anther/tissue culture process.

CONCLUSIONS

The in vitro culture of numerous crop plants may be used for the recovery of unique and important genetic variability which cannot be easily derived by standard breeding practices. Certain types of variability will be useful, other types will be deleterious. Somaclonal variation ultimately must be understood and controlled to be used effectively by plant genetic engineers and cellular geneticists. If properly controlled, this source of variability will become a major adjunct to standard breeding practices in the future.

LITERATURE CITED

Aviv, D., and E. Galun. 1977. *Isolation of tobacco protoplasts in the presence of isopropyl-N-phenylcarbamate and their culture and regeneration into plants.* Z. Pflanzenphysiol. 83:267–73.

Beach, K. H., and R. R. Smith. 1979. *Plant regeneration from callus of red clover and crimson clover.* Plant Sci. Lett. 16:231–28.

Behnke, M. 1979. *General resistance to late blight of Solanum tuberosum plants regenerated from callus.* Theor. Appl. Genet. 56:151–52.

Ben-Jaacov, J., and R. W. Langhans. 1972. *Rapid multiplication of Chrysanthemum plants by stem-tip proliferation.* Hort. Sci. 7:289–90.

Bottino, P. J. 1975. *The potential of genetic manipulation in plant cell cultures for plant breeding.* Rad. Bot. 15:1–16.

Bourgin, J. P. 1978. *Valine-resistant plants from in vitro selected tobacco cells.* Mol. Gen. Genet. 161:225–30.

Brettell, R. I. S., and D. S. Ingram. 1979. *Tissue culture in the production of novel disease-resistant crop plants.* Biol. Rev. 54:329–45.

Bright, S. W. J., B. J. Miflin, and S. E. Rognes. 1982. *Threonine accumulation in the seeds of a barley mutant with an altered aspartate kinase.* Biochem. Genet. (in press).

Bright, S. W., J., L. C. Featherstone, and B. J. Miflin. 1979. *Lysine metabolism in a barley mutant resistant to S(2-aminoethyl)-cysteine.* Planta 146:629–33.

Brock, R. D., E. A. Friederick, and J. B. Langridge. 1973. *The modification of amino acid composition of higher plants by mutation and selection.* Pages 329–38 in *Nuclear Techniques for Seed Protein Improvement.* IAEA, Vienna.

Burk, L. G., J. F. Chaplin, G. V. Gooding, and N. T. Powell. 1979. *Quantity production of anther-derived haploids from a multiple disease resistant tobacco hybrid. 1. Frequency of plants with resistance or susceptibility to tobacco mosaic virus (TMV), potato virus (PYV) and root knot (RK).* Euphytica 28:201–208.

Burk, L. G., and D. F. Matzinger. 1976. *Variation among anther-derived doubled haploids from an inbred line of tobacco.* J. Hered. 67:381–84.

Carlson, P. S. 1973. *Methionine sulfoximine-resistant mutants of tobacco.* Science 180:1366–68.

Carlson, J., and J. Widholm. 1978. *Separation of two forms of anthranilate synthetase from 5-methyltryptophan susceptible and resistant cultured Solanum tuberosum cells.* Physiol. Plant. 44:251–55.

Chaleff, R. S., and P. S. Carlson. 1974. *Somatic cell genetics of higher plants. Ann. Rev. Genet.* 8:267–78.

Chaleff, R. S., and M. F. Parsons. 1978a. *Direct selection in vitro for herbicide-resistant mutants of Nicotiana tabacum*. Proc. Nat. Acad. Sci. USA 75:5704–5707.

———. 1978b. *Isolation of a glycerol-utilizing mutant of Nicotiana tabacum*. Genetics 89:723–28.

Chen, C. M., and C. C. Chen. 1979. *Selection and regeneration of 5-methyltryptophan resistant rice plants from pollen callus*. Natl. Sci. Counc. Monthly ROC 7(4):378–82.

Coffman, W. R., and B. O. Juliano. 1979. *Seed protein improvement in rice*. In *Seed Protein Improvement in Cereals and Grain Legumes*. IAEA, Vienna.

Cummings, D. P., C. E. Green, and D. D. Stuthman. 1976. *Callus induction and plant regeneration in oats*. Crop Sci. 16:465–70.

Day, P. R. 1977. *Plant genetics: Increasing crop yield*. Science 197:1334–39.

Deambrogio, E., and P. J. Dale. 1980. *Effect of 2,4-D on the frequency of regenerated plants in barley and on genetic variability between them*. Cereal Res. Comm. 8:417–23.

Devreux, M., and V. Laneri. 1974. *Anther culture, haploid plants, isogenic line and breeding research in Nicotiana tabacum* L. Pages 101–107 in *FAO/IAEA, Polyploidy and Induced Mutations in Plant Breeding*. IAEA, Vienna.

Gamborg, O. L., J. P. Shyluk, D. S. Brar, and F. Constabel. 1977. *Morphogenesis and plant regeneration from callus of immature embryos of sorghum*. Plant Sci. Lett. 10:67–74.

Gengenbach, B. G., and C. E. Green. 1975. *Selection of T-cytoplasm maize callus cultures resistant to Helminthosporium maydis Race T pathotoxin*. Crop Sci. 15:645–49.

Gengenbach, B. G., C. E. Green, and C. M. Donovan. 1977. *Inheritance of selected pathotoxin resistance in maize plants regenerated from cell cultures*. Proc. Nat. Acad. Sci. USA 74:5113–17.

Green, C. E. 1977. *Prospects for crop improvement in the field of cell culture*. Hort Sci. 12:7–10.

Green, C. E., and C. M. Donovan. 1980. *Effect of aspartate-derived amino acids and aminoethyl cysteine on growth of excised mature embryos of maize*. Crop Sci. 20:358–62.

Gresshoff, P. M. 1979. *Cycloheximide resistance in Daucus carota cell cultures*. Theor. Appl. Genet. 54:141–42.

Grout, B. W. W., and P. Crisp. 1980. *The origin and nature of shoots propagated from cauliflower roots*. J. Hort. Sci. 55:65–70.

Hackett, W. P., and J. M. Anderson. 1967. *Aseptic multiplication and maintenance of differentiated carnation shoot tissue derived from shoot apices*. Proc. Amer. Soc. Hort. Sci. 90:365–69.

Halsall, O., R. D. Brock, and J. B. Langridge. 1972. *Selection for high lysine mutants*. Genetics Report, CSIRO Division of Plant Industry. P. 31.

Heinz, D. J. 1973. *Sugarcane improvement through induced mutations using vegetable propagules and cell culture techniques*. Pages 53–59 in *Induced Mutations in Vegetatively Propagated Plants*. Proc. of a panel Sept. 11–15, 1972. IAEA, Vienna.

Heinz, D. J., M. Krishnamurthi, L. G. Nickell, and A. Maretzki. 1977. *Cell, tissue and organ culture in sugarcane improvement*. Pages 3–17 in J. Reinert and Y. P. S. Bajaj, eds., *Applied and Fundamental Aspects of Plant Cell, Tissue and Organ Culture*. Springer, Berlin.

Heinz, D. J., and G. W. P. Mee. 1971. *Morphologic, cytogenetic, and enzymatic variation in Saccharum species hybrid clones derived from callus tissue*. Amer. J. Bot. 58:257–62.

Henke, R. R., M. A. Mansur, and M. J. Constantin. 1978. *Organogenesis and planlet formation from organ- and seedling-derived calli of rice (Oryza sativa)*. Physiol. Plant. 44:11–14.

Hibberd, K. A., and C. E. Green. 1982. *Inheritance and expression of lysine plus threonine resistance selected in maize tissue culture*. Proc. Nat. Acad. Sci. USA 79:559–63.

Hibberd, K. A., T. Walter, C. E. Green, and B. G. Gengenbach. 1980. *Selection and characterization of a feedback-insensitive tissue culture of maize*. Planta 148:183–87.

Ibrahim, R. K. 1969. *Normal and abnormal plants from carrot root tissue cultures*. Can. J. Bot. 47:825–26.

Kao, K. W., R. N. Miller, O. L. Gamborg, and B. L. Harvey. 1970. *Variations in chromosome number and structure in plant cells grown in suspension cultures*. Can. J. Genet. Cytol. 12:297–301.

Kleinhofs, A., and R. Behki. 1977. *Prospects for plant genome modification by non-conventional methods*. Ann. Rev. Genet. 11:79–101.

Larkin, P. J., and L. R. Scowcroft. 1981. *Somaclonal Variation—A novel source of variability from cell cultures for plant improvement*. Theor. Appl. Genet. 60:197–214.

Lawyer, A. L., M. B. Berlyn, and I. Zelitch. 1980. *Isolation and characterization of glycine hydroxamate-resistant cell lines of Nicotiana tabacum*. Plant Physiol. 66:334–41.

Liu, M.-C., K. C. Shang, W. H. Chen, and S. C. Shih. 1977. *Tissue and cell culture as aids to sugarcane beeeding. Aneuploid cells and plants induced by treatment of cell suspension cultures with colchicine. Proc. 16th Cong. Int. Soc. Sugarcane Technol.* I.:29–41.

Maliga, P. 1978. *Resistant mutants and their use in genetic manipulation*. Pages 381–92 in T. A. Thorpe, ed., *Frontiers of Plant Tissue Culture*. International Association for Plant Tissue Culture, Calgary, Alberta.

———. 1981. *Streptomycin resistance is inherited as a recessive Mendelian trait in Nicotiana sylvestris line*. Theor. Appl. Genet. 60:1–3.

Maliga, P., A. SZ. Breznovits, and L. Marton. 1973. *Streptomycin-resistant plants from callus culture of haploid tobacco*. Nature New Biology 244:29–30.

Maliga, P., A. SZ. Breznovits, L. Marton, and F. Job. 1975. *Non-Mendelian streptomycin-resistant tobacco mutant with altered chloroplasts and mitochondria*. Nature (Lond.) 224:401–402.

Maliga, P., G. Lazar, Z. Svab, and F. Nagy. 1976. *Transient cycloheximide resistance in tobacco cell line*. Mol. Gen. Genet. 149:267–71.

Maliga, P., L. Marton and A. SZ. Breznovits. 1973. *5-bromodeoxyuridine-resistant cell lines from haploid tobacco*. Plant Sci. Lett. 1:119–21.

Mastrangelo, I. A., and H. H. Smith. 1977. *Selection and differentiation of aminopterin resistant cells of Datura innoxia. Plant Sci. Lett. 10:171–79.*

Matern, U., G. Strobel, and J. Shepard. 1978. *Reaction to phytotoxins in a potato population derived from mesophyll protoplasts*. Proc. Nat. Acad. Sci. USA 75:4935–39.

Müller, A., and R. Grafe. 1978. *Isolation and characterization of cell lines of Nicotiana tabacum lacking nitrate reductase*. Mol. Gen. Genet. 161:67–76.

Nabors, M. W. 1976. *Using spontaneously occurring and induced mutations to obtain agriculturally useful plants*. BioScience 26:761–68.

Nabors, M. W., S. E. Gibbs, C. S. Bernstein, and M. E. Meis. 1980. *NaCl-tolerant tobacco plants from cultured cells*. Z. Pflanzenphysiol. 97:13–17.

Nelson, O. E. 1977. *The applicability of plant cell and tissue culture techniques to plant improvement*. Pages 67–76 in I. Rubenstein, R. Phillips, and C. E. Green, eds., *Molecular Genetic Modification of Eukaryotes*. Academic Press, New York.

Nickell, L. G., and D. J. Heinz. 1973. *Potential of cell and tissue culture techniques as aids in economic plant improvement*. Pages 109–28 in A. M. Srb, ed. *Genes, Enzymes and Populations*. Plenum Press, New York.

———. 1977. *Crop improvement in sugarcane: Studies using in vitro methods*. Crop Sci. 17:717.

Nishi, T., Y. Yamada, and E. Takahashi. 1968. *Organ redifferentiation and plant restoration in rice callus*. Nature (Lond.) 219:508–509.

Oono, K. 1975. *Production of haploid plants of rice (Oryza sativa) by anther culture and their use for breeding*. Bull. Nat. Inst. Agric. Sci. D26:139–222.

———. 1978a. *High frequency mutations in rice plants regenerated from seed callus*, page 52. (abst.). Fourth Int. Congr. Plant Tissue Cell Culture, Calgary, Canada.

———. 1978b. *Test tube breeding of rice by tissue culture*. Trop. Agric. Res. Series 11:109–23.

Polacco, J. C., and M. L. Polacco. 1977. *Inducing and selecting valuable mutation in plant cell culture: A tobacco mutant resistant to carboxin*. Ann. N.Y. Acad. Sci. 287:385–400.

Radin, D. N., and P. S. Carlson. 1978. *Herbicide-tolerant tobacco mutants selected in situ and recovered via regeneration from cell culture*. Genet. Res. 32:85–9.

Rutger, J. N., M. L. Peterson, and C. H. Hu. 1977. *Registration of Calrose 76*. Crop Sci. 17:978.

Schaeffer, G. W. 1979. *Selection of biochemical variants from cell culture*. Pages 74–78 in R. Durbin, ed., *Nicotiana: Procedures for Experimental Use*. USDA Tech. Bull. No. 1586.

———. 1982. *Recovery of heritable variability in anther-derived double-haploid rice*. Crop Sci. 22:1160–64.

Schaeffer, G. W., and F. T. Sharpe Jr. 1981. *Lysine in seed protein from S-aminoethyl-L-cysteine resistant anther derived tissue cultures of rice*. In Vitro 17:345–52.

Scowcroft, W. R. 1977. *Somatic cell genetics and plant improvement*. Pages 39–81 in N. C. Brady, ed., *Advances in Agronomy,* vol. 29. Academic Press, New York.

Shepard, J. F., D. Bidney, and E. Shahin. 1980. *Potato protoplasts in crop improvement*. Science 28:17–24.

Siegemund, F. 1981. *Selection von Resistenzmutanten in pflanzlichen zellkulturen-eine Ubersicht*. Biol. Zbl. 100:155–65.

Singh, R., and D. Axtell. 1973. *High lysine mutant gene (hl) that improves protein quality and biological value of grain sorghum*. Crop Sci. 13:535–39.

Skirvin, R. M., & J. Janick. 1976a. *Tissue culture-induced variation in scented Pelargonium spp*. J. Amer. Soc. Hort. Sci. 101:281–90.

———. 1976b. *'Velvet Rose' Pelargonium, a scented geranium*. Hort Sci. 11:61–62.

Stimart, D. P., P. D. Ascher, and J. S. Zagorski. 1980. *Plants from callus of the interspecific hybrid Lilium 'Black Beauty'*. Hort Sci. 15:313–15.

Sung, Z. R., and S. Jacques. 1980. *5-fluorouracil resistance in carrot cultures. Its use in studying interactions of the pyrimidine and arginine pathways*. Planta 148:389–96.

Thomas, E., P. J. King, and I. Potrykus. 1979. *Improvement of crop plants via single cells in vitro—an assessment*. Z. Pflanzenzuchtg. 82:1–30.

Wenzel, G., O. Schieder, T. Przewozny, S. K. Sopory, and G. Melchers. 1979. *Comparison of single cell culture derived Solanum tuberosum L. plants and a model for their application in breeding programs*. Theor. Appl. Genet. 55:49–55.

Widholm, J. M. 1972. *Anthranilate synthetase from 5-methyltryptophan-susceptible and resistant cultured Daucus carota cells*. Biochim. Biophys. Acta. 279:48–57.

———. 1974a. *Selection and characterization of biochemical mutants of cultured plant cells*. Pages 287–89 in H. E. Street, ed., *Tissue Culture and Plant Sciences*. Blackwell, Oxford.

———. 1974b. *Cultured carrot cell mutants: 5-methyltryptophan-resistant trait carried from cell to plant and back*. Plant Sci. Lett. 3:323–30.

———. 1977a. *Selection and characterization of amino-acid analog resistant plant cell cultures*. Crop Sci. 17:597–600.

———. 1977b. *Selection and characterization of biochemical mutants*. Pages 112–22 in W. Barz, E. Reinhard, and M. M. Zenk, eds., *Plant Tissue Culture and Its Biotechno-*

logical Applications. Springer-Verlag, Berlin-Heidelberg-New York.
———. 1978a. *Selection and characterization of a Daucus carota L. cell line resistant to four amino acid analogues*. J. Exp. Bot. 29:1111–16.
———. 1978b. *The selection of agriculturally desirable traits with cultured plant cells*. Pages 189–99 in K. W. Hughes, R. Henke, and M. Constantin, eds. *Propagation of Higher Plants Through Tissue Culture. A Bridge Between Research and Application*. Proc. Intern. Symp., University of Tennessee, Knoxville.

six

NEW PLANTS VIA TISSUE CULTURE

18] Hybrid and Cybrid Production via Protoplast Fusion

by EDWARD C. COCKING*

ABSTRACT

Protoplasts can be fused by use of a range of chemical fusogens, and also by electrical procedures; and the resultant heterokaryons can sometimes be suitably cultured to form hybrid plants. To determine factors influencing the degree of nuclear hybridity we have developed a simple procedure for the mechanical isolation of heterokaryons, using both bright field and fluorescence procedures. We are also utilizing a nitrate reductase deficient mutant, coupled with fusion with irradiated protoplasts, to regulate the degree of nuclear hybridity. Cybrid production by protoplast fusion is also being assessed by use of both irradiation procedures and enucleate subcellular units, such as microplasts. Somatic hybrid and cybrid plants produced by these unconventional genetic manipulation procedures will be of importance in plant improvement programs. The range of agricultural applications from protoplast fusions is extensive, and this aspect of plant genetic manipulations can now be developed, provided the fusion products can be adequately cultured.

INTRODUCTION

Agriculture is man's oldest industry, and it is to be expected that studies on the production of new plants via tissue culture will sooner or later have an impact on traditional agricultural and horticultural practice. Many questions are arising in this respect, and the challenge is to provide realistic answers. Will the aspirations to produce custom-designed crops be realized, and will this represent a second green revolution? When will this new technology arrive, and will it affect international commerce in foodstuffs? Will genetic engineering in plants live up to the hopes of cell and molecular biologists, corporate conglomerates, and world-hunger specialists? The answers to most of these

*Plant Genetic Manipulation Group, Department of Botany, University of Nottingham, Nottingham NG7 2RD, U.K.

GENETIC ENGINEERING: APPLICATIONS TO AGRICULTURE (Beltsville Symposium 7—Lowell D. Owens, ed.) Rowman & Allanheld, Totowa

questions are to be found in a fuller understanding of the advantages, and limitations, of these new, unconventional breeding procedures, coupled with an appreciation of the historical development of agriculture during the past few thousand years. It may be easy to identify objectives, but the lack of adequate basic knowledge of factors influencing cell division and plant regeneration from in vitro cultures, and an insufficient knowledge of the molecular biology of plant cells, frequently impedes progress. Two plus two does not make four until somebody actually does the addition! Also, as recently emphasized by Hanson (1979), it should not be forgotten that scientists do not produce food—farmers do.

Because the cell wall is absent in somatic protoplasts, they can be fused with other protoplasts, thereby providing opportunities for somatic hybridizations (Cocking 1981a). Both nuclear and cytoplasmic interactions will be involved. Heterokaryosis and fusion of nuclei will result in the development of hybrid cells after mitosis. The cytoplasmic mix is novel, and cybrids (cytoplasmic hybrids) will sometimes arise from such somatic fusions. If these cellular manipulations are combined with adequate in vitro culture methodology, hybrid plants and cybrid plants can be regenerated. The challenge is to exploit these novel hybrids for agricultural and horticultrual improvements.

SOMATIC HYBRIDIZATION

The ready availability of enzymatically isolated protoplasts opened up the possibility of inducing such protoplasts to fuse and to produce heterokaryons (Power, Cummins, and Cocking 1970), and during the past decade there has been extensive work with the objective of producing somatic hybrid plants from these heterokaryons. From the earliest of these experiments it was clear that adequate procedures for the selection of heterokaryons would be required if these somatic fusions were to be successfully utilized for various programs of plant improvement. Several procedures, often based on principles of microbiological complementation selection (Cocking 1978), have been employed for selection; and studies within the *Petunia* genus produced amphidiploid somatic hybrid plants by using a complementation selection procedure which involved fusing wild-type leaf protoplasts of one species with albino protoplasts from cell suspension cultures of the other species. From these protoplast fusions flowering plants with 28 chromosomes of the somatic hybrid *P. parodii* ($2n = 14$) ⊗ *P. hybrida* ($2n = 14$), and of *P. parodii* ($2n = 14$) ⊗ *P. inflata* ($2n = 14$) were produced between these sexually compatible species. The consequences of fusion of sexually incompatible *P. parodii* leaf protoplasts with albino protoplasts from cell suspensions of *P. parviflora* (Power et al. 1980) enabled the production of amphidiploid somatic hybrid plants with 32 chromosomes, *P. parodii* ($2n=14$) ⊗ *P. parviflora* ($2n=18$) (Cocking 1981b). This novel hybrid could be of considerable horticultural interest, particularly if further hybridizations, either sexually or somatically, could extend the range of floral types of this "hanging basket" type of petunia. Indeed, there are many possible applications of somatic hybridization to the improvement of horticultural species.

As far as crop species are concerned, there have been many extensive reviews, and reference should be made to these for general background information (Sharp et al. 1979). The power of conventional plant breeding is becoming increasingly appreciated, and probing questions are being asked of the new technology associated with plant somatic and molecular genetics—particularly concerning what this new technology offers that conventional plant breeding does not. These very pertinent questions have been comprehensively discussed by Cocking et al. (1981), and it was concluded that although plant breeding by well-established, conventional methods will continue to make the key contribution to meet the demands of increased food production, multiple gene transfer mediated by somatic hybridization will, during the next few decades, assume increasing significance in the improvement of forage legumes, vegetables, and fruit crops. It was also concluded that many of the techniques associated with this new technology are still at the experimental stage, and that several major drawbacks will have to be overcome before they can be applied to agricultural practice, but the effort should be well worthwhile. It was also suggested that during the next decade it will be advantageous to attempt to incorporate into the recipient crop species some limited, perhaps single, genetic attribute of a donor species via a transformation procedure to eliminate the need for backcrossing and recurrent selection.

While, as we have seen, significant advances have been made in our ability to produce somatic hybrids by protoplast fusions, one of the limitations in further advancement has been our lack of knowledge of the details of many of the stages involved in this multistage process. The successful outcome of somatic hybridization, whether it is to produce nuclear hybrids, limited gene transfer, or cybrids will depend on a basic knowledge of the many steps in this procedure. These steps involve isolation of protoplasts, division of cells regenerated from protoplasts, controlled fusion of protoplasts, and the identification and selection of heterokaryons, hybrid and cybrid cells, and the regeneration of plants. This further development of the subject is from a baseline of experience mainly with tobaccos and petunias, and often some of the difficulties encountered are related to special cultural features of the crop species.

Isolation and culture of protoplasts. Detailed descriptions of protoplast isolation and culture have been described by Evans and Cocking (1978). Recently it has been established that the use of Bio-Gel purification of cell wall degrading enzymes considerably increases plating efficiency for several crop species, and a simple procedure has been described (Patnaik, Wilson, and Cocking 1981). The use of purified enzyme, with its associated enhancement of plating efficiency, is particularly important when attempting to culture mechanically isolated heterokaryons (Patnaik et al. 1982). An even greater enhancement of reproducible yield and plating efficiency can often be obtained with the use of a new enzyme mixture (Patnaik and Cocking 1982). The use of protoplasts isolated from seedling roots (Xu, Davey, and Cocking 1981, 1982) and from seedling cotyledons (Lu, Pental, and Cocking 1982) could also be particularly advantageous when somatic hybridizations are being undertaken. The use of germinating seeds overcomes the difficulty encountered with the production of leaf material, and an adequate supply of experimental material can be produced rapidly with a minimum of resources. It has also been suggested for somatic hybridization assessments that protoplasts from roots and etiolated cotyledons could be used, instead of cell suspension protoplasts, for fusion with green cotyledon leaf material or leaf mesophyll protoplasts (Lu, Pental and Cocking 1982).

Regeneration of plants from protoplasts. The regeneration of plants from protoplasts is central to the utilization of new developments in somatic cell genetics for plant-breeding programs, and the recent successes in regenerating plants from seedling root and cotyledon protoplasts is likely to enhance greatly the spectrum of species in which regeneration from protoplasts is possible. It is likely that during the next decade regeneration of plants from protoplasts of the major crops will be accomplished. The regeneration of tomato plants from tomato leaf protoplasts is a step toward this goal, an accomplishment that took more than three years' work to perfect (Wiscombe and Cocking 1982). It should be noted that there is evidence that regeneration capability is usually dominant; consequently a regeneration capability in both species being fused is not essential, although of course it may be advantageous. Loss of regeneration capability is frequently encountered when biochemical mutants are used for selection, since long periods of in vitro culture are often needed to isolate a suitable mutant. Sometimes regeneration from protoplasts of such auxotrophic mutants is, however, possible (Pental et al. 1982).

Fusion of plant protoplasts. As fully discussed by Power, Evans, and Cocking (1978), a balance has to be established between fusion frequency and subsequent protoplast viability. In general, the greater the extent of induced fusion the lower will be the viability of the resultant heterokaryons. Protoplast systems also vary greatly in their sensitivity to the various fusion agents, so that most of the fusion methods have to be adapted to the specific systems under examination. Jaworski (1978) has highlighted some of these cultural and developmental problems following fusion. He questioned whether there is a fundamental lack of knowledge of the early events following protoplast fusion, and whether there are more incompatibility barriers than presently envisioned.

While electrically induced protoplast fusion may indeed enhance fusion frequency and reproducibility, its effect on cultural capability is not yet clear (Zimmermann 1982). Figure 18.1 summarizes the effect of three different chemical fusogens on the frequency of somatic hybrid plant formation (Cocking 1980). The use of PEG as a fusogen, however, may have to be reinvestigated, since there is some evidence that purification of commercial-grade PEG leads to loss of fusogenicity, and that antioxidants such as α-tocopherol added to commercial-grade PEG probably promote fusion in the co-presence of PEG (Honda et al. 1981).

Selection for somatic hybridization. Whether hybrids or cybrids are desired, a general scheme of selecting somatic hybrids between any two (or more) wild-type species is desirable for plant-breeding programs. Some successes have

THE EFFECT OF FUSION METHODS ON THE FREQUENCY OF SOMATIC HYBRID FORMATION

FUSION PROCEDURE	% FUSION	NUMBER OF SOMATIC HYBRIDS*
Polyethylene glycol (PEG)	4%	6
High pH/Ca^{++} ions	4%	440
$NaNO_3$	<0.01%	1
Mixture control (no fusion)	0	0

* This represents the number of separate events (per $4x10^6$ protoplasts) recorded, following the fusion of P. parodii protoplasts with those of albino P. hybrida.

Figure 18.1. The influence of different fusogens on the frequency of somatic hybrid formation.

been achieved using biochemical-type selection procedures (Cocking 1978; Flores, Kaur-Sawhney, and Galston 1981) but, as previously emphasized, one of the major problems presently encountered is a lack of suitable mutants for complementation selection.

We have recently developed a simple procedure for the aseptic manual isolation of individual heterokaryons. Heterokaryons were identified with bright field illumination using an inverted microscope, and were isolated by means of a micromanipulator and capillary pipette coupled to a specially constructed syringe. When cell suspension protoplasts were labeled with fluorescein isothiocyanate and fused with mesophyll protoplasts, the heterokaryons exhibited an apple-green cytoplasmic fluorescence (from cell suspension protoplasts) and a red chloroplast fluorescence (from mesophyll protoplasts). Using an inverted microscope with a suitable fluorescence attachment, we have also observed that the differential fluorescence from chloroplasts and fluorescein isothiocyanate in individual heterokaryons persists in the cells of small colonies developed from such heterokaryons. Manual- or fluorescence-activated cell-sorting procedures may not therefore be required, since selection of individual heterokaryons is not then required (Patnaik et al. 1982). Such methods are also readily applicable to the selection of cybrids.

Fusion products, hybrid and cybrid cells. Central to the development of protoplast fusion products is the stage at which fusion of heterospecific nuclei occurs. Earlier studies have been comprehensively reviewed by Constabel (1978), and it was concluded that the development of fusion products, formation of heterokaryons, and growth of hybrid cells, in particular the behavior of nuclei, paralleled the earlier experiments on somatic hybridization performed with animal cell lines. The data available, however, has been somewhat limited for plant heterokaryons. In particular there has been no detailed comparative analysis of the behavior of nuclei after fusion between the protoplasts of two species either sexually compatible or incompatible. Recently we have attempted such a comparative assessment of nuclear behavior following fusion between protoplasts of *Nicotiana rustica* and those of *N. tabacum* (sexually compatible) and *P. parodii* (sexually incompatible). In both instances it was clearly established that nuclear fusion, while somewhat delayed in the case of tobacco-petunia fusions, always preceded any mitotic division in the hybrid cells. It was also established that tobacco/petunia hybrid cell division was significantly delayed in comparison with the onset of cell division in the parental species (Patnaik 1982). Differences in mitotic cycles could influence the direction of chromosome elimination in such hybrid cells; and with further study it may be possible to control the direction of chromosome elimination in such hybrids by suitable combinations between interphase nuclei and mitotic nuclei. From the earlier studies of Power et al. (1975) on the consequences of fusion of petunia and *Parthenocissus* protoplasts, the extent to which the asymmetry of mitotic cycle times and phases between genera being studied for their somatic hybridization potential would result in selective chromosome elimination was not clear. It was suggested that careful matching of genera for similar mitotic cycles might be necessary if amphiploids were desired.

Recent studies in which a fluorescence marker has been employed for the analysis of heterokaryons have revealed that fusion results in the formation of three main types of heterokaryon. One type has an approximately equal amount of the cytoplasm of each of the parental species, one type mainly the cytoplasm of only one of the parental species (somewhat equating to the situation holding in maternal cytoplasmic inheritance in sexual hybridization), and the other type mainly the cytoplasm of the other parental species (again somewhat equating to the situation holding in maternal cytoplasmic inheritance in sexual hybridization, when the other species is the female). This spectrum of heterokaryon types has not previously been detected, largely because of the absence of suitable cytoplasmic markers; most previous workers have assumed that only the type of heterokaryon with an approximately equal amount of the cytoplasm of each of the parental species is produced. The significance for the adequate selection and characterization, both nuclear and cytoplasmic, of resultant somatic hybrids and cybrids is likely to be profound (for a detailed discussion of this recent work see Patnaik 1982).

Protoplast fusions provide novel opportunities for obtaining heterozygosity of extra-chromosomal genes. As far as chloroplasts are concerned, the characterization of chloroplast organelles indicates that the two types of parental chloroplast in the somatic hybrid eventually sort out for one or the other, or in some cases only one, of the parental types. The sorting-out process of the parental chloroplasts in the somatic hybrid population is likely to be greatly influenced by many factors encountered during protoplast fusion and subsequent selection and regeneration of somatic hybrids. For instance, the analysis of the chloroplast DNA (cpDNA) by various restriction endonucleases of the three somatic hybrids previously produced by the fusion of protoplasts from wild-type mesophyll cells of *P. parodii* with protoplasts of *P. hybrida, P. inflata,* and *P. parviflora* showed the presence of only one cpDNA of *P. parodii* in the somatic hybrids (Kumar et al. 1982). It was suggested that this unidirectional sorting out phenomenon, in favor of *P. parodii* chloroplasts, was most probably the result of the strong selections in favor of *P. parodii* used in the production of these somatic hybrids. It was not possible to determine the amount of chloroplast heterozygosity from the Fraction 1 profiles because the large subunit polypeptides (coded by cpDNA) of the Fraction 1 protein are identical in these four *Petunia* species (Kumar, Wilson, and Cocking 1981). No recombination between cpDNAs was detected in any of the three hybrids, indicating that cpDNA recombination must be an infrequent event, if it occurs at all. Opportunities for increasing cytoplasmic variability by protoplast fusion may be greater with mitochondria than with chloroplasts. Less detailed analysis has been carried out on mitochondrial DNA. Nevertheless, there is some evidence for mitochondrial recombination in cytoplasmic hybrids of *N. tabacum* obtained by protoplast fusion (Belliard, Vedel, and Pelletier 1979). The heterogeneity of most mitochondrial DNAs isolated from *Nicotianas* is complex, and recent work suggests that those from some *Brassica* species are less complex, making *Brassicas* preferable for this type of analysis (Lebacq and Vedel 1981).

Cocking (1981a) discusses disadvantages often encountered in attempting to combine the total genetic structures of two species and greater advantage in attempting to incorporate into the recipient crop species some limited, perhaps single, genetic attribute of a donor species. It may be possible to achieve this objective using suitable somatic fusions. For instance, leaf protoplasts of *N. rustica* fused with irradiated protoplasts may yield plants with genetic traits similar to those obtained as a result of sexual transformation (for a detailed discussion, see Cocking 1981). It may also be possible, using such somatic transformation, to transfer nitrate reductase genes between different species. Protoplasts from a nitrate reductase-minus mutant of tobacco (Pental et al. 1982) can be fused with x-ray–irradiated legume protoplasts, followed by selection for nitrate reductase proficiency among the somatic cells: in future work this selection for a limited gene transfer by fusion could result in a legume or cereal with improved nitrate reductase activity, and perhaps improved yield (for a detailed discussion, see Cocking 1981a).

Protoplast fusion has made it possible to produce hybrids containing different types of cytoplasms which otherwise would never occur through sexual crosses. Using cytoplasmic male sterile *N. tabacum*, Belliard et al. (1978) and Gleba et al. (1979) were able to obtain new cytoplasmic hybrids (cybrids) which contained nuclear information from the same species. As discussed previously (see Cocking 1981a), if fusion of protoplasts is coupled with procedures for the inactivation of the nuclear genome of one of the species, and with suitable selection procedures, a range of novel cybrids can theoretically be obtained (Fig. 18.2). Recently a new somatic hybrid plant containing two different cytoplasms and a mixture of nuclear genomes of two tobacco species, *N. tabacum* and *N. glutinosa*, was obtained following fusion of protoplasts derived from suspension culture cells of cytoplasmic male sterile *N. tabacum* (*N. debneyi* cytoplasm) and of fertile *N. glutinosa* (Uchimiya 1982). Rather than irradiating protoplasts to inactivate nuclei, fractionation of protoplasts into enucleate subprotoplasts (Lörz 1981; Archer, Landgren, and

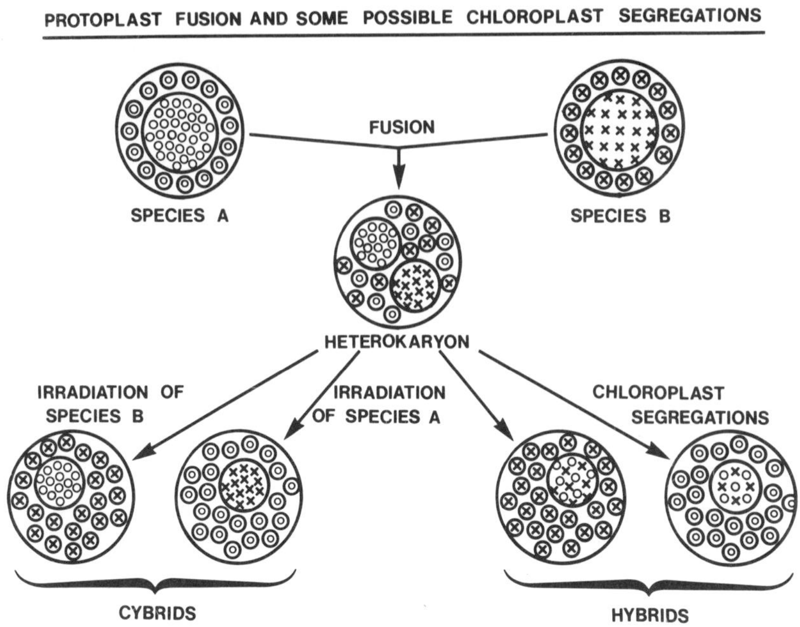

Figure 18.2. Protoplast fusion and some possible chloroplast segregation. (Cocking 1981.)

Bonnett 1982) or enucleate microplasts (Bilkey, Davey, and Cocking 1982) may provide suitable enucleate units for fusions to produce a range of cybrids, thereby avoiding any irradiation effects on the cytoplasm and the possibility that irradiation has not always completely eliminated the nuclear contribution of irradiated protoplasts (Fig. 18.3). Transfer of cytoplasmically based herbicide resistance may also be possible. Indeed, the opportunities for the plant breeder are considerable in relation to cybrid plant production by protoplast fusions, since cytoplasmic influences are generally quite important in crop yield.

APPLICATIONS OF SOMATIC HYBRIDIZATION IN AGRICULTURE

Past discussion on the use of protoplast fusion technology has mainly concentrated on problems and potentials (Bhojwani, Evans, and Cocking 1977). Recently, however, the resurgence of the successful culture of several of the forage legumes, with regeneration of plants from protoplasts, is providing a basis for actual assessments of the role that protoplast fusion technology will play in forage legume improvements by exploring extraspecific genetic variation (Razdan and Cocking 1981).

For instance, white clover (*Trifolium repens*) is an important forage crop in

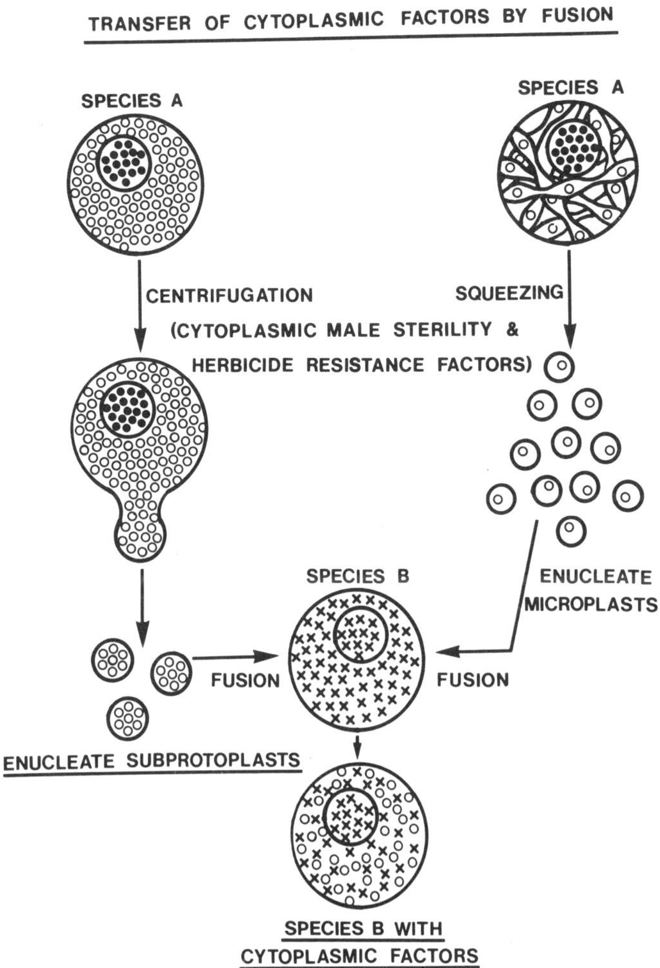

Figure 18.3. Transfer of cytoplasmic factors by fusion. (Cocking 1981.)

temperate regions but is difficult to establish, and there is a relationship between white clover and bloat in ruminant animals (Rodgers 1975). Sexual incompatability within the genus *Trifolium,* and between *Trifolium* and other forage legumes, prevents the introduction of desirable characters by conventional means. One such forage legume is sainfoin (*Onobrychis viciifolia*), which combines persistency, drought resistance, and high palatability with valuable protein and nonbloating properties (Goplen 1969). Conditions for the isolation, culture, and regeneration of mesophyll protoplasts of white clover and sainfoin have now been established; protoplast fusion for somatic hybridization, coupled with heterokaryon isolation procedures, can now be realistically undertaken for somatic hybridization assessments (Ahuja, Cocking, and Davey 1982).

In the *Medicago* genus, which contains the forage legumes most cultivated in temperate and subtropical climates, somatic hybridization provides a means of overcoming sexual incompatability and of transforming characters from *M. coerulea* to *M. sativa* (alfalfa, lucerne) and to *M. glutinosa*. Protoplasts of *M. sativa* have been developed for use in genetic manipulations, with successful plant regeneration being obtained from mesophyll protoplasts by three independent laboratories (Kao and Michayluk 1980; Santos et al. 1980; Johnson et al. 1981). Realistic assessment of gene transfers by somatic hybridization can now be undertaken, since conditions for protoplast isolation, culture, and somatic embryogenesis have been established for *M. coerulea* and *M. glutinosa,* as well (Arcioni et al. 1982).

CONCLUSIONS

Hybrid and cybrid production via protoplast fusion is clearly a realistic objective for the plant breeder who needs to explore extraspecific genetic variation. As we have seen, such objectives are presently restricted for several important crop species, mainly due to a lack of reproducible regeneration of plants from protoplasts; and more basic work is required on heterokaryon selection and hybrid nuclear behavior so that gene transfers can be manipulated more specifically. Nevertheless, it is to be expected that progress, while often slow, will be steady and that somatic hybridization will progressively add its contribution to crop improvement by more conventional procedures, particularly in legumes, vegetables, fruit crops, and forestry.

LITERATURE CITED

Ahuja, P.S., E.C. Cocking, and M.R. Davey. 1982. *An assessment of the cultural capabilities of Trifolium repens (White Clover) and Onobrychis viciifolia (Sainfoin) mesophyll protoplasts as a basis for somatic hybridisation.* Crop Sci. (in press).

Archer, E.K., C.R. Landgren, and H.T. Bonnett. 1982. *Cytoplast formation and enrichment from mesophyll tissues of Nicotiana spp.* Plant Sci. Lett. (in press).

Arcioni, S., M.R. Davey, A.V.P. dos Santos, and E.C. Cocking. 1982. *Somatic embryogenesis in tissues from mesophyll cell suspension protoplasts of Medicago coerulea and M. glutinosa.* Z. Pflanzenphysiol. 106:105–10.

Belliard, G., G. Pelletier, F. Vedel, and F. Quetier. 1978. *Morphological characteristics and chloroplast DNA distribution in different cytoplasmic parasexual hybrids of Nicotiana tabacum.* Mol. Gen. Genet. 165:231–37.

Belliard, G., F. Vedel, and G. Pelletier. 1979. *Mitochondrial recombination in cytoplasmic hybrids of Nicotiana tabacum by protoplast fusion.* Nature (Lond.) 281:401–403

Bhojwani, S.S., P.K. Evans, and E.C. Cocking. 1977. *Protoplast technology in relation to crop plants: Progress and problems.* Euphytica 26:343–60.

Bilkey, P., M.R. Davey, and E.C. Cocking. 1982. *Isolation, origin and properties of enucleate plant microplasts.* Protoplasma 110:147–51.

Cocking, E.C. 1978. *Selection and somatic hybridisation*. Pages 151–58 in T.A. Thorpe, ed. *Frontiers of Plant Tissue Culture 1978*. Int. Assoc. Plant Tissue Culture.

———. 1980. *Concluding remarks*. Pages 419–25 in F. Sala et al. eds., *Plant Cell Culture: Results and Perspectives*. Elsevier, Amsterdam.

———. 1981a. *Opportunities from the use of protoplasts*. Phil. Trans. Roy. Soc. Lond. B. 292:557–68.

———. 1981b. *Plant cell hybrids and somatic hybrid plants*. Pages 206–11 in *Chromosomes Today, vol. 7*. George Allen & Unwin, London.

Cocking, E.C., M.R. Davey, D. Pental, and J.B. Power. 1981. *Aspects of plant genetic manipulation*. Nature (Lond.) 293:265–70.

Constabel, F. 1978. *Development of protoplast fusion products, heterokaryotes and hybrid cells*. Pages 141–49 in T.A. Thorpe, ed., *Frontiers of Plant Tissue Culture 1978*. Int. Assoc. Plant Tissue Culture.

Evans, P.K., and E.C. Cocking. 1978. *Isolated plant protoplasts*. Pages 103–35 in H.E. Street, ed., *Plant Tissue and Cell Culture*. Blackwell Scientific Publications, Oxford.

Flores, H.E., R. Kaur-Sawhney, and A.W. Galston. 1981. *Protoplasts as vehicles for plant propagation and improvement*. Pages 241–79 in K. Maramorosch, ed., *Advances in Cell Culture,* vol. 1. Academic Press, New York.

Gleba, Y.Y., N.M. Piven, I.K. Komarnitskii, and A.K.M. Sytnik 1978. *Parasexual cytoplasmic Nicotiana tabacum + N. debneyi hybrids (cybrids) obtained by protoplast fusion*. Doklady Academii Nauk SSSR 240:1223–26.

Goplen, B.P. 1969 *Legume Breeding*. Canadian Forage Crops Sympos. 225–59. Modern Press, Saskatoon.

Hanson, H. 1979. Conf. Agricultural Production: *Research and Development Strategies for the 1980s*. Bonn (Rockefeller Foundation, New York).

Honda, K., Y. Maeda, S. Sasakawa, H. Ohno, and E. Tsuchida. 1981. *The components contained in polyethylene glycol of commercial grade (PEG–6000) as cell fusogen*. Biochem. Biophys. Res. Comm. 101:165–71.

Jaworski, E.G. 1978. *Propagation of higher plants through tissue culture—a bridge between research and application*. Pages 1–10 in K.W. Hughes et al., eds., *Propagation of Higher Plants Through Tissue Culture*. Int. Symposium, University of Tennessee, Knoxville.

Johnson, L.B., D.L. Stuteville, R.K. Higgins, and D.Z. Skinner 1981. *Regeneration of alfalfa plants from selected regen S clones*. Plant Sci. Lett. 20:297–304.

Kao K.N., and M.R. Michayluk. 1980. *Plant regeneration from mesophyll protoplasts of Alfalfa*. Z. Pflanzenphysiol. 96:135–41.

Kumar, A., D. Wilson, and E.C. Cocking. 1981. *Polypeptide composition of Fraction 1 protein of the somatic hybrid between Petunia parodii and P. parviflora*. Biochem. Genet. 19:255–61.

Kumar, A., E.C. Cocking, W.A. Bovenberg, and A.J. Kool. 1982. *Restriction endonuclease analysis of chloroplast DNA in interspecies somatic hybrids of Petunia*. Theor. Appl. Genet. 62:377–84.

Lebacq, P., and F. Vedel. 1981. *Sal 1 restriction enzyme analysis of chloroplast and mitochondrial DNAs in the genus Brassica*. Plant Sci. Lett. 23:1–8.

Lörz, H., J. Paszhowski, C. Dierks-Ventling, and I. Potrykus. 1981. *Isolation and characterisation of cytoplasts and miniprotoplasts derived from protoplasts of cultured cells*. Physiol. Plant. 53:386–91.

Lu, D.Y., D. Pental, and E.C. Cocking. 1982. *Plant regeneration from seedling cotyledon protoplasts*. Z. Pflanzenphysiol. 107:59–63.

Patnaik, G. 1982. *Factors affecting isolation and division of plant heterokaryons*. Ph.D., Disser. The University of Nottingham.

Patnaik, G., and E. C. Cocking. 1982. *A new enzyme mixture for the isolation of leaf protoplasts*. Z. Pflanzenphysiol. 107:41–5.

Patnaik, G., E.C. Cocking, J. Hamill, and D. Pental. 1982. *A simple procedure for the manual isolation and identification of plant heterokaryons*. Plant Sci. Lett. 24:105–10.

Patnaik, G., D. Wilson, and E.C. Cocking. 1981. *Importance of enzyme purification for increased plating efficiency and plant regeneration from single protoplasts of Petunia parodii*. Z. Pflanzenphysiol. 102:199–205.

Pental, D., S. Cooper-Bland, K. Harding, E.C. Cocking, and A.J. Müller. 1982. *Cultural studies on nitrate reductase deficient Nicotiana tabacum mutant protoplasts*. Z. Pflanzenphysiol. 105:219–27.

Power, J.B., S.F. Berry, J.V. Chapman, and E.C. Cocking. 1980. *Somatic hybridisation of sexually incompatible Petunias: Petunia parodii, Petunia parviflora*. Theor. Appl. Genet. 57:1–4.

Power, J.B., S.E. Cummins, and E.C. Cocking. 1970. *Fusion of isolated plant protoplasts*. Nature (Lond.) 225:1016–18.

Power, J.B., P.K. Evans, and E.C. Cocking. 1978. *Fusion of plant protoplasts*. Pages 370–81 in G. Poste and G.L. Nicholson, eds. *Membrane Fusions* (Cell Surface Reviews, vol. 5). Elselvier North Holland, New York.

Power, J.B., E.M. Frearson, C. Hayward, and E.C. Cocking. 1975. *Some consequences of the fusion and selective culture of Petunia and Parthenocissus protoplasts*. Plant Sci. Lett. 5:197–207.

Razdan, M.K., and E.C. Cocking. 1981. *Improvement of legumes by exploring extra-specific genetic variation*. Euphytica 30:819–33.

Rodgers, H.H. 1975. *Forage legumes (with particular reference to Lucerne and Red Clover)*. Plant Breeding Institute Annual Report 22–57.

Santos, A.V.P. dos, D.E. Outka, E.C. Cocking, and M.R. Davey. 1980. *Organogenesis and somatic embryogenesis in tissues derived from leaf protoplasts and leaf explants of Medicago sativa*. Z. Pflanzenphysiol. 99:261–70.

Sharp, W.K., P.O. Larsen, E.F. Paddock, and V. Raghavan. 1979. *Plant Cell and Tissue Culture: Principles and Applications*. Ohio State Press, Columbus.

Uchimiya, H. 1982. *Somatic hybridisation between male sterile Nicotiana tabacum and N. glutinosa through protoplast fusion*. Theoret. Appl. Genet. 61:69–72.

Wiscombe, A., and E.C. Cocking. 1982. *Regeneration of plants from tomato leaf protoplasts*. Z. Pflanzenphysiol. 106:97–104.

Xu, Z–H., M.R. Davey, and E.C. Cocking. 1981. *Isolation and sustained division of Phaseolus aureus (Mung Bean) root protoplasts*. Z. Pflanzenphysiol. 104:289–98.

———. 1982. *Plant regeneration from root protoplasts of Brassica*. Plant Sci. Lett. 24:117–21.

Zimmermann, U. 1982. *Electric field-mediated fusion and related electrical phenomena*. Biophys. Biochim. Acta. Biomembranes. (in press).

19] Dihaploids via Anthers Cultured in Vitro

by P. S. BAENZIGER and G. W. SCHAEFFER*

ABSTRACT

General methods have been developed for the recovery of dihaploids. Anther culture is the most recent method and can be used to recover large numbers of haploid plants in several crops. Techniques for developing dihaploids via anther culture, the breeding theory behind their use, how dihaploids are being used in breeding programs, and their future uses are discussed. The main breeding advantages of dihaploidy are the rapid derivation of homozygous lines and the consequent ability to select among those lines in the absence of dominance variation. The main disadvantage is the need for large evaluation nurseries. Most of the anther culture–derived dihaploids used in breeding programs have been derived from tobacco, though good progress has been made in wheat and rice. In tobacco, the technique may induce variation which is usually deleterious. Dihaploids derived from interspecific hybridization (barley and preliminarily tobacco) do not exhibit as much deleterious variation as observed with anther culture, indicating the variation may be caused by the technique for developing dihaploids and not dihaploidy itself. The expectation is that useful variation may also be recovered. The potential for dihaploids in biochemical selection, mutation breeding, and gene transfer systems is also discussed.

INTRODUCTION

The initial discovery of Guha and Maheshwari (1964) that anthers from *Datura innoxia* Mill. could be cultured and haploid plants regenerated was made almost two decades ago. This process, which was later confirmed by Nitsch and Nitsch (1969) in tobacco *(Nicotiana tabacum* L.), has, in the succeeding years, been extended to numerous plant species (Schaeffer, Baenziger, and

*Field Crops Laboratory, and Tissue Culture and Molecular Genetics Laboratory, Agricultural Research Service, U.S. Department of Agriculture, Beltsville, Maryland 20705.

GENETIC ENGINEERING: APPLICATIONS TO AGRICULTURE (Beltsville Symposium 7—Lowell D. Owens, ed.) Rowman & Allanheld, Totowa

Worley 1979; Collins and Genovesi 1982). Haploidy, anther culture techniques, and their eventual applications to plant breeding and related sciences have been reviewed by many authors (Nitzsche and Wenzel 1977; Reinert and Bajaj 1977; Sink and Padmanabhan 1977; Kihara 1979; Maheshwari, Tyagi, and Malhotra 1980; Collins and Genovesi 1982). With this in mind, the purpose of this chapter is not to review extensively the literature with regards to existing methodologies, but rather to present a plant breeder's perspective of how the products of these techniques are being and will be used in crop improvement. As such, this report has been divided into four main areas: (1) techniques for successful anther culture and dihaploid plant production, (2) dihaploid breeding theory, (3) present use of dihaploids in breeding programs, and (4) future use of dihaploids in breeding programs.

ANTHER CULTURE TECHNIQUES

Many factors affect the production of haploid plants from anther culture. Among the most important are the composition of the medium upon which the anthers are placed after excision, health of the donor plant, plant growth or developmental stage, plant genotype, and the absence or presence of anther or whole plant pretreatment. The importance of the medium has been reviewed by Collins and Genovesi (1982). Often one medium is used for the initial callus formation, a different one for callus growth and maintenance, and a third for plant regeneration. Alterations in either inorganic salt levels, sucrose concentrations, nitrate-to-ammonium-nitrogen ratio, plant hormone levels, presence of activated charcoal, amino acid concentrations, and complex compounds (such as casein hydrolysates, lactalbumin hydrolysates, coconut milk, and potato extracts) have been found to be necessary in regenerating plantlets from anthers of various species. Androgenesis in barley (*Hordeum vulgare* L.) has been improved by using media "conditioned" with barley anthers (Xu, Huang, and Sunderland 1981). These conditioning anthers are "bicellular" (binucleate), whereas the anthers cultured for haploid callus production are "unicelluar" (uninucleate). In general, plants grown under constant and optimal conditions are the best anther donors, and the microspore stage is normally of critical importance for successful culturing. For instance, Tomes and Collins (1976) found binucleate microspores were generally best for anther culture of tobacco and its related species. In wheat (*Triticum aestivum* L. em Thell), however, the mid-uninucleate stage is preferred (Ouyang et al. 1973). Though the cellular basis for the specificities in microspore optimum for culturing are not yet fully defined or understood, in many cases either the single premitotic or the vegetative nucleus is believed to be the primary origin of haploid callus. Thus, variations in growth conditions which affect microspore development, including the differentiation of the vegetative and generative nucleus development (Bennett 1976), may explain many of the seasonal response variations for culturability recorded when donor plants are grown in greenhouses (Dunwell 1976; Bernard 1977; Picard, De Buyser, and Henry 1978; Durr and Fleck 1980). Anther culturability is also thought to have a genetic basis (Picard and

De Buyser 1977; Picard, De Buyser, and Henry 1978; Bullock et al. 1982), and some of the differing reports regarding the best media can probably be explained by the various researchers using different genotypes. For example, in anther culture of corn (Miao et al. 1978), oats (H. W. Rines, personal communication), rye (Wenzel, Hoffman, and Thomas 1977), potato (Jacobsen and Sopory 1978), and wheat (Picard and De Buyser 1977; Schaeffer et al. 1979; Bullock et al. 1982), some genotypes have been found to be more responsive than others. Finally, pretreatments and after-excision treatments such as thermal shock have been useful (Picard et al. 1978; Schaeffer et al. 1979; Sunderland 1979). With tobacco anthers culturability was also enhanced by starving the intact donor plants for nitrogen (Sunderland 1978).

Once haploid callus is produced, plants may be regenerated by transferring the callus from a callus initiation medium to a differentiation medium. The differentiation medium often has lower levels of growth hormones, particularly auxins, than the initiation medium. After haploid plants have been regenerated, three main doubling techniques have been used; spontaneous doubling which often occurs during callus development, regeneration from pith callus as done with tobacco (Kasperbauer and Collins 1972), and treatment with colchicine (Jensen 1974).

DIHAPLOID BREEDING THEORY

The main interest and advantage in dihaploid breeding has been its ability to create homozygous lines more quickly than most plant-breeding systems. In self-pollinated crops, homozygous or "pure-breeding" lines are marketed to commercial growers. Among cross-pollinated crops, homozygous or inbred lines are used in hybrid production. A dihaploid breeding program is compared with a conventional winter wheat breeding program in Table 19.1. Four years after the initial cross has been made, the dihaploid preliminary yield trial has been harvested, whereas in a conventional program the preliminary yield trial is not harvested until six years after the cross has been made. Of course, each breeder organizes his program differently and the time comparisons will differ somewhat with the breeding method, but on the average, dihaploid breeding will shorten the time needed for inbreeding by at least two years. It should be noted that dihaploid breeding is but one method of obtaining homozygous lines. Regardless of the method used, a careful period of evaluation must follow before the line is released for commercial sale. This evaluation period, involving many years and locations for testing, is almost independent of the breeding system. Hence, dihaploid breeding decreases only the period needed to obtain homozygous lines. It does not shorten the period required for evaluation.

Dihaploids can be used very effectively in selection programs. Table 19.2 compares the proportion of a dihaploid and F_2 breeding population that will have the desired phenotype and genotype when only three genes are being selected. If three recessive traits are being selected, 1/8 of the dihaploid population will be selected, whereas only 1/64 of the F_2 population will be

Table 19.1 Comparison of Dihaploid and Conventional Breeding Methods for Winter Wheat

Time[a]	Dihaploid	Conventional
5-7/0	Select parents	
8-9/1	Vernalize seed, transplant to greenhouse	
11/1	Make crosses to produce F_1 seed	
12/1	Harvest seed, vernalize seed	
2-3/1	Transplant to greenhouse	Transplant to field
4/1	Excise anthers	
7/1	Calli forming	Harvest F_2 seed
9/2	Double and regenerate plants	
10/2	Transplant to field	Plant F_2 seed
7/2	Harvest dihaploids	Harvest F_3 seed
7/3	Select and harvest dihaploid rows	Harvest F_4 seed
7/4	Harvest preliminary dihaploid yield trial	Harvest F_5 seed
7/5	Harvest advanced dihaploid yield trial	Select and harvest F_6 head rows
7/6	Harvest advanced dihaploid yield trial	Harvest preliminary yield trial

[a]Month/year. Year 1 starts with vernalizing parental seed for crossing.

selected. Selected lines in both cases will be true breeding. If three dominant traits are being selected, 1/8 of the dihaploid population will be selected, whereas 27/64 of the F_2 population will be selected. All of the dihaploid selections will be true breeding, but only 1/27 of the F_2 selections will be true breeding. Breeders would select the 27/64 and then grow those selections in the F_3 generation for a second evaluation to determine which 1/27 was true breeding and which 26/27 were segregating. If a segregating family was selected in the F_3 generation, further evaluations would be needed in the F_4 and later generations until the desired true breeding types could be found. In the dihaploid populations, all selections are true breeding, and no evaluations will be required after the initial one.

With the development of methods for obtaining doubled haploids, new breeding theories were also developed. Doubled haploid breeding is most similar to single-seed descent (SSD) breeding (Grafius 1965; Brim 1966) in that both methods involve rapid inbreeding without selection to achieve homozygous ("pure breeding") lines, followed by yield testing and selection. Riggs and Snape (1977), using a computer simulation, concluded that there are no differences between dihaploids and SSD when linkage is not present. When genes are in the coupling phase, dihaploids tend to have a higher population mean and variance. When genes are in the repulsion phase, dihaploids will tend to have a lower population mean and variance. They concluded that the

Table 19.2 Comparison of Dihaploid and F_2 Generation Selection for a Trigenic Trait

Mode of inheritance	Dihaploid	F_2 generation
3 Recessive Genes		
Proportion of population	$(1/2)^3$	$(1/4)^3$
Phenotypic selection	1/8	1/64
True breeding	1/8	1/64
3 Dominant Genes		
Proportion of population	$(1/2)^3$	$(3/4)^3$
Phenotypic selection	1/8	27/64
True breeding	1/8	1/64

higher recombination of SSD may be advantageous. As most breeders choose parents so that genes will be in repulsion, SSD may be the better method. Dihaploid breeding may be more suitable than SSD for species having large numbers of chromosomes, as linkage should be less important. Snape and Simpson (1981) determined that if genes in the repulsion phase are important, F_2 plants rather than F_1 plants should be used for dihaploid breeding. The additional recombination may be worth the additional time delay. Little advantage was found by waiting until the F_3 generation or by intermating F_2s before extracting dihaploids.

Many breeders have argued that early generation selection in which the population is still segregating is the most efficient way of breeding, because it allows one to select lines within relatively small populations that contain all of the desired genes (Sneep 1977; Nass 1979). Other breeders have argued, however, that while all the desired genes may be present in those early generation materials, the presence of dominance variation (variation due to the deviation of the heterozygote from the mid-parental value which will not be present in a homozygous line) and the lack of evaluation techniques for early generation testing that is sufficiently accurate to identify superior materials having those desirable genes (Knott 1972; Knott and Kumar 1975) hinder effective selection. The theoretical advantages of doubled haploid breeding have been summarized in two major papers.

The first paper (Griffing 1975) showed that the phenotypic variances for diploid and dihaploid populations are:

$$\text{Diploid: } \sigma_P^2\,(D) = \sigma_A^2 + \sigma_D^2 + \sigma_E^2$$

$$\text{Dihaploid: } \sigma_P^2\,(H) = 2\sigma_A^2 + \sigma_E^2$$

where $\sigma_{\bar{P}}{}^2$ is the phenotypic variance (variation due to differences among lines), $\sigma_{\bar{A}}{}^2$ is the additive genetic variance (variation due to the differences between homozygotes at a single locus), $\sigma_{\bar{D}}{}^2$ is the dominance genetic variance (described above), and σ_E^2 is the environmental variance (variation due to environmental effects). Griffing also compared recurrent selection methods, methods used to concentrate favorable alleles from many individuals by selection in each generation among the progeny produced by intermating the selected individuals or their progeny of the previous generation (Allard 1960) using diploid and/or dihaploid individuals, clones, and general combining ability (line performance as determined by the average performance of its F_1s in crosses with other lines). In comparisons among the diploid methods he found that the advantages of clonal and general combining ability over individual selection, particularly for traits of low heritability, largely disappear when the population size is restricted, assuming equal selection cycle lengths (time required to intermate lines and undertake the appropriate evaluation for recurrent selection). Yet the very considerable advantages of haploid versus diploid selection schemes, measured on a per cycle basis, do not disappear when total plant numbers are restricted, as would be the case in conventional plant-breeding programs, again assuming equal selection cycle lengths. The relative cycle length is quite important. Hence, the key to the successful inclusion of dihaploid breeding methods in breeding programs are techniques for the rapid extraction of dihaploids.

The second important paper (Scowcroft 1978) extended the work of Griffing and considered how Griffing's estimated efficiencies would change by having relatively few doubled haploids among which to select, but having a population of 1000 lines that were conventionally derived. This extension is important because many important crops have dihaploid production systems which can produce only limited numbers of dihaploids. Even when only 20 dihaploids are available and five are selected, compared to selecting 50 lines from 1000 conventionally derived lines, the dihaploid system is 0.84 to 3.55 times more efficient than the diploid selection system, assuming equal selection cycle lengths. Dihaploid breeding is most advantageous when selecting traits having low heritability caused by large dominance variations. Dihaploid breeding is least advantageous for traits of high heritability (heritability equalling 1). From this, Scowcroft concluded that before deciding to use dihaploid selection one should know the: (a) length of selection cycle for dihaploid and diploid cycles, (b) number of fertile dihaploids which can be conveniently produced and evaluated per selection cycle, and (c) mode of gene action (additive or dominance) and magnitude of environmental effects.

Dihaploids also are useful in statistical genetic studies. Diallel (crossing in all possible combinations a series of genotypes [Allard 1960]) experiments using dihaploids can estimate additive and additive-by-additive epistatic variation and can test for additive epistasis, linkage, and gene association. Baker (1978) described the limitations of conventional diallels as being the inability (a) to determine whether genes are independently distributed in the parents (which is critical for interpreting the results) and (b) to assume no epistasis. Researchers using dihaploid diallels can test to determine whether the assumptions are met (Choo and Reinbergs 1979). Besides using dihaploids to estimate genetic variances, they can be used in population improvement (Choo, Christie, and

Reinbergs 1979) for self-pollinated crops using the selective diallel mating system (N. F. Jensen 1970). Lines can be selected on line or on line and cross performance. As the dihaploids are pure breeding, any high-yielding lines identified in the population improvement program can be increased and released for commercial use.

Most of this discussion has considered the advantages and disadvantages of dihaploid breeding in self-pollinated crops in which the inbred line would be sold commercially. Dihaploids can also be used in population improvement of cross-pollinated crops. Choo and Kannenberg (1978), using a computer simulation, found that the response to dihaploid mass selection was superior to diploid mass selection and was equivalent to S_1 selection. Dihaploid mass selection would be faster than S_1 selection if replicated yield trials are used. Nevertheless, they found more desirable genes were lost with dihaploid mass selection than in S_1 selection, which they speculated may have been associated with the small sample sizes used in the simulation.

From theoretical considerations, the advantages of dihaploid breeding for many breeding objectives outweigh the disadvantages. Many of the disadvantages can be avoided with careful planning.

PRESENT USES OF DIHAPLOIDS IN BREEDING

In discussing the use of dihaploids in plant-breeding programs, it should be recognized that there are many methods of developing dihaploids. Besides anther culture, haploids and dihaploids are produced spontaneously by natural and genetic induction. The indeterminant gametophyte gene in maize (Kermicle 1969) and the haploid initiator gene in barley (Hagberg and Hagberg 1980) both genetically induce spontaneous haploidy. Semigamy can produce paternal and maternal haploids in cotton (Turcotte and Feaster 1974). Interspecific hybridization has led to the formation of maternal haploids in both barley (Kasha and Kao 1970) and tobacco (Burk, Gerstel, and Wernsman 1979). Twin seedlings and chemical or physical shock also produce haploids (Lacadena 1974). Doubled haploid breeding methodology is well developed in both barley (using interspecific hybridization) and tobacco (using anther culture and some interspecific hybridization).

In barley, doubled haploids were compared to single-seed descent and pedigree-derived lines. No differences were found between the lines derived by the three methods for grain yield, heading date, and plant height (Park et al. 1976). In a second analysis of this study, Choo, Reinbergs, and Park (1982) compared the frequency distributions between the dihaploid and SSD populations to determine if gene linkage were sufficiently large to cause SSD to be the preferred breeding method, as Riggs and Snape (1977) suggested. The frequency distributions for both methods were not significantly different, indicating that linkage is not important for these traits in barley (a crop with a low chromosome number, 2n = 14). In 1978, Song et al. compared dihaploid-derived lines to bulk-derived lines and determined that the bulk-derived lines had a higher mean and less variation than the dihaploid lines. This result was explained by natural selection removing the low-yielding lines from the bulk population. The best dihaploid and bulk-derived lines, however, were not different for yield. Reinbergs, Park, and Song (1976) also showed that superior barley crosses could be identified by testing as few as 20 dihaploid lines derived from the cross. Reinbergs et al. (1978) determined that the yield stability, as determined by response to the environment (Eberhart and Russell 1966), for dihaploid lines were not different from conventionally derived check cultivars. These studies indicate that in barley, dihaploidy is a useful breeding method and that complete homozygosity is not detrimental. As further proof of the usefulness of the dihaploid breeding method, many breeding stations have dihaploid production facilities. Ciba-Geigy Seeds Ltd. of Canada breeds barley exclusively by the doubled haploid breeding method. K. M. Ho (personal communication) believes that he can effectively evaluate four times as many lines because he is evaluating them on a plot basis, rather than on an individual plant basis. The company began using the method in 1974 and by 1979, after two years of testing in regional performance trial, licensed "Mingo" barley. Mingo was the highest-yielding barley in Ontario in 1980 (K. M. Ho, personal communication; Ho and Jones 1980; Kasha and Reinbergs 1981).

Dihaploid breeding using anther culture is most developed in tobacco, but here the advantages and uses of dihaploid breeding are not as clear as they are in barley. The ambiguity stems from the differing results that have been reported in flue-cured tobacco (grown in North Carolina) and in burley tobacco (grown in Kentucky). In 1972, Burk, Gwynn, and Chaplin reported reduced vigor in flue-cured tobacco dihaploids, whereas Collins, Legg, and Litton (1974) found no substantial reduced vigor in burley tobacco dihaploids. Burk and Matzinger (1976), using a highly inbred, hence homozygous line, found anther culture induced variation which was usually deleterious. Dihaploid parents, their F_1s and F_2s yielded less than the equivalent conventionally derived parents, F_1s, and F_2s in the flue-cured tobacco (Arcia, Wernsman, and Burk 1978). This study indicated that the reduced vigor in the dihaploid parents may not have been caused by complete homozygosity resulting in inbreeding depression, since the dihaploids hybrids, which should have heterozygosity equal to the conventionally derived hybrids, still yielded less. This result indicated that homozygosity alone did not explain the dihaploid yield depression. Burk, Gerstel, and Wernsman (1979) found disparate genetic ratios for two of three known monogenically controlled disease resistances in flue-cured tobacco, which indicated that the gametic array of the F_1 anther source was not accurately represented in the dihaploid progenies. Burk and Chaplin (1980) obtained some dihaploids that were superior to either parent in flue-cured tobacco, indicating that dihaploid breeding can produce useful materials. In comparing dihaploids to SSD lines in flue-cured tobacco, however, Schnell, Wernsman, and Burk (1980) found SSD lines had a higher mean and lower variance than dihaploid lines. They concluded that SSD would be the preferred breeding method. Recent work in the burley tobaccos using a diallel found no vigor reduction, and the results were similar to a conventional line diallel (Deaton, Legg, and Collins 1982). The vigor reduction found in the flue-cured tobaccos has been attributed to the loss of residual heterozygosity and to the anther and tissue culture techniques inducing variation (probably through mutagenesis). Little evidence showing the presence of residual heterozygosity has been presented, however, that would not invalidate the theory. Some evidence which tends to support the belief that the techniques themselves induce variation, generally deleterious, are that F_1 synthetics of dihaploids from a cultivar had a lower yield than the flue-cured tobacco cultivar from which they were derived. If the dihaploid vigor reduction were solely caused by the loss of residual heterozygosity, the F_1 synthetic which reestablished heterozygosity should not also exhibit a vigor reduction (Brown and Wernsman 1982). Also, De Paepe, Bleton, and Grangbe (1981) and Javier (1981) have found continued reductions with succeeding anther culture cycles. After the first anther culture cycle, the dihaploids should be completely homozygous and there should be no residual heterozygosity to remove. Maternal dihaploids developed through use of interspecific hybridization do not show as large a vigor reduction as do anther-derived dihaploids, and the smaller vigor reduction may be caused by the tissue culture doubling procedure (Wernsman, personal communication). Hence, it appears that tobacco tissue and anther culture do induce variation and may preferentially select for deleterious traits, though the magnitude of the deleterious effects may vary with the tobacco type, flue-cured or burley.

In trying to put the conflicting tobacco research in perspective, three points should be remembered. Tissue culture techniques have often been reported to induce variation. Larkin and Scowcroft (1981) have recently reviewed tissue culture–induced variation, which they call "somaclonal variation." They believe somaclonal variation may be very useful in breeding programs. Second, even in flue-cured tobacco, dihaploid lines with useful traits and improved yields have been selected. Hence the induced variation and vigor reductions are not so severe as to negate the efficacy of dihaploid breeding via anther culture. Finally, tobacco yields are measured as leaf yields, which is similar to forage yields. Grain crop breeders should therefore avoid extrapolating too readily from tobacco research findings. For example, while semidwarf tobacco may reduce leaf yield, Schaeffer (1982) has developed a short-statured rice from anther callus of "Calrose 76". It may be a potentially useful, new source of dwarfing.

Another crop in which anther culture is being developed for use in dihaploid breeding programs is wheat. As early as 1972, Picard and De Buyser produced anther callus with little plantlet regeneration. Cultivars and F_1s were anther-cultured and plants regenerated by Ouyang et al. (1973) and by Wang et al. (1973). The first major medium improvement was the Chinese potato medium (Research Group 301, 1976). Schaeffer et al. (1979) anther-cultured wheats adapted to the United States and found large cultivar differences for androgenesis, indicating the importance of genotype in successful anther culture. For breeders to effectively utilize dihaploid breeding, the techniques should work for most cultivars and the diverse group of F_1s found in a breeding program. Bullock et al. (1982) determined that the F_1s of cultivars having good androgenic potential and cultivars having poor androgenic potential were anther culturable. Hence, androgenic potential can be transferred from a parent to its F_1, which implies that by careful parent selection, using existing technologies, diverse F_1s can be cultured. No reciprocal differences for androgenesis were found, indicating that the inheritance of androgenesis was controlled by nuclear genes. The regenerated haploid plants were tested for their resistance to powdery mildew (*Erysiphe graminis* f. sp. *tritici*) and examined for the presence of awns. Although an inadequate number of plants were regenerated to allow a statistical evaluation of the gametic array, the haploid plants did express the parental classes, and powdery mildew screening at the hemizygous level appeared to be possible. Hence, haploids could be screened for important disease resistances prior to doubling and field transplanting, if either procedure is costly. Picard, De Buyser, and Henry (1978) found that species cytoplasms in an alloplasmic line series and the plant environment can affect androgenesis.

They also found that anther culture techniques can induce variation. We have found variants among our anther culture–derived genotypes, including an awned variant from an awnletted source, a short family of plants from a normal height cultivar, a tall variant from a normal height cultivar, and a vigorous, but extremely late heading and maturing, plant. Many of these variants appear to be chromosomal aberrations, because semisterility is often found among these aberrant plants. The cytological basis for our variants is being studied by Drs. Moshe Feldman, Gordon Kimber, and Esra Galun. Most of our variation is very obvious "off-types," however. Wheat anther culture has been reported to produce aneuploids and other chromosomal changes (Hu et al. 1980; 1981). In very preliminary yield trials we have not found subtle variation, which would be considerably more difficult for plant breeders to discern and would potentially make dihaploid breeding less useful, particularly if the variation were preferentially deleterious. Large yield trials have been planted this year at Aberdeen, Idaho, in cooperation with Dr. D. M. Wesenberg. These should give us our first extensive test to determine whether our techniques are inducing subtle variation.

Anther culture has been reported to preferentially select lines that have better androgenic potential. Dihaploids and their F_1s were superior for anther culture to the source cultivars and their F_1s (Picard and De Buyser 1977). Rives and Picard (1977) found high levels of abnormal pollen in dihaploids and believed that the abnormal pollen may be the source of anther callus, which would explain the improved androgenetic potential of the dihaploids. In our laboratory and with our genotypes, however, anther culture does not appear to select for improved androgensis (M. Lazar, personal communication). In summary, wheat anther culture has progressed rapidly in the last ten years. Still needed are an improved medium (particularly one that is not genotype specific), improved doubling techniques, and extensive field testing to determine whether the gametic array is accurately represented and whether the technique induces variation.

FUTURE USE OF DIHAPLOIDS IN PLANT BREEDING

One of the main future uses of dihaploids will be the development of new cultivars. The new cultivars will be the products of recurrent selection schemes, superior crosses identified by their dihaploids, and by better evaluation at the haploid level of gene pyramids for disease or stress resistance. One advantage of cultivar development via dihaploidy is that no inadvertant field selection is associated with the inbreeding process. Obviously, cultivars are released on the basis of their response to local environments. During the segregating generations, however, natural selection may be harmful. For example, the F_1 of a winter by spring growth habit wheat cross will have a spring growth habit. The F_2 and each succeeding generation will have proportionately more spring growth habit plants than will the final population when homozygosity is achieved. Growing this segregating population at any time in winter conditions will lose both the homozygous and the heterozygous spring growth habit types. Similar results will be found in every cross segregating for a dominant and recessive gene when the recessive trait is desired, and the dominant trait will be selected against. There is no assurance that the desired plant types will be competitive in a bulk (a population of segregating lines from a cross in a self-pollinated crop that are grown together in a "bulk"). Semidwarf wheats, while highly desirable, are less competitive than tall plants in a bulk (Khalifa and Qualset 1975). Hence, field selection in early generations can be harmful. Inbreeding without field selection will also be useful in instances where the techniques for developing dihaploid plants are either too difficult for most breeders to use at their location or do not readily lend themselves to large-scale mass production. Breeders could send their materials to central laboratories, which would produce dihaploids, increase the seed, and return the true breeding lines to the breeder. An example of a dihaploid developed in one area and released in another is Gwylan, a recent dihaploid barley developed by the Welsh Plant Breeding Station, currently being released in New Zealand (Kasha and Reinbergs 1981). Many of the international germplasm enhancement and breeding centers, such as CIMMYT, may find breeding without field selection useful.

A second future use of haploidy and dihaploidy will be in biochemical selection experiments and mutational breeding. Selection and mutation at the haploid level have two major advantages over selection and mutation at the diploid level: (1) because every gene is hemizygous and will appear as being pseudo-dominant, selection can be done not only for dominant genes, but also for recessive genes, and (2) once the cell type is selected, chromosome doubling will make the trait and any other genetic change homozygous. Chromosome doubling is important in that it saves one generation of selfing, which would be required if diploid selection schemes were used where only one allele at a locus was altered. Also, mutational events are rarely well directed. Hence, many genes in addition to the desired gene may be altered. In a dihaploid all of the desired and undesired changes would be expressed, enabling the breeder to better determine which is the best breeding method to use to develop an agronomic type having the desired trait. Undesirable genes in the heterozygous condition are lost very slowly from a population.

A third important use of dihaploids will be in basic genetic and physiological mechanism studies. Homozygous lines can be developed for genes that would be impossible to obtain using conventional breeding techniques. Two examples would be lines homozygous for the dominant male sterile gene in wheat (because the gene is dominant, it cannot be selfed or crossed to other lines having the gene to obtain homozygosity) and self-incompatibility alleles (again, heterozygotes carrying the alleles cannot be selfed or crossed to other lines having the gene). In physiological studies, one of the sources of unexplained variation is the variation within a cultivar. Dihaploids should be the most uniform material available and would immediately remove one of the sources of unexplained variation, thereby improving the precision of the test. Researching the physiology of winterhardiness, an area wherein most of the research methods are inexact, is an example of the potential usefulness of truly homozygous lines in keeping experimental error to a minimum and more accurately measuring experiment-to-experiment variation.

Finally, dihaploids will be useful in the development of new chromosomal and cytoplasmic variation (Hu et al. 1981). Haploidy has a long history in the development of useful cytogenetical stocks. Many aneuploids stocks were developed from haploid by diploid hybridization (Sears 1954). The preliminary studies of wheat anther culture indicate that chromosomal aberrations may be developed through anther culture techniques (Feldman, personal communication). It is too early to tell if these will be useful or not, but variation is being created. As the cytoplasm of the microspore does not participate genetically in normal fertilization, it may not be well conserved genetically and could be a potentially useful source of cytoplasmic variation. Further, as gene or chromosome transfer systems are developed, haploids and haploid aneuploid cell lines would be excellent receiver lines for the genes. Once the gene or chromosome is transferred, the recipient cells would be doubled to attain homozygosity, and even transferred recessive genes would be expressed.

CONCLUSIONS

Dihaploid breeding in some crops is already a reality and is proving itself to be a useful breeding method. In the majority of crops, further improvements in techniques are needed. As breeding requires large populations for selection, the anther culture system, with its potential of having every microspore within an anther form calli from which plants are regenerated, is potentially the best system for generating the numbers of dihaploid plants that are needed. Initial research will concentrate on developing non-genotype–specific methods that will produce haploid plants efficiently. Doubling efficiency will also need to be improved in that every generation the seed needs to be increased is one more generation that could be used for evaluation, the critical phase. Finally, extensive field tests will need to be undertaken to determine how accurately the gametic array is expressed and whether the techniques introduce useful or deleterious variation. In addition, further breeding theory should be developed. As an adjunct to existing breeding systems alone, dihaploidy deserves to be thoroughly researched and developed. Dihaploidy will also have an important role in biochemical selection and mutation breeding, basic physiology and genetics, and in gene transfer systems and cytogenetics. It may become one of the most powerful plant-breeding and genetic tools available in crop improvement.

ACKNOWLEDGMENTS

The authors wish to acknowledge the help of Drs. L. G. Burk, G. B. Collins, K. M. Ho, K. J. Kasha, H. W. Rines, and E. A. Wernsman, who provided additional information of their current research for use in this paper. The authors also wish to acknowledge the assistance of Drs. M. D. Lazar and D. T. Kudirka, and Mr. W. P. Bullock for critical review of the manuscript. The research was in part supported by Competitive Grant No. 59–2246–0–1–436–0 and BARD Project No. US 234–80.

LITERATURE CITED

Allard, R. W. 1960. *Principles of plant breeding*. John Wiley & Sons, New York. 485 pp.

Arcia, M. A., E. A. Wernsman, and L. G. Burk. 1978. *Performance of anther-derived dihaploids and their conventionally inbred parents as lines, in F_1 hybrids, and in F_2 generations*. Crop Sci. 18:413–8.

Baker, R. J. 1978. *Issues in diallel analysis*. Crop Sci. 18:533–36.

Bennett, M. D. 1976. *The cell in sporogenesis and spore development*. In M. M. Yoeman, ed., *Cell Division in Higher Plants*. Academic Press, New York.

Bernard, S. 1977. *Study of some factors controlling in vitro androgenesis of hexaploid Triticale*. Ann. Amelior. Plantes 27(6):639–55.

Brim, C. A. 1966. *A modified pedigree method of selection in soybeans*. Crop Sci. 6:220.

Brown, J. S., and E. A. Wernsman. 1982. *Nature of reduced productivity of anther-derived dihaploid lines of flue-cured tobacco*. Crop Sci. 22:1–5.

Bullock, W. P., P. S. Baenziger, G. W. Schaeffer, and P. J. Bottino. 1982. *Anther culture of wheat F_1's and their reciprocal crosses*. Theor. Appl. Genet. 62:155–59.

Burk, L. G., and J. F. Chaplin. 1980. *Variation among anther-derived haploids from a multiple disease-resistant tobacco hybrid*. Crop Sci. 20:334–38.

Burk, L. G., J. F. Chaplin, G. V. Gooding, and N. T. Powell. 1979. *Quantity production of anther-derived haploids from a multiple disease resistant tobacco hybrid. I. Frequency of plants with resistance or susceptibility to tobacco mosaic virus (TMV), potato virus Y (PVY), and root knot (RK)*. Euphytica 28:201–08.

Burk, L. G., D. U. Gerstel, and E. A. Wernsman. 1979. *Maternal haploids of Nicotiana tabacum L. from seed*. Science 206:585.

Burk, L. G., G. R. Gwynn, and J. F. Chaplin. 1972. *Diploidized haploids from aseptically cultured anthers of Nicotiana tabacum*. J. Hered. 63:355–60.

Burk, L. G., and D. F. Matzinger. 1976. *Variation among doubled haploid lines obtained from anthers of Nicotian tabacum L*. J. Hered. 67:381–84.

Choo, T. M., B. R. Christie, and E. Reinbergs. 1979. *Doubled haploids for estimating genetic variances and a scheme for population improvement in self-pollinating crops.* Theor. Appl. Genet. 54:267–71.

Choo, T. M., and L. W. Kannenberg. 1978. *The efficiency of using doubled haploids in a recurrent selection program in a diploid, cross-fertilized species.* Can. J. Genet. Cytol. 505–11.

Choo, T. M., and E. Reinbergs. 1979. *Doubled haploids for estimating genetic variance in the presence of linkage and gene association.* Theor. Appl. Genet. 55:129–32.

Choo, T. M., E. Reinbergs, and S. J. Park. 1982. *Comparison of frequency distribution of doubled haploid and single seed descent lines in barley.* Theor. Appl. Genet. 61:215–8.

Collins, G. B., and A. D. Genovesi. 1982. *Anther culture and its application to crop improvement.* Pages 1–24 in D. T. Tomes, B. E. Ellis, P. M. Harney, K. J. Kasha, and R. L. Peterson, eds., *Applications of Plant Cell and Tissue Culture to Agriculture and Industry.* Univ. of Guelph.

Collins, G. B., P. D. Legg, and C. C. Litton. 1974. *The use of anther-derived haploids in Nicotiana. II. Comparison of doubled haploid lines with lines obtained by conventional breeding methods.* Tob. Sci. 18:40–2.

Deaton, W. R., P. D. Legg, and G. B. Collins. 1982. *A comparison of burley tobacco doubled-haploid lines with their source inbred cultivars.* Theor. Appl. Genet. 62:69–74.

De Paepe, R., D. Bleton, and F. Grangbe. 1981. *Basis and extent of genetic variability among doubled haploid plants obtained by pollen culture in Nicotiana sylvestris.* Theor. Appl. Genet. 59:177–84.

Dunwell, J. M. 1976. *A comparative study of environmental and developmental factors which influence embryo induction and growth in cultured anthers of Nicotiana tabacum.* Envir. and Exp. Bot. 16:109–18.

Durr, A., and J. Fleck. 1980. *Production of haploid plants of Nicotiana langsdorffii.* Plant Sci. Lett. 18:75–79.

Eberhart, S. A., and W. R. Russell. 1966. *Stability parameters for comparing varieties.* Crop Sci. 6:36–40.

Grafius, J. E. 1965. *Short cuts in plant breeding.* Crop Sci. 5:377.

Griffing, B. 1975. *Efficiency changes due to use of doubled haploids in recurrent selection methods.* Theor. Appl. Genet. 46:367–86.

Guha, S., and S. C. Maheshwari. 1964. *In vitro production of embryos from anthers of Datura.* Nature (Lond.) 204:497.

Hagberg, A., and G. Hagberg. 1980. *High frequency of spontaneous haploids in the progeny of an induced mutation in barley.* Hereditas 93:341–43.

Ho, K. M., and G. E. Jones. 1980. *Mingo Barley.* Can. J. Plant Sci. 60:279–80.

Hu Han, Z. Y. Xi, J. Jing, and X. Wang. 1981. *Production of pollen-derived wheat aneuploid plants through anther culture.* Pages 767–78 in *Cell and Tissue Culture Techniques for Improvement of Cereal Crops.* Beijing, China.

Hu Han, Z. Y. Xi, J. W. Ouyang, Hao Shui, M. Y. He, Z. Y. Xu, and M. Z. Zou. 1980. *Chromosome variation of pollen mother cell of pollen-derived plants in wheat (Triticum aestivum L.)* Sci. Sinica 23 (7):905–14.

Jacobsen, E., and S. K. Sopory. 1978. *The influence and possible recombination of genotypes on the production of microspore embryoids in anther cultures of Solanum tuberosum and dihaploid hybrids.* Theor. Appl. Genet. 52:119–23.

Javier, E. L. 1981. *Variability among nonconventionally derived diploid lines of Nicotiana tabacum L.* Ph.D. dissertation, N.C. State University, Raleigh.

Jensen, C. J. 1974. *Chromosome doubling techniques in haploids.* Pages 153–90 in K. J. Kasha, ed., *Haploids and Higher Plants. Proc. First Internat. Symp. June 10–14. Univ. Guelph Press, Ontario.*

Jensen, N. F. 1970. *A diallel selective mating system for cereal breeding.* Crop Sci. 10:629–35.

Kasha, K. J., and K. N. Kao. 1970. *High frequency haploid production in barley (Hordeum vulgare L.)* Nature (Lond.) 225:874–76.

Kasha, K. J., and E. Reinbergs. 1981. *Recent developments in the production and utilization of haploids in barley*. Pages 655–65 in M. J. C. Asher, ed., *Barley Genetics IV*. Proc. 4th Int. Barley Genet. Symp. Edinburgh, U. K.

Kasperbauer, M. J., and G. B. Collins. 1972. *Reconstitution of diploids from leaf tissue of anther-derived haploid in tobacco*. Crop Sci. 12:98–101.

Kermicle, J. L. 1969. *Androgenesis conditioned by mutation in maize*. Science 166:1422–24.

Khalifa, M. A., and C. O. Qualset. 1975. *Intergenotypic competition between tall and dwarf wheats. II. In hybrid bulks*. Crop Sci. 15:640–44.

Kihara, H. 1979. *Artificially raised haploids and their uses in plant breeding*. Seiken Ziko 27–28: 14–29.

Knott, D. R. 1972. *Effects of selection for F_2 plant yield on subsequent generations in wheat*. Can. J. Plant Sci. 52:721–26.

Knott, D. R., and J. Kumar. 1975. *Comparison of early generation yield testing and a single seed descent procedure in wheat breeding*. Crop Sci. 15:295–99.

Lacadena, J. R. 1974. *Spontaneous and induced parthenogenesis and androgenesis*. Pages 13–32 *in* K. J. Kasha, ed., *Haploids and Higher Plants*. Proc. First Internat. Symp. University of Guelph, Canada.

Larkin, P. J., and W. R. Scowcroft. 1981. *Somaclonal variation—a novel source of variability from cell cultures for plant improvement*. Theor. Appl. Genet. 60:197–214.

Maheswari, S. C., A. K. Tyagi, and K. Malhotra. 1980. *Induction of haploidy from pollen grains in Angiosperms—the current status*. Theor. Appl. Genet. 58:193–206.

Miao, S. H., C. S. Kuo, Y. L. Kwei, A. T. Sun, S. Y. Ku, W. L. Lu, Y. Y. Wang, M. L. Chen, M. K. Wu, and L. Hang. 1978. *Induction of pollen plants of maize and observation on their progeny. Pages 23–34 in Proc. Symp. Plant Tissue Culture,* May 25–30. Science Press, Peking.

Nass, H. G. 1979. *Selecting superior spring wheat crosses in early generations*. Euphytica 28:161–67.

Nitsch, J. P., and C. Nitsch. 1969. *Haploid plants from pollen grains*. Science 163:85–87.

Nitzsche, W., and G. Wenzel. 1977. *Haploids in Plant Breeding*. Verlag Paul Parey, Berlin and Hamburg, 101 pp.

Ouyang, T. W., H. Hu, C. C. Chuang, and C. C. Tseng. 1973. *Induction of pollen plants from anthers of Triticum aestivum L. cultured in vitro*. Sci. Sin. 16:79–95.

Park, S. J., E. J. Walsh, E. Reinbergs, L. S. P. Song, and K. J. Kasha. 1976. *Field performance of doubled haploid barley lines in comparison with lines developed by the pedigree and single seed descent methods*. Can. J. Plant Sci. 56:467–74.

Picard, E., and J. De Buyser. 1972. *Obtention de plantules haploides de Triticum aestivum L. a partes d'antheras in vitro*. C. R. Acad. Sci Seri D 277:1463–66.

———. 1977. *High production of embyroids in anther culture of pollen derived homozygous spring wheats*. Ann. Amel. Plantes. 27:483–88.

Picard, E., J. De Buyser, and Y. Henry. 1978. *Technique de production d'haploides de blé par culture d'antheres in vitro*. La Selectionneur Français. 26:25–37.

Reinbergs, E., S. J. Park, and L. S. P. Song. 1976. *Early identification of superior barley crosses by the doubled haploid technique*. Z. Pflanzenzuchtg. 76:215–24.

Reinbergs, E., L. S. P. Song, T. M. Choo, and K. J. Kasha. 1978. *Yield stability of doubled haploid lines of barley*. Can. J. Plant Sci. 58:929–33.

Reinert, J., and Y. P. S. Bajaj. 1977. *Anther culture: Haploid production and its significance*. Pages 251–67 in J. Reinert and Y. P. S. Bajaj, ed., *Applied and Fundamental Aspects of Plant Cell, Tissue, and Organ Culture*. Springer-Verlag, Berlin.

Research Group 301. 1976. *A sharp increase of the frequency of pollen-plant induction in wheat with potato medium*. Acta Genet. Sin. 3:30–1.

Riggs, T. J., and J. W. Snape. 1977. *Effects of linkage and interaction in a comparison of theoretical populations derived by diploidized haploid and single seed descent methods*. Theor. Appl. Genet. 49:111–15.

Rives, M., and E. Picard. 1977. *A case of genetic assimilation: Selection through androgenesis or parthenogenesis of haploid producing systems (an hypothesis)*. Ann. Amel. Plantes. 27:489–91.

Schaeffer, G. W. 1982. *Recovery of heritable variability in anther-derived doubled-haploid rice*. Crop Sci. 22:1160–64.

Schaeffer, G. W., P. S. Baenziger, and J. Worley. 1979. *Haploid plant development from anthers and in vitro embryo culture of wheat*. Crop Sci. 19:697–702.

Schnell, R. J., II, E. A. Wernsman, and L. G. Burk. 1980. *Efficiency of single-seed-descent vs. anther-derived dihaploid breeding methods*. Crop Sci. 20:619–22.

Scowcroft, W. R. 1978. *Aspects of plant cell culture and their role in plant improvement*. Pages 181–98. in *Proceedings of a Symposium on Plant Tissue Culture,* May 25–30. Science Press, Peking.

Sears, E. 1954. *The aneuploids of common wheat*. Missouri Agr. Exp. Sta. Res. Bull. 572:59.

Sink, K. C., Jr., and V. Padmanabhan. 1977. *Anther and pollen culture to produce haploids: progress and application for the plant breeder*. Hort. Sci. 12:143–48.

Snape, J. W., and E. Simpson. 1981. *The genetical expectations of doubled haploid lines derived from different filial generations*. Theor Appl. Genet. 60:123–28.

Sneep, J. 1977. *Selection for yield in early generations of self-fertilizing crops*. Euphytica 26:27–30

Song, L. S. P., S. J. Park, E. Reinbergs, T. M. Choo, and K. J. Kasha. 1978. *Doubled haploid vs. the Bulk Plot method for production of homozygous lines in barley*. Z. Pflanzenzuchtg. 81:271–80.

Sunderland, N. 1978. *Strategies in the improvement of yields in anther culture*. Pages 65–86 in *Proceedings of Symposium on Plant Tissue Culture*. Science Press, Peking.

———. 1979. *Comparative studies of anther and pollen culture*. Pages 203–19 in W. R. Sharp, P. O. Larsen, E. F. Paddock, and V. Raghavan, eds., *Plant Cell and Tissue Cultures: Principles and Applications*. Ohio State Univ. Press, Columbus.

Tomes, D. T., and G. C. Collins. 1976. *Factors affecting haploid plant production from in vitro cultures of Nicotiana species*. Crop Sci. 16:837–40.

Turcotte, E. L., and C. V. Feaster. 1974. *Semigametic production of cotton haploids*. Pages 53–64 in K. J. Kasha, ed., *Haploids and Higher Plants*. Proc. First Internat. Symp. University of Guelph, Canada.

Wang, C. C., C. C. Chu, C. S. Sun, S. H. Wu, K. C. Yin, and C. Hsu. 1973. *The androgenesis in wheat (Triticum aestivum) anthers cultured in vitro*. Sci. Sinica 16(2):218–25.

Wenzel, G., F. Hoffman, and E. Thomas. 1977. *Increased induction and chromosome doubling of androgenetic haploid rye*. Theor. Appl. Genet. 51:81–86.

Xu, Z. H., B. Huang, and N. Sunderland. 1981. *Culture of barley anthers in conditioned media*. J. Exp. Botany 32(129):767–78.

20] Morphogenesis and Regeneration in Tissue Culture

by TREVOR A. THORPE*

ABSTRACT

The capacity to regenerate plantlets from cultured tissues is a requirement for successful application of tissue culture technology to agriculture. Regeneration can occur via somatic embryogenesis or primordium formation, but most plant species follow the second route. It appears that manipulation of the nutrient medium and the culture environment, etc., allows competent cells to demonstrate their intrinsic capacity for organized development, which is ultimately the reflection of selective gene activity. This manifests itself through biochemical, biophysical, physiological, and structural changes in the cultured tissues. This chapter describes some aspects of these events mainly in relation to shoot primordium initiation in tobacco callus. The behavior of subcultured cells is briefly discussed. Then the structural and physiological bases of organized development are outlined. This is followed by a review on the role of endogenous phytohormones, in general, and ethylene, in particular, in primordium formation. Research on carbohydrate utilization and metabolism, as well as our initial studies on nitrogen metabolism, is then described. These studies indicate that shoot primordium formation has high requirements for energy and reducing power. In addition, a part of the carbohydrate is used osmotically. Nitrogen assimilation and aromatic amino acid metabolism are enhanced during shoot formation. These results are in agreement with the hypothesis that the initiation of organized development involves a shift in metabolism.

INTRODUCTION

The application of plant tissue culture technology to the improvement of agricultural crops and horticultural and forest tree species requires that plantlets be regenerated from the selected material. In many cases this material

*Department of Biology, University of Calgary, Calgary, Alberta T2N 1N4, Canada.

GENETIC ENGINEERING: APPLICATIONS TO AGRICULTURE (Beltsville Symposium 7—Lowell D. Owens, ed.) Rowman & Allanheld, Totowa

will have been in culture for a period of time, during which genetic and epigenetic modification of the cells have been taking place. Thus, immediate direct regeneration of plantlets from explants is not possible under these conditions. One of the major problems facing researchers using in vitro modification techniques is that by the time modified cells have been obtained, the tissue will no longer form plantlets.

Plantlet formation from cultured cells and tissues occurs via two routes. One is by the production of somatic embryos, in which cells undergo structural and organizational changes similar to those which occur during zygotic embryogenesis, i.e., a rudimentary plant with a root/shoot axis is formed. The second route involves the production of primordia which subsequently undergo organogenesis. In most cases shoot primordia followed by leafy vegetative shoots are formed. These then become rooted via root primordium formation and subsequent root organogenesis. Essentially, therefore, either bipolar or unipolar structures are initially formed. Most plants follow the unipolar route, i.e., separate primordia and then organs are initiated; thus plantlet formation requires several steps.

Controlled organ formation in vitro was first reported by White (1939), who obtained shoots in a tobacco hybrid, and by Nobécourt (1939), who obtained roots in carrot callus. Subsequent work in the 40s and 50s showed that numerous species could form callus (wound parenchyma tissue) and shoots and roots *de novo*. These studies culminated in the now-classical finding of Skoog and Miller (1957) that a basic regulatory mechanism underlying organogenesis involved a balance between auxin and cytokinin; a relatively high level of auxin to cytokinin favored root formation, the reverse favored shoot formation, and intermediate ratios favored callus proliferation. The first report of somatic embryogenesis appeared later, when Reinert (1958) demonstrated the phenomenon using carrot callus. Subsequent studies using this carrot system led to the firm establishment of the concept of totipotency (Steward et al. 1964).

These early studies, using a variety of explants and numerous plant species, showed that successful culture and organized development required providing the sterilized explant or cultured tissue an appropriate nutrient medium and proper culture conditions (Murashige 1974, 1977). Five classes of compounds, namely (1) inorganic macro- and micro-nutrients, (2) carbon and energy source, (3) vitamins, (4) reduced nitrogen, and (5) phytohormones, are usually sufficient for media for most plant species. Physical factors such as the form of the medium (liquid or semisolid), pH, light, temperature, and humidity are all important in the manipulation of organized development. Thus, much of the work done in tissue culture has concerned itself with optimizing these conditions to achieve organ or embryo formation.

Although a very large number of species respond in culture to variations of phytohormones in the presence of other media components by producing shoots, roots, and embryos, many species fail to do so. It would appear, therefore, that these external manipulations set up conditions so that cells in the right physiological state can respond and undergo organized development. Thus, the process of primordium or embryo formation can be viewed as being regulated at various distinct levels. The manipulation of the culture environment brings about changes in the cultured cells which can be described histologically and ultrastructurally. These structural changes are themselves the result of preceding biochemical and biophysical events which ultimately reflect selective gene activity (Thorpe 1980, 1982; Thorpe and Biondi 1981). While virtually nothing is known about organized development at the molecular level, a fair amount of information has been gleaned in the last decade concerning the regulation of organized development at the structural, physiological, and biochemical levels. This article will review studies dealing mainly with organogenesis, and will discuss in greater detail some of the current work from my laboratory. These studies deal with shoot primordium initiation in tobacco callus. For recent reviews dealing with somatic embryogenesis, reference can be made to Kohlenbach (1978), Wetherell (1979), Sharp et al. (1980), and Thorpe (1982).

BEHAVIOR OF SUBCULTURED CELLS

Callus can be initiated from virtually any part of the plant. The initial behavior of a cultured tissue may vary with the overall physiological status of the original plant. After one or two subcultures, the "typical" callus for that species, under the culture conditions used, will be obtained. It is important to recognize however, that the culture conditions act in a selective manner, and they favor certain cells which are best suited to that particular culture environment. These favored cells relatively quickly become dominant, especially under liquid culture conditions. The type of callus or cell suspension obtained can therefore vary, and the ideal material for regeneration must often be determined empirically. For tobacco callus under the experimental conditions I use, the tissue is initiated and subcultured at 4-week intervals in the dark, and only smooth, white, compact callus is used as inoculum for subculture or experimentation. For the latter, approximately 5-week-old callus subcultured at least three times and not older than a year is used.

Cells of most plant cultures are not uniform: variations occur in the size, degree of vacuolation and cytoplasmic content, cell wall characteristics and shape, especially under liquid culture conditions. Thus, the typical callus cell is a highly vacuolated parenchyma cell having a thin peripheral layer of cytoplasm and a faintly staining nucleus which is usually adpressed against the cell wall. Callus cells can be considered the least differentiated of all the mature cell types of the plant, but they are definitely not undifferentiated. It is preferable to refer to callus as being unorganized (Torrey 1966). It should also be pointed out that tracheary elements often are found in subcultured callus.

With continued subculture of callus certain changes occur. One change is the loss of a phytohormone requirement for growth. Such tissue may be auxin-independent or cytokinin-independent, or both, and is said to be habituated. In terms of morphogenesis, such habituated tissues, apparently committed to rapid cell division, rarely differentiate without some prior manipulative treatment. Another change that takes place is the loss of morphogenetic potential.

This generally occurs more quickly in relation to rooting than to shoot formation in tobacco. This reduction in organogenetic and embryogenetic potential has been observed in many tissues and is caused by both genetic and epigenetic factors. A third change often observed is the formation of tissue with altered texture, usually increased friability, which may reduce the morphogenetic capacity of the tissue. In tobacco such tissues do not readily produce shoots under our experimental conditions. These problems are discussed in greater detail by Murashige (1974) and Thorpe (1982).

Even when cells appear cytologically similar, they may differ greatly in their biochemical competency, a fact now being used for cell selection for product formation (Zenk 1978). Furthermore, it is this variation in genetic or epigenetic features of cultured cells that is being exploited in the selection of somaclonal variants (Larkin and Scowcroft 1981).

The importance of cell selection for regeneration has been illustrated by the work of Rice, Reid, and Gordon (1979), who were attempting to manipulate competent cell types. They found that they could select cells from a 17-year-old tobacco cell line that had lost its capacity to regenerate plants, by a process of repeated subculture on a shoot-forming medium, and by selecting sectors of the culture mass that appeared more viable, greener, and more compact. By this process normal shoots ultimately were obtained which could be rooted. This means that many of the cells which have become modified in culture and possess desirable culture traits may prove recalcitrant in terms of their regenerative capacity. Such cells may require prior manipulative treatments before they can respond to the inductive conditions which lead to primordium or embryo formation.

MANIPULATION OF ORGANIZED DEVELOPMENT

In manipulating *primordium formation,* many growth-active substances, phytohormones as well as other types of compounds, have been included in the medium. As already indicated, a large number of plant species respond to a suitable auxin/cytokinin balance by forming shoots and roots. Evans, Sharp and Flick (1981) found that for 75% of the species that form shoots, kinetin or benzylaminopurine was used in a concentration range of 0.05 to 46 μM. Auxins such as indoleacetic acid (IAA) and naphthaleneacetic acid (NAA) were used in concentrations of 0.06 to 27 μM. Graminaceous species tend to have a lower cytokinin requirement for shoot formation than other species. Indolebutyric acid and NAA are the most commonly used auxins for rooting.

In a number of cases, however, the permissive phytohormonal balance leads to the induction of organogenetic tissue only. This tissue will then develop into organs in a medium with an altered phytohormonal balance. In addition, in some cases, only exogenous auxin or cytokinin may be sufficient to bring about organogenesis. Finally, apparently anomalous situations have been reported, e.g., by Walker et al. (1978). They found that a high level of auxin and a low level of cytokinin resulted in shoot formation in alfalfa callus, while the reverse favored root formation. Such findings can be reconciled easily, when one realizes that the endogenous auxin/cytokinin balance is important for the initiation of organogenesis.

In place of cytokinins, a variety of substituted purines, pyrimidines, and ureas have been used successfully to bring about organogenesis. Similarly, various auxin-like compounds can satisfy the auxin requirements in vitro. Other phytohormones added to the medium have been shown to play a role in organogenesis. These include gibberellins and abscisic acid. No generalizations can be made with respect to the effects of these substances in shoot and root formation, as they have been shown variously to repress, enhance, or be without effect on organ formation in different plant species (Thorpe 1980).

In addition to phytohormones, other metabolites have been shown to stimulate organogenesis in different species. Such metabolites include adenine, guanine, uracil, uridine, amino acids, various phenolic acids, and nicotine (see Thorpe 1980). Thus, many classes of compounds can play a role in organogenesis. The interaction between these different substances is in agreement with the basic ideas of Skoog and Miller (1957), a concept which guides most of the research performed on in vitro organogenesis.

In manipulating *embryogenesis,* two media components, in particular, play crucial roles. These are auxin and nitrogen (Kohlenbach 1978). The importance of auxin was first recognized by Halperin and Wetherell (1964). Further studies showed that the process of somatic embryogenesis normally takes place in two stages; first, the induction of cells with embryogenic competence (referred to as embryogenic masses or clumps, proembryos, proembryonic tissue) in the presence of high concentrations of auxin; and second, the development of the embryogenic masses into embryos in the absence of or in the presence of a lowered concentration of auxin. 2,4-Dichlorophenoxyacetic acid is the most commonly used auxin. Reduced nitrogen in the form of NH_4^+ is also apparently required (see Kohlenbach 1978; Wetherell 1979).

STRUCTURAL AND PHYSIOLOGICAL BASES OF ORGANIZED DEVELOPMENT

Studies carried out mainly with tobacco callus and carrot cell suspensions have given a general idea of some of the events associated with primordium formation and somatic embryogenesis, respectively (see Thorpe 1982).

In brief, the induction of organogenesis or embryogenesis leads to distinct structural patterns in the tissue. After varying lengths of time in culture (2–8 days), random cell division activity in organogenetic tissue occurs and produces regions of high mitotic activity. These regions produce meristematic centers or meristemoids containing densely plasmatic cells, which are initially apolar but rapidly show directional divisional activity to form unipolar primordia (see Thorpe 1980). The induction of proembryonic tissue involves superficial cell division activity producing groups of densely plasmatic cells. On transfer to auxin-free medium these cells divide and go through a series of morphological changes that are comparable to zygotic embryogenesis, since structures similar to globular, heart, and torpedo stage embryos can be found (see Wetherell 1979).

Both types of morphogenic phenomena begin with changes in single cells. This does not mean, however, that the entire primordium or embyro arises from the derivatives of that activated cell. In organogenetic tissues, surrounding cells are often induced to divide and thus may become incorporated in the new structure (e.g., see Chlyah 1974; Smith and Thorpe 1975).

In shoot-forming tobacco callus, zones of preferential cell division activity are observed in the lower half of the tissue by day 8 in culture (Thorpe and Murashige 1970; Ross, Thorpe, and Costerton 1973; Maeda and Thorpe 1979). Within these zones meristemoids and then primordia are formed. Meristemoids consist of spherical masses of small isodiametric meristem-like cells with dense cytoplasm and a high nucleo-cytoplasmic ratio. These cells are microvacuolated (Ross, Thorpe, and Costerton 1973), they contain a higher content of all organelles, and the nuclei contain more nucleolar material (Asbell 1977). Furthermore, meristemoid cells stain densely for RNA and protein; and the organ-forming tissue contains large amounts of stored starch, which disappears during meristemoid and primordium formation (Thorpe and Murashige 1970). The process is not synchronous, so meristemoids can be observed in 8-to-14-day-old tissue, primordia in 10-day-old and older tissue, and the emergence of the primordia from the base of the callus as early as day 12 (Ross, Thorpe, and Costerton 1973; Thorpe 1979). Thus during the 6–12 day period in culture, the critical events associated with primordium formation occur.

What are the physiological requirements for the initiation of this process? Concentration or diffusion gradients of materials from the medium into the tissue have been implicated in determining the loci at which primordium initiation begins (Ross and Thorpe 1973). This mechanism, however, cannot explain how individual cells become activated and induced to undergo an altered pathway of differentiation. This process has at least three requirements: (1) cell dedifferentiation. (2) cell interaction, and (3) reaction to specific signals (Thorpe 1980; Thorpe and Biondi 1981). The net result is specific changes in the tissue which precede and are, presumably, causative to organized development. In particular, I have been concerned with the role of phytohormones and carbohydrates in this process and more recently with nitrogen assimilation and amino acid metabolism.

ROLE OF ENDOGENOUS PHYTOHORMONES

The importance of phytohormones in initiating and regulating organized development has already been indicated. For these to be effective in organized development, the critical balance and/or concentration must be within the tissue, and at specific loci. Yet to date, we have virtually no data on the concentration of or changes in endogenous phytohormones in callus tissue during differentiation.

Indirect studies based on changes in cathodic isoperoxidases have been used to indicate probable changes in endogenous auxin in shoot and root initiation (see Thorpe 1978, 1982).

Tobacco callus contains gibberellic acid (GA)-like substances and is capable of metabolizing exogenous GAs (Lance, Reid, and Thorpe 1976; Lance et al. 1976). As a matter of fact, the level as well as the spectrum of endogenous GA-like substances change during shoot formation. It seems that GAs are involved in normal tissue growth and differentiation, but that tobacco callus synthesizes enough for the organogenetic process, and that inhibition of organ formation by exogenous GAs results from the supraoptimum levels imposed on the cells. Thus, the stimulation of organogenesis in vitro by GAs (e.g., in *Arabidopsis* callus, Negrutiu Jacobs and Cachita 1978), can be interpreted as indicative of low endogenous GA levels (Thorpe 1978).

More recently, we have been examining the role of endogenous ethylene in differentation of shoots and roots. For rooting studies, we used excised tomato leaf-discs which require exogenous auxin for root formation (Coleman and Greyson 1977). We found that applied ethylene or ethephon did not stimulate rooting in the leaf discs. In the presence of optimum levels of IAA these substances significantly inhibited root formation. Ethylene production was positively correlated with increased IAA concentrations at various times during the culture period and, as a consequence, with the rooting response. Yet separate testing of equimolar concentrations of different auxins and auxin-like compounds showed no positive correlation between the rate of ethylene production and subsequent rooting response. Aeration of gas-tight flasks containing leaf discs and absorption of ethylene evolved from the discs (by mercuric perchlorate in gas-tight flasks) or pretreatment of leaf discs with $AgNO_3$ (an inhibitor of ethylene action) significantly enhanced IAA-induced root regeneration. Thus, these studies indicated that ethylene was not a rooting hormone per se. Furthermore, ethylene (whether applied externally or synthesized by the tissue) did not appear to account for the ability of auxin to stimulate rooting (Coleman et al. 1980).

Shoot-forming tobacco callus was found to produce less endogenous ethylene than non-shoot-forming tissue cultured in the light (16 h photoperiod) or the dark. In shoot-forming tissue more ethylene was produced early in culture (days 0–5) than later. Also dark-grown tissue produced much more ethylene than light-grown tissue. On the basis of experiments in which (a) gaseous ethylene was added to or (b) CO_2 removed from the flasks, or (c) Ethrel (an ethylene releasing agent) or (d) 1-aminocyclopropane 1-carboxylic acid (an ethylene precursor) was added to the medium, it was determined that this gaseous phytohormone had two contrary effects on shoot primordium formation. Early in culture (days 0–5) endogenous or exogenous ethylene inhibited organogenesis, but later (days 5–10) exogenous ethylene or increased endogenous ethylene production speeded up primordium formation (Huxter, Thorpe, and Reid 1981). Thus, it is conceivable that if the endogenous ethylene production could be reduced early in culture and stimulated during meristemoid and primordium formation, these structures would be produced earlier and proceed faster and more synchronously to shoot development than normally occurs. Here again, the changes associated with ethylene production and action are apparently independent of auxin action, as judged by changes in specific cathodic isoperoxidases (Thorpe and Gaspar 1978; Thorpe, Tran Thanh Van, and Gaspar 1978).

We have pointed out the difficulty in designating a specific role for ethylene in growth and organized development, since this phytohormone is gaseous and diffuses easily through the tissue. In addition, endogenous ethylene is measured after it has passed through the cells, and therefore could have affected them prior to being released. Thus, there is no direct method of separating cause and effect relationships. Nevertheless, we have postulated that, although most of the ethylene released is a by-product of general metabolism, the initial endogenous amounts produced could act as a stimulant to growth and differentiation (Huxter, Reid and Thorpe 1979).

METABOLISM AND ORGANOGENESIS

Histo- and cyto-chemical approaches, as well as regular biochemical techniques, have been used to glean some data on the metabolic aspects of differentiation. Most of these studies are fragmentary, since a thematic approach to examining basic aspects of differentiation generally has not been undertaken. Nevertheless, direct and indirect evidence has shown that DNA, RNA, and protein synthesis occur during, and are probably necessary for, organ formation (see Thorpe 1980; Thorpe and Biondi 1981). Higher contents of RNA and protein have been found in organ-forming tissues. In addition, evidence for the synthesis of special low molecular weight shoot-forming proteins has also been obtained (Hasegawa, Yasuda and Cheng 1979; Yasuda, Hasegawa and Cheng 1980). These proteins presumably serve structural, but perhaps mainly enzymatic, purposes. Two aspects of our work will be discussed; (a) carbohydrate utilization and metabolism and (b) nitrogen assimilation and amino acid metabolism during primordium formation.

Our approach has been to compare shoot-forming tobacco callus with non-shoot-forming tissues which is achieved by either reducing the level of cytokinin in the medium (non-shoot-forming tissue) or by the addition of GA_3 to the shoot-forming medium. This allows us to separate aspects of metabolism apparently important in differentiation from those involved only in growth.

Carbohydrate utilization and metabolism. Changes in the starch content of organ-forming tissues have been observed in several tissues, since the initial observation of Thorpe and Murashige (1968) (see also Thorpe 1980). In shoot-forming tobacco callus the peak of starch accumulation occurred just prior to the formation of meristemoids (Thorpe and Murashige 1970; Thorpe and Meier 1972). Examination of the activities of enzymes involved in starch metabolism revealed that the accumulation of starch resulted from increased synthetic activity, while the reduction during meristemoid and primordium formation involved enhanced rates of degradation (Thorpe and Meier 1974). A continuous supply of free sugars from the medium was required also for shoot formation (Thorpe 1974). While there was no difference in the uptake of ^{14}C-sucrose into shoot-forming and non-shoot-forming tobacco callus, there was a steady and linear incorporation of ^{14}C into starch in the former (D. Brown and T. Thorpe, unpublished).

One probable role of the starch (and free sugars) is to serve as a readily available reserve of energy for the organogenetic process. Compared to non-organ-forming tissue, shoot-forming callus had a higher rate of respiration (Thorpe and Meier 1972; Ross and Thorpe 1973), increased activities of enzymes of both the Embden-Meyerhof glycolytic and pentose phosphate pathways, and enhanced ^{14}C-glucose oxidation rates (Thorpe and Laishley 1973). Shoot-forming tissues also had higher levels of total adenosine phosphates and NAD^+ and a lower energy charge (Brown and Thorpe 1980a). More recent studies using isolated mitochondria added weight to the above conclusions (Brown and Thorpe 1982). Although no differences in the capacity of mitochondrial enzymes were observed between isolated shoot-forming and non-shoot-forming mitochondria, there was a trend toward higher and more efficient respiration. During meristemoid and primordium formation there was a decrease in the relative level of the alternate cyanide-insensitive pathway and an increase in respiration via the normal cyanide-sensitive pathway, which produces more ATP per molecule oxidized.

One possible function of the enhancement of the pentose phosphate pathway (PPP) during shoot formation is the production of reducing power (NADPH) for reductive biosynthesis. This role was supported when $^{14}CO_2$ incorporation into acid-stable compounds and activities of enzymes involved in malate metabolism in dark-grown cultures were examined (Plumb-Dhindsa, Dhindsa, and Thorpe 1979). These enzymes had greater activities in shoot-forming tissues and, therefore, malate metabolism seems to be important in the production of NADPH via the NADP-linked malate enzyme. Measurement of the NADPH and $NADP^+$ pools revealed a faster decline, more complete utilization of NADPH and a greater build up of $NADP^+$ levels in shoot-forming tissue, compared to non-shoot-forming tissue (Brown and Thorpe 1980a). This variation in the NADPH/NADP ratios could be involved in in vivo regulation of the PPP. While the variation dropped to <0.05 in shoot-forming callus, it remained relatively constant in non-shoot-forming tissue. Glucose-6-P dehydrogenase (G6PD) was fully inhibited by NADPH/NADP ratios >2.0, whereas 6-phosphogluconate dehydrogenase (6PGD) required a ratio of >5.0 (Thompson and Thorpe 1980). Since the level of G6PD is 2 to 3 times that of 6PGD in shoot-forming tissue, it appears that regulation of the pathway could occur by (a) coarse control of the amount of G6PD present and (b) by fine control due to variation in the NADPH/NADP ratios.

We also found that there is a specific osmotic component for shoot formation (Brown, Leung, and Thorpe 1979). One-third of the sucrose incorporated into the medium fulfills this requirement in tobacco callus, since we could replace one-third of the medium sucrose by mannitol. This sugar alcohol is considered non-metabolizable in most plant species and, therefore, could act in a colligative fashion intracellularly. An, as yet, incomplete study on mannitol metabolism in tobacco cell suspensions has indicated that cells absorb less than 5% of the fed mannitol in a 14-hour period (also true for carrot; and ca. 10% for radiata pine cotyledons). About 50% of the mannitol taken up was metabolized, ca. 4% was released as CO_2, ca. 30% was converted into an unidentified, low molecular weight sugar or other neutral compounds, and a further ca. 15% was converted into basic, acidic, and insoluble compounds (Thompson and Thorpe 1981). Thus, the osmotic effect of mannitol apparently takes place from outside the cell, i.e., from mannitol in the cell wall, intercellular spaces, etc., i.e., in the apoplast. In any case, much of the sucrose placed in the medium also remains outside of the cells. Whether or not the osmotic agent is outside the plasmalemma or inside the tonoplast, its osmotic effect on organelles in the cytoplasm is likely to be the same. The efficiency of mannitol as an osmotic agent thus, apparently, is a consequence of its low uptake into the cells.

The finding of an osmotic component to shoot formation prompted us to examine the water relations of the cultured tissues (Brown and Thorpe 1980b). We found that shoot-forming tobacco callus grown in a medium with the same water potential as non-shoot-forming tissue maintained greater (i.e., more negative) water and osmotic potentials, as well as greater (more positive) pressure potentials than non-shoot-forming tissue. These differences were observed by day 2 in culture, which is prior to any visible histological changes in the tissue, and were maximum at day 6, which is at the time of the first visible changes leading to organized development. These differences were maintained throughout the culture period. They did not result from differences in the uptake of sucrose from the medium. In addition, fresh weight/dry weight ratios and water content of the tissues were of the same order.

We believe that the greater water, osmotic, and pressure potentials in shoot-forming tissue could be maintained in part by (a) the accumulation of malate early in culture (Plumb-Dhindsa, Dhindsa, and Thorpe 1979), (b) the accumulation of free sugars from the medium throughout the culture period (Thorpe 1974), and (c) the degradation products of starch at the time of meristemoid and primordium formation (Thorpe and Meier 1974). In addition, we have recently found that shoot-forming tobacco callus has high levels of threonine-serine and proline (Table 20.1, L. Hardy and T. Thorpe, unpublished). Thus, it appears that the morphogenic system uses a variety of metabolites colligatively, and furthermore, different metabolites contribute osmotically at different stages of the process.

The significance of the osmotic adjustment for growth and organized development remains to be determined. It is reasonable to assume, however, that biophysical events play as important a role as do biochemical ones in the process. The biophysical events may involve changes in membrane properties. One possible consequence of the increased osmotic potential of shoot-forming tissue is the enhancement of the activity of the mitochondria (D. Brown and T. Thorpe, unpublished). Figure 20.1 shows that a range of osmotic potential of 100 and 200 milliosmoles was found in non-shoot-forming (callus) and shoot-forming (shoots) tissue, with a difference of about 300 milliosmoles at the time of shoot primordium formation. Increasing osmolarities enhanced mitochondrial activity. Thus, the capacity of the isolated mitochondria to utilize substrate (state 3 respiration) increased 23%, the coupling of oxidative phos-

Table 20.1 Amino Acid Composition (m moles/g FW) of Shoot-Forming (SF) and Non-Shoot-Forming (NSF) Tobacco Callus

Amino acid	Tissue	Days in culture 7	10	16
Lysine	SF	0.18	0.19	0.28
	NSF	0.17	0.11	0.24
Histidine	SF	0.14	0.21	0.31
	NSF	0.09	0.09	0.14
Aspartic acid	SF	0.35	0.05	0.67
	NSF	0.18	0.23	0.42
Threonine and Serine	SF	11.52	7.10	12.50
	NSF	2.55	2.27	3.35
Glutamate	SF	0.39	0.15	0.10
	NSF	0.29	0.39	0.08
Proline	SF	T	4.69	7.80
	NSF	T	0.29	0.48
Glycine	SF	0.58	0.48	0.63
	NSF	0.32	0.37	0.49
Alanine	SF	1.00	0.65	0.63
	NSF	1.15	1.59	1.30
Valine	SF	T	T	0.20
	NSF	T	T	0.18
Methionine	SF	0.21	0.17	0.29
	NSF	0.19	0.16	0.27
Leucine	SF	0.24	0.23	0.38
	NSF	0.17	0.16	0.43
Tyrosine	SF	1.76	0.57	0.14
	NSF	0.01	0.04	0.26
Phenylalanine	SF	4.61	7.50	7.50
	NSF	3.38	5.61	6.42

Note: T = trace.

phorylation (RCR, or state 3 to state 4 ratio) increased by 34%, and the efficiency of ATP production (ADP/O) ratio increased a striking 60% in going from the osmotic potential found in growing tissue to that in the shoot-forming tobacco. Therefore, the early osmotic adjustment observable by day 2 in shoot-forming tissue could be a mechanism to increase the metabolic capacity of mitochondria to produce the energy required for primordium formation.

Nitrogen assimilation and amino acid metabolism. We have recently begun to study these topics in shoot-forming tobacco callus. As an initial examination of nitrogen metabolism (Table 20.2), we found that shoot-forming tissue maintained higher levels of total-N, protein-N, and nitrite-N, but similar levels of

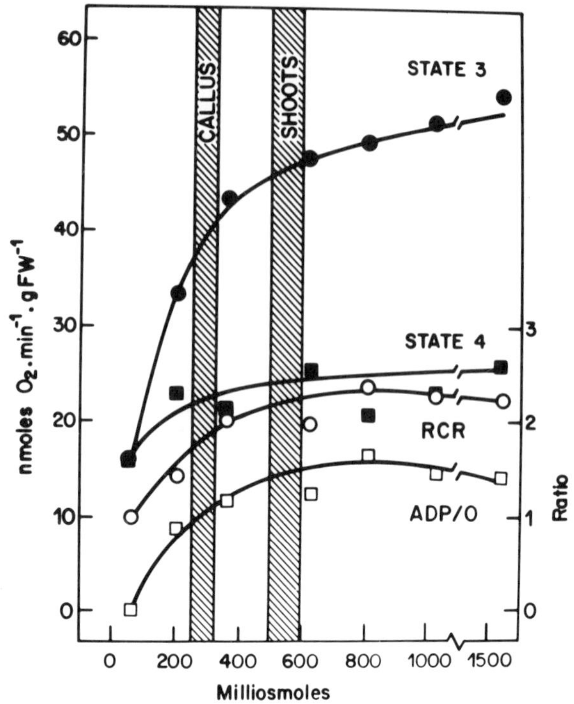

Figure 20.1. Effect of varying osmolarities, adjusted with mannitol, on the functioning of mitochondria isolated from 17-day-old shoot-forming tobacco callus. Oxygen consumption is shown in the closed symbols, and the respiratory control ratio (RCR) and ADP/O ratios in open symbols. State 3 is the active or ADP-dependent state, while state 4 is the resting state. Succinate was the substrate. Striped bars indicate the native osmolarity range found in non-shoot-forming (callus) and shoot-forming (shoots) tissue during primordium formation. Points represent the data means from two separate experiments.

amino-N, in comparison to non-shoot-forming tissues (Hardy and Thorpe 1981). Nitrate reductase activity was higher in shoot-forming tissue, but the activity of nitrite reductase was similar in both shoot-forming and non-shoot-forming tissues (L. Hardy and T. Thorpe, unpublished). Preliminary amino acid analyses (Table 20.1) of three stages of organogenesis showed that amino acid pool sizes changed during culture, and that, in general, shoot-forming tissue had higher levels of amino acids (L. Hardy and T. Thorpe, unpublished).

As indicated earlier, increased activities of both the pentose phosphate and glycolytic pathways are observed during shoot formation. An important function of these pathways is to provide erythrose-4-phosphate and phosphoenol pyruvate, respectively, for aromatic amino acid biosynthesis via the shikimate pathway. We have looked at the capacities of key enzymes in this pathway

Table 20.2 Content of Nitrogen-Containing Metabolites in Shoot-Forming (SF) and Non-Shoot-Forming (NSF) Tobacco Callus

Metabolite	Tissue	Days in culture					
		0	4	7	10	13	16
Total nitrogen	SF	1.3	3.7	3.8	3.9	3.4	3.9
	NSF	1.3	2.1	2.4	2.5	2.2	2.2
Protein	SF	0.4	2.4	4.1	4.3	4.7	5.1
	NSF	0.4	1.4	2.5	3.7	3.9	4.1
Amino-N $\times 10^4$	SF	0.3	0.4	0.5	1.0	1.2	1.1
	NSF	0.3	0.4	0.5	0.7	0.8	1.0
Nitrite-N $\times 10^3$	SF	1.6	2.2	2.4	3.4	2.9	3.9
	NSF	1.6	1.7	1.8	2.0	2.0	2.1

Note: Values in mg/g FW are the average of two independent experiments.

(Beaudoin-Eagan and Thorpe 1981). We found that 2 to 10-fold higher levels of activity of DAHP synthetase (the coupling enzyme), shikimate kinase (rate limiting step in pathway), chorismate mutase and anthranilate synthetase (branch points enzymes) were observed in shoot-forming tissues, compared to non-shoot-forming tissues (Table 20.3). Differences were evident by day 6 in culture and reached a maximum between days 12 to 15, which is coincident with meristemoid and primordium formation.

When the tissues were fed ^{14}C-glucose, labeled shikimate and quinate were found (L. Beaudoin-Eagan and T. Thorpe. unpublished). More label was incorporated into quinate than shikimate (Table 20.4), and the incorporation

Table 20.3 Specific Activity of Key Enzymes of the Shikimate Pathway in Shoot-Forming (SF) and Non-Shoot-Forming (NSF) Tobacco Callus

Enzyme	Tissue	Days in culture					
		0	3	6	9	12	15
DAHP[a] synthetase $\times 10$	SF	0.5	2.5	6.8	12.5	22.0	20.5
	NSF	0.5	0.5	2.1	2.4	5.7	5.8
Shikimate kinase $\times 10^4$	SF	0.3	0.3	1.8	2.5	4.0	3.7
	NSF	0.3	0.3	0.4	0.4	0.4	0.4
Chorismate mutase $\times 10$	SF	3.0	3.2	3.5	5.0	6.6	10.0
	NSF	3.0	3.1	3.0	3.2	3.1	3.1
Anthranilate synthetase $\times 10$	SF	7.5	9.5	9.5	13.0	16.0	20.7
	NSF	7.5	10.0	10.0	10.0	8.0	10.0

[a] 3-Deoxyarabinoheptulosonate-7-phosphate.
Note: Data average of two experiments which showed similar trends. Specific activity—μmoles/mg Protein/min.

Table 20.4 Incorporation of Radioactivity from ^{14}C-Glucose into Shikimate and Quinate in Shoot-Forming (SF) and Non-Shoot-Forming (NSF) Tobacco Callus

		Days in culture					
Metabolite	Tissue	0	3	6	9	12	15
Shikimate	SF	876	1,938	3,450	3,817	3,490	3,184
	NSF	574	576	936	948	880	873
Quinate	SF	3,690	12,919	13,226	15,751	13,200	12,751
	NSF	2,195	4,916	6,137	8,320	8,180	7,713

Note: Tissue was incubated for 2 h in appropriate liquid medium. Values in DPM/g FW are the average of two independent experiments.

into both metabolites was greater in shoot-forming tissue than in non-shoot-forming tissue. Similarly, when the tissues were fed ^{14}C-shikimate, the end products of the pathway (phenylalanine, tyrosine and tryptophan) were all labeled (L. Beaudoin-Eagan and T. Thorpe, unpublished). As can be seen in Table 20.5, the level of incorporation into the aromatic amino acids in shoot-forming tissue was higher than non-shoot-forming tissue. Furthermore, the incorporation into tyrosine in shoot-forming tissue was greater than into the other two amino acids. This finding is interesting since the shoot-forming medium contains tyrosine for optimum shoot formation. The turnover of these compounds as well as the metabolism of the aromatic amino acids is being investigated.

These latter studies will take us into phenylpropanoid metabolism. Earlier studies by Tryon (1956) and Skoog and his co-workers (Sargent and Skoog 1960; Skoog and Montaldi 1961) have implicated scopoletin and its glucoside,

Table 20.5 Incorporation of Radioactivity from ^{14}C-Shikimate into Aromatic Amino Acids in Shoot-Forming (SF) and Non-Shoot-Forming (NSF) Tobacco Callus

		Days in culture					
Amino acid	Tissue	0	3	6	9	12	15
Phenylalanine	SF	15,655	17,680	24,460	42,805	38,590	39,665
	NSF	14,368	16,490	16,635	16,880	16,795	21,480
Tyrosine	SF	16,665	17,655	25,310	128,355	122,505	119,860
	NSF	14,938	16,770	17,700	17,605	17,655	21,755
Tryptophan	SF	14,050	17,990	23,675	49,650	39,280	41,030
	NSF	14,110	16,810	16,245	16,775	16,585	21,990

Note: Tissue was incubated for 1 h in appropriate liquid medium. Values in DPM/g FW are the average of two independent experiments.

scopolin, in tobacco tissue growth and differentiation. The effectiveness of tyrosine and other substituted phenols in stimulating organ formation in tobacco callus was attributed in part to its involvement in lignin synthesis (Lee 1962). Evidence in favor of this view has been contradictory (Dougall and Shimbayashi 1960; Dougall 1962), and even the more recent study by Hasegawa, Murashige, and Mudd (1977) has not significantly increased our understanding of the role of tyrosine in organogenesis. They found that ^{14}C-tyrosine was incorporated into polymeric substances including proteins (both cytoplasmic and cell wall), polysaccharides, and possibly lignin.

During organ formation the measured capacities, and presumably in vivo activities, of many enzymes change. Some of these enzyme changes can be correlated with pool sizes of metabolites, etc., while others cannot. Nevertheless, these fragmentary reports indicate that many areas of metabolism, not unexpectedly, may be involved in organ formation (see Thorpe 1980; Thorpe and Biondi 1981).

CONCLUSIONS

The initiation of organized development is a complex morphogenetic phenomenon, in which extrinsic and intrinsic factors play a role. Organized development can be regulated through manipulation of culture medium, the culture environment, and by judicious selection of the inoculum. Manipulation of these factors allows cells that are quiescent or committed only to cell division to undergo a transition, which at the molecular level involves selective gene activity. This is reflected by biochemical, biophysical, and physiological changes leading to structural organization which can be monitored in the differentiating system. Much is known about the manipulation of factors regulating organized development, but virtually nothing is known about regulation at the molecular level. Certainly, the activation of cells leading to differential gene action must include the cellular events of replication, transcription, processing, and translation. Furthermore, intercellular communication is apparently necessary, and phytohormones seem to play a controlling role in the complex processes leading to primordium or embryo formation (see Thorpe 1980; Thorpe and Biondi 1981).

The physiological and biochemical events associated with organogenesis have been my concern. Using tobacco callus as the main experimental material, it has been possible to describe the developmental process leading to shoot primordium formation and to examine physiological and biochemical events in relation to this sequence. The key histological events leading to shoot primordium formation occur during the 6- 12-day culture period. Changes in the levels of endogenous phytohormones seem important, and our recent studies on ethylene support the complex involvement of phytohormones in differentiation. Examination of the role of carbohydrate in the process has given a clearer picture. Carbohydrate has a dual role in providing both energy (ATP) and osmotic adjustment. Various biochemical measures have supported the idea that primordium formation is a high energy-requiring process in which accumulated starch and free sugars from the medium are utilized. The process is metabolically very demanding as there is also a high requirement for reducing power (NADPH). Studies initiated recently on nitrogen assimilation and amino acid metabolism also indicate, not unexpectedly, that N metabolism is also important to the differentiation process.

These findings are in agreement with the hypothesis that organized development involves a shift in metabolism that leads to changes in the content and spectrum of both structural and enzymatic proteins. The metabolic changes we have oserved precede and are coincidental with organized development. Thus, as Bonner (1965) pointed out, such changes in which new enzymes originally absent are synthesized or enzymes which are present show increased synthesis and/or activity *a priori* must be a cause rather than the result of differentiation. It is obvious that information on the biochemical pathways involved in differentiation is very meager, and much more work needs to be done on this subject. In the final analysis, regulation of organized development will be understood only when molecular biology is able to provide answers. Without a better understanding of the regulation of organized development, many of the possible advances to be made in plant improvement, via tissue culture methods, will remain unrealized, since at present we lack the capacity to regenerate plantlets from cultured tissues of many agriculturally important plants.

ACKNOWLEDGMENTS

The author acknowledges with gratitude the contribution of colleagues, associates, and students to the personal research reported in this article. In particular, I am grateful to Dan Brown, Lynda Beaudoin-Eagan, Lorraine Hardy, and Michael Thompson, who have contributed as yet unpublished material. Financial support from the Natural Sciences and Engineering Research Council (formerly the National Research Council) of Canada and the University of Calgary is gratefully acknowledged.

LITERATURE CITED

Asbell, C. W. 1977. *Ultrastructural modifications during shoot formation in vitro.* In Vitro 13:180.

Beaudoin-Eagan, L., and T. A. Thorpe. 1981. *Shikimic acid pathway activity in shoot-forming tobacco callus.* Plant Physiol. suppl. 67:154.

Bonner, J. A. 1965. *Development.* Pages 850–66 in J. A. Bonner and J. E. Varner, eds., *Plant Biochemistry,* 2nd edition. Academic Press, New York.

Brown, D. C. W., D. W. M. Leung, and T. A. Thorpe. 1979. *Osmotic requirement for shoot formation in tobacco callus.* Physiol. Plant. 46:36–41.

Brown, D. C. W., and T. A. Thorpe. 1980a. *Adenosine phosphate and nicotinamide adentine dinucleotide pool sizes during shoot initiation in tobacco callus.* Plant Physiol. 65:587–90.

———. 1980b. *Changes in water potential and its components during shoot formation in tobacco callus.* Physiol. Plant. 49:83–7.

———. 1982. *Mitochondrial activity during shoot formation and growth in tobacco callus*. Physiol. Plant. 54:125–30.

Chlyah, H. 1974. *Formation and propagation of cell division centers in the epidermal layers of internodal segments of Torenia fournieri grown in vitro*. Can. J. Bot. 52:867–72.

Coleman, W. K., and R. I. Greyson. 1977. *Analysis of root formation in leaf discs of Lycopersicon esculentum Mill. cultured in vitro*. Ann. Bot. (Lond.) 41:307–20.

Coleman, W. K., T. J. Huxter, D. M. Reid, and T. A. Thorpe. 1980. *Ethylene as an endogenous inhibitor of root regeneration in tomato leaf discs in vitro*. Physiol. Plant. 48:519–25.

Dougall, D. K. 1962. *On the fate of tyrosine in tobacco callus tissue*. Aust. J. Biol. Sci. 15:619–22.

Dougall, D. K., and K. Shimbayashi. 1960. *Factors affecting growth of tobacco callus tissue and its incorporation of tyrosine*. Plant Physiol. 35:396–404.

Evans, D. A., W. R. Sharp, and C. E. Flick. 1981. *Growth and behavior of cell cultures*. Pages 45–113 in T. A. Thorpe, ed., *Plant Tissue Culture—Methods and Applications in Agriculture*. Academic Press, New York.

Halperin, W., and D. F. Wetherell. 1964. *Adventive embryony in tissue cultures of the wild carrot, Daucus carota,* Amer. J. Bot. 51:274-83.

Hardy, E. L., and T. A. Thorpe. 1981. *Nitrogen metabolism in shoot-forming tobacco callus*. Plant Physiol. suppl. 67:7.

Hasegawa, P. M., T. Murashige, and J. B. Mudd. 1977. *The fate of L-tyrosine-UL-^{14}C in shoot forming tobacco callus*. Physiol. Plant. 41:223–30.

Hasegawa, P. M., T. Yasuda, and T. -Y. Cheng. 1979. *Effect of auxin and cytokinin on newly synthesized proteins of cultured Douglas fir cotyledons*. Physiol. Plant. 46:211–17.

Huxter, T. J., D. M. Reid, and T. A. Thorpe. 1979. *Ethylene production by tobacco (Nicotiana tabacum) callus*. Physiol. Plant. 46:374–80.

Huxter, T. J., T. A. Thorpe, and D. M. Reid. 1981. *Shoot initiation in light- and dark-grown tobacco callus: The role of ethylene*. Physiol. Plant. 53:319–26.

Kohlenbach, H. W. 1978. *Comparative somatic embryogenesis*. Pages 59–66 in T. A. Thorpe, ed., *Frontiers of Plant Tissue Culture 1978*. University of Calgary Press, Canada.

Lance, B., R. C. Durley, D. M. Reid, T. A. Thorpe, and R. P. Pharis. 1976. *The metabolism of [^3H]-gibberellin A_{20} in light- and dark-grown tobacco callus cultures*. Plant Physiol. 58:387–92.

Lance, B., D. M. Reid, and T. A. Thorpe. 1976. *Endogenous gibberellins and growth of tobacco callus cultures*. Physiol. Plant. 36:287–92.

Larkin, P. J., and W. C. Scowcroft. 1981. *Somaclonal variation—a novel source of variability from cell cultures for plant improvement*. Theor. Appl. Genet. 60:197–214.

Lee, T. T. 1962. *Effects of substituted phenols on growth, bud formation and IAA oxidase activity*. Ph.D. dissertation, University of Wisconsin, Madison.

Maeda, E., and T. A. Thorpe. 1979. *Shoot histogenesis in tobacco callus cultures*. In Vitro 15:415–24.

Murashige, T. 1974. *Plant propagation through tissue culture*. Ann. Rev. Plant Physiol. 25:135–66.

———. 1977. *Clonal crops through tissue culture*. Pages 392–403 in W. Barz, E. Reinhard, and M. H. Zenk, eds., *Plant Tissue Culture and Its Bio-technological Application*. Springer-Verlag, Berlin, Heidelberg, New York.

Negrutiu, I., M. Jacobs, and D. Cachita. 1978. *Some factors controlling in vitro morphogenesis of Arabidopsis thaliana*. Z. Pflanzenphysiol. 86:113–24.

Nobécourt, P. 1939. *Sur les radicelles naissant des cultures de tissus végétaux*. C. R. Séanc. Soc. Biol. 130:1271.

Plumb-Dhindsa, P. L., R. S. Dhindsa, and T. A. Thorpe. 1979. *Non-autotrophic CO_2 fixation during shoot formation in tobacco callus*. J. Exp. Bot. 30:759–67.

Reinert, J. 1958. *Untersuchungen über die Morphogenese an Gewebekulturen*. Ber dt. Bot. Ges. 71:15.

Rice, T. B., R. K. Reid, and P. N. Gordon. 1979. *Morphogenesis in Field Crops* Pages 262–78 in K. W. Hughes, R. Henke, and M. Constantin, eds., *Propagation of Higher Plants Through Tissue Culture—A Bridge Between Research and Application*. Conf-780411. U.S. Tech. Inf. Serv., Springfield, Virgina.

Ross, M. K., and T. A. Thorpe. 1973. *Physiological gradients and shoot initiation in tobacco callus cultures*. Plant Cell Physiol. 14:473–80.

Ross, M. K., T. A. Thorpe, and J. W. Costerton. 1973. *Ultrastructural aspects of shoot initiation in tobacco callus cultures*. Amer. J. Bot. 60:788–95.

Sargent, J. A., and F. Skoog. 1960. *Effects of indoleacetic acid and kinetin on scopoletin and scopolin levels in relation to growth of tobacco tissues in vitro*. Plant Physiol. 35:934–41.

Sharp, W. R., M. R. Sondahl, L. S. Caldas, and S. B. Maraffa. 1980. *The physiology of in vitro asexual embryogenesis*. Pages 268–310 in J. Janick, ed., *Horticultural Reviews* vol. 2. AVI Publ., Wesport, Conn.

Skoog, F., and C. O. Miller. 1957. *Chemical regulation of growth and organ formation in plant tissues cultured in vitro*. Symp. Soc. Exp. Biol. 11:118–30.

Skoog, F., and E. Montaldi. 1961. *Auxin-kinetin interaction regulating the scopoletin and scopolin levels in tobacco tissue cultures*. Proc. Nat. Acad. Sci. USA 47:36–49.

Smith, D. R., and T. A. Thorpe. 1975. *Root initiation in cuttings of Pinus radiata seedlings. I. Developmental sequence*. J. Exp. Bot. 26:184–92.

Steward, F. C., M. O. Mapes, A. E. Kent, and R. D. Holsten. 1964. *Growth and development of cultured plant cells*. Science (N.Y.) 163:20–27.

Thompson, M. R., and T. A. Thorpe. 1980. *Control of hexose monophosphate oxidation in shoot-forming tobacco callus*. Plant Physiol. suppl. 65:89.

———. 1981. *Mannitol metabolism in cell cultures*. Plant Physiol. suppl. 67:27.

Thorpe, T. A. 1974. *Carbohydrate availability and shoot formation in tobacco callus cultures*. Physiol. Plant. 30:77–81.

———. 1978. *Physiological and biochemical aspects of organogenesis in vitro*. Pages 49–58 in T. A. Thorpe, ed., *Frontiers of Plant Tissue Culture 1978*. University of Calgary Press, Canada.

———. 1979. *Regulation of organogenesis in vitro*. Pages 87–101 in K. W. Hughes, R. Henke, and M. Constantin, eds., *Propagation of Higher Plants Through Tissue Culture—A Bridge Between Research and Application*. Conf-780411. U.S. Tech. Inf. Serv., Springfield, Virginia.

———. 1980. *Organogenesis in vitro: Structural, physiological, and biochemical aspects*. Int. Rev. Cytol. suppl. 11A:71–111.

———. 1982. *Callus organization and de novo formation of shoots, roots and embryos in vitro*. Pages 115–38 in D. T. Tomes, B. E. Ellis, P. M. Harney, K. J. Kasha, and L. R. Peterson, eds., Applications of Plant Cell and Tissue Culture to Agriculture and Industry. University of Guelph Press, Ontario.

Thorpe, T. A., and S. Biondi. 1981. *Regulation of plant organogenesis*. Adv. Cell Culture 1:213–39.

Thorpe, T. A., and T. Gaspar. 1978. *Changes in isoperoxidases during shoot formation in tobacco callus*. In Vitro 14:522–26.

Thorpe, T. A., and E. J. Laishley. 1973. *Glucose oxidation during shoot initiation in tobacco callus cultures*. J. Exp. Bot. 24:1082–89.

Thorpe, T. A., and D. D. Meier. 1972. *Starch metabolism, respiration, and shoot formation in tobacco callus cultures*. Physiol. Plant. 27:365–69.

———. 1974. *Starch metabolism in shoot-forming tobacco callus*. J. Exp. Bot. 25:288–94.

Thorpe, T. A., and T. Murashige. 1968. *Starch accumulation in shoot-forming tobacco callus culture*. Science 160:421–22.

———. 1970. *Some histochemical changes underlying shoot initiation in tobacco callus culture*. Can. J. Bot. 48: 277–85.

Thorpe, T. A., M. Tran Thanh Van, and T. Gaspar. 1978. *Isoperoxidases in epidermal layers of tobacco and changes during organ formation in vitro*. Physiol. Plant. 44:388–94.

Torrey, J. G. 1966. *The initiation of organized development in plants*. Adv. Morphogen. 5:39–91.

Tryon, K. 1956. *Scopoletin in differentiating and nondifferentiating cultured tobacco tissue*. Science 123:590

Walker, K. A., P. C. Yu, S. J. Sato, and E. G. Jaworski. 1978. *The hormonal control of organ formation in callus of Medicago sativa L. cultured in vitro*. Amer. J. Bot. 65:654–59.

Wetherell, D. F. 1979. *In vitro embryoid formation in cells derived from somatic plant tissues*. Pages 102–24 in K. W. Hughes, R. Henke, M. Constantin, eds., *Propagation of Higher Plants Through Tissue Culture—A Bridge Between Research and Application*. Conf-780411 U.S. Tech. Inf. Serv., Springfield, Virginia.

White, P. R. 1939. *Controlled differentiation in a plant tissue culture*. Bull. Torrey Bot. Club 66:507–13.

Yasuda, T., P. M. Hasegawa, and T. -Y. Cheng. 1980. *Analysis of newly synthesized proteins during differentiation of cultured Douglas fir cotyledons*. Physiol. Plant. 48:83–87.

Zenk, M. H. 1978. *The impact of plant cell culture on industry*. Pages 1–13 in T. A. Thorpe, ed., *Frontiers of Plant Tissue Culture 1978*. University of Calgary Press, Canada.

21] Rapid Clonal Propagation by Tissue Culture

by RICHARD H. ZIMMERMAN*

ABSTRACT

The rapid large-scale increase of any plant clone, whether developed through genetic engineering or not, is possible only through the use of an in vitro propagation system. Such a system can be based on regeneration of plants from existing meristems, *de novo* formation of organ primordia from callus, or somatic embryogenesis. Each system utilizes the steps of explant establishment, propagule multiplication, plantlet regeneration, and acclimatization to greenhouse or field conditions. The principles of rapid clonal propagation are illustrated using examples drawn from research on apples and other fruit crops. Several factors that have significant impact on rapid clonal propagation are discussed. They include juvenility, latent bacterial contaminants, tissue vitrification, differential response among clones, and phenotypic stability of regenerated plants.

INTRODUCTION

Rapid clonal propagation is not only the final step necessary in order to utilize vegetatively propagated plants that have been genetically engineered, but it is also widely used for increasing the quantity of any selected clone, no matter how developed or selected. The term "clone" is used here in its classical meaning to denote a group of genetically uniform individuals that have originated from a single plant and have been reproduced only by vegetative (a-sexual) means, in contrast to its recent usage to denote a plant derived from a single cell.

As the means to carry out rapid clonal propagation have been developed, the techniques have been adopted, with surprising quickness in some instances, for use by producers of a wide range of plants. Production on a commercial

*Fruit Laboratory, Agricultural Research Service, U.S. Department of Agriculture, Beltsville, Maryland 20705.

GENETIC ENGINEERING: APPLICATIONS TO AGRICULTURE (Beltsville Symposium 7—Lowell D. Owens, ed.) Rowman & Allanheld, Totowa

scale was first accomplished with orchids (Murashige 1974) and soon afterwards with an ever-increasing number of ornamental foliage and flower crops (Holdgate 1977). Within the past ten years, techniques have been developed for a number of fruit, woody ornamental, tree, and specialty crops, and commercial production of many of these crops is now well under way (Boxus, Quoirin, and Laine 1977; Zimmerman 1979, Levy 1981; Evans and Sharp 1982; Harney 1982; Lane 1982). Seed-propagated agronomic and vegetable crops have made limited use of rapid clonal propagation technology, but this situation could well change as efficient and reliable methods for somatic embryogenesis are developed.

The general principles of rapid clonal propagation, which apply to all crops, have been reviewed numerous times in recent years (Murashige 1974; Vasil and Vasil 1980; Harney 1982; Lane 1982; Zimmerman 1982a,b). These techniques have been applied to a wide variety of crops, but I will limit this discussion mainly to fruit crops, primarily apple (*Malus sylvestris* Mill.). Furthermore, I will not attempt to cover related topics such as the use of in vitro culture to free plants of specific disease organisms, e.g., viruses. It must be reemphasized, however, that rapid clonal propagation by itself does not ensure that the plants produced will be free of disease organisms (J. B. Jones 1979; Smith and Oglevee-O'Donovan 1979).

POTENTIAL CULTURE SYSTEMS

Three main types of culture systems can be used for rapid clonal propagation: (a) the use of a differentiated organ as an explant for direct production of buds, shoots, bulbs, corms, etc.; (b) the regeneration of differentiated organs by organogenesis from callus; and (c) somatic embryogenesis. Factors affecting both organogenesis and somatic embryogenesis in vitro have been discussed by Thorpe in the preceding chapter and elsewhere (Thorpe 1982).

The first of these three systems is the one most commonly used at present and will be emphasized in the following discussion. Although it has the lowest potential for multiplication of the three, very large numbers of shoots per explant can still be produced in one year (Levy 1981). Advantages of this technique are its adaptability to a wide range of plants, its relative simplicity, and its increased likelihood of producing phenotypically uniform, genetically stable plants. Shoot-tips or meristem-tips are the usual explant for many crops, including all the deciduous fruit crops. Proliferation can occur by stimulation of axillary buds or by formation of adventitious buds on stems or leaves. Other potential explants include bulb scales of lilies and other bulb crops from which adventitious bulbs are formed (see Murashige 1974).

The second system involves obtaining callus from the original explant, which can be any of a wide variety of organs including leaf, stem, root, flower bud, and inflorescence, among others. From the callus, organogenesis is induced, which usually results in differentiation of adventitious buds or shoots that are multiplied and then rooted, as in the first system. This callus system is effective for a wide range of plants, many of them ornamental (Murashige

1974), but not yet for deciduous fruit trees. Since the plants are regenerated from callus, concern about the phenotypic stability and uniformity of the regenerated plants has limited the application of this method, although it is used commercially for some ornamental plants.

The third system, somatic embryogenesis, has been described for numerous plants (Tisserat, Esan, and Murashige 1979; Sharp et al. 1980), but it is not yet in general use as a method of rapid clonal propagation. While the potential of the technique is great, numerous obstacles remain to be overcome, including the problem of insuring the genetic stability and phenotypic uniformity of the plants produced from somatic embryos.

CULTURE SYSTEM STEPS

Propagating a plant using tissue culture techniques requires that the organs, tissues, or cells be carried through a sequence of steps in which different environmental and cultural conditions are used. These have been defined in terms of different stages (Murashige 1974), with Stage I being explant establishment, Stage II being multiplication of the propagules, and Stage III being preparation for reestablishing the plants in soil. This definition of Stage III is very broad and covers several procedures requiring different environmental and cultural conditions. Thus, Stage III is often considered to be the one in which whole plants are regenerated from the proliferated cultures, usually by induction of rooting on shoots. Stage IV is added for acclimatizing the newly regenerated plants to greenouse of field conditions.

Explant establishment. The first step in the culture system is to initiate growth of an explant in vitro. Explants can be obtained from either actively growing or fully dormant plants, depending upon the plant species, source of stock plants, time of year, and culture methods to be used. Explants obtained from plants grown in a greenhouse are less likely to be contaminated than from field material, particularly if plants are maintained as a relatively low humidity for several weeks prior to explant collection and if water is kept off the shoots during greenhouse care. Actively growing shoot tips taken from chilled plants that have been forced into growth are good sources of explants for apples (O. P. Jones 1979) and peaches (Hammerschlag 1982b). Explants 1 to 3 cm long are prepared from such actively growing shoots. Short meristem-tips (0.1 to 1.5 mm long) can be dissected from axillary buds on actively growing shoots or from dormant buds. For the latter, meristem-tips will grow regardless of whether the buds have received the cold treatment normally required to break dormancy.

When explants are prepared from actively growing shoot tips, the leaves are removed and the shoot tip is washed in a dilute detergent solution, treated with dilute sodium or calcium hypochlorite, and finally rinsed in sterile distilled water to remove any surface-borne contaminants. An initial wash in running tap water for 30 to 90 minutes is often useful in this regard (Jones, Pontikis, and Hopgood 1979). Duration of treatment and concentration of hypochlorite solution varies with species, tissue source, and growth stage of the plant. More rigorous procedures can be used with dormant buds since the meristem-tip is well protected by bud scales.

Various media formulations have been tried, but that of Murashige and Skoog (1962) has been effective with a wide array of fruit crops, other woody plants, and numerous herbaceous crops. Other media used effectively with both fruit crops and woody plants are those of Lepoivre (Quoirin and Lepoivre 1977) and the woody plant medium (WPM) of McCown (Lloyd and McCown 1980). An agar-solidified medium is used most often, but a liquid medium has been more effective in some cases, e.g., blackberry (Broome and Zimmerman 1978), peach (Hammerschlag 1982b), and apple (Zimmerman and Broome 1980).

Growth regulator requirements during the establishment stage vary depending upon the plant being cultured and the explant being used. Meristem-tips, particularly very small ones, seem to respond better to lower concentrations of growth regulators than do the larger shoot tip explants. Benzyladenine (BA) has been the most effective cytokinin with a few exceptions, such as blueberry, for which 2-isopentenyladenine (2iP) has been more efficacious. For shoot tips, 1 mg/l BA is a usual concentration, and for meristem-tips a reduction to 0.1 mg/l is common. Either indolebutyric acid (IBA) or naphthaleneacetic acid (NAA) is used as the auxin at a concentration equimolar to or less than that of the cytokinin.

Little has been done experimentally to determine the optimum environmental conditions for explant establishment but, typically, constant temperatures of 25° to 27°C and a photosynthetic photon flux density (PPFD) of 30 to 100 μmol $sec^{-1}m^{-2}$ for 16 or 24 hr per day are used. After 3 to 6 weeks, explants have developed enough to be transferred to a multiplication medium.

Propagule multiplication. Established explants are recultured on a medium similar or identical to that used for establishment but often containing a higher concentration of growth regulators, particularly cytokinin. Although BA concentrations of 1 to 3 mg/l are used typically, concentrations as high as 10 mg/l have been used successfully. The resulting shoots are characterized by very short internodes and deeply lobed leaves. When transferred to medium with lower cytokinin concentration, these shoots elongate normally and new leaves have a more typical appearance. With additional time in culture, many axillary buds will begin to grow. The tips of the shoots from these buds are then subcultured for additional proliferation and the bases are recultured to permit regrowth of shoots from buds that had not grown previously. Some production of adventitious buds may occur from the recultured shoot bases or callus that occasionally grow around them (Nasir and Miles 1981).

Culturing the proliferating tissue in a liquid medium can increase the multiplication rate significantly. When the tissues are left too long in liquid medium, however, they become vitreous, i.e., water-soaked in appearance with translucent leaf blades sometimes accompanied by distortion of the blade and petiole. Alternating between liquid and solid medium enhances shoot proliferation without production of abnormal shoots (Snir and Erez 1980).

Reculturing and subculturing more frequently can also increase the rate of multiplication, and this method has been used to build up the number of shoots more rapidly with some crops, e.g. strawberry.

Proliferation of some woody plants, including apple, is rather slow at first, and this response has been linked to the necessity for adapting the explants to the culture conditions. This delay is more noticeable when explants are derived from mature, fruiting trees than from seedlings. The appearance of the culture also reflects the rate of adaptation since early leaves produced in culture are more likely to be entire with little lobing. As the number of subcultures increase, lobing of the leaves, typical of the juvenile seedling form, increases. Since increased levels of cytokinin also seem to increase lobing of the leaves, this may indicate a role for cytokinin in the normal phase change of these plants.

Environmental conditions used for the proliferation stage are basically the same as those used for explant establishment.

Plantlet regeneration. After the proliferating cultures have produced an adequate number of propagules, some or all of these propagules can be taken to produce whole plants. Proliferating cultures that produce shoots must be induced to form adventitious roots at the base of excised shoots. The ease with which this can be done varies greatly depending upon a number of factors, which include the species of plant being cultured, the cultivar within the species, the time elapsed since the culture was initiated, and the development of the culture since the last preceding subculture.

For root induction on shoots the composition of the medium is modified. The concentration of mineral salts and sugar is normally reduced to half-strength or less of that used for proliferation, cytokinin is omitted, and the auxin concentration is adjusted, often to a higher level. Gibberellic acid has been found to enhance rooting in "Calita" Japanese plum *(Prunus salicina)* under certain conditions (Rosati, Marino, and Swierczewski 1980), but it is generally not used in the rooting medium.

Rooting of shoots can be done in vitro on solid or in liquid medium or by direct rooting under non-aseptic conditions using intermittent mist or high humidity to maintain the turgidity of the cuttings. When this latter method is used, the rooting and acclimatization steps are combined. Rooting in vitro, on agar medium is the usual method. The agar provides support for the cutting, so the resulting plant has a straight stem and can be handled easily. Shoots can also be rooted in either stationary or agitated liquid medium. When cuttings are rooted in an agitated liquid culture, geotropism is overcome to some extent and the roots on the resulting plants are difficult to insert into a soil mix for the acclimatization stage. For stationary liquid cultures, support for the shoot is provided by a filter paper bridge or by a material such as perlite or vermiculite. Unsupported cuttings in stationary liquid medium develop undesirable crooked stems. Roots developed in liquid medium will sometimes initiate with less auxin and tend to elongate more rapidly than those in agar medium.

Few data are available concerning the optimum temperature and light intensity, quality, and duration for plantlet regeneration. Lane (1978) reported that dropping the temperature from 28°C to 23° or 18°C reduced the percentage of apple shoots forming roots and the number of roots formed per shoot. Increasing the PPFD from 100 to 400 μmol $sec^{-1}m^{-2}$ reduced rooting of 'Jonathan' apple shoots in vitro, as did dropping the temperature from 26°C to 21°C, and the effects appeared to be additive (R. H. Zimmerman, unpublished). Rooting of Myrobalan plum *(Prunus cerasifera)* was more rapid at 26° C than at 21°C, although the final rooting percentages were similar (Hammerschlag 1982a). In contrast, 'Calita' Japanese plum rooted better at 21°C than at 26°C or 30°C unless GA_3 was added to the medium (Rosati, Marino, and Swierczewski 1980).

Both auxin concentration and type can affect root initiation. The most effective concentration can vary with species, cultivar, and culture conditions. With apple cultivars, liquid medium over an opaque support (e.g., vermiculite) induced better rooting than did agar medium when the concentration (0.01 mg/l) of auxin was very low, but these differences disappeared when auxin concentrations were increased 10 to 100 times (Zimmerman 1981). Both IBA and NAA have been equally effective in rooting shoots of apple cultivars in vitro (Zimmerman 1981).

Addition of various phenolic compounds to the rooting medium has given mixed results with fruit crops. Phloroglucinol was first reported to stimulate rooting of apple shoots by Jones (1976), and these results were confirmed and extended to additional crops (James 1979; James and Thurbon 1979b, 1981; Jones, Pontikis and Hopgood 1979; James, Knight, and Thurbon 1980). In contrast, little or no effect from phloroglucinol was found for many apple cultivars and rootstocks (Snir and Erez 1980; Zimmerman and Broome 1981) or blackberry (Broome and Zimmerman 1978). Chlorogenic acid stimulated rooting in Myrobalan plum (Hammerschlag 1982a), and quercitin, rutin, and phloridzin were effective on peach (Mosella Ch. and Macheix 1979).

Acclimatization. Once complete plants have been regenerated in vitro, they must be conditioned for return to the environment outside the culture vessel. Since the plants have grown continuously in conditions of high humidity and low light, the leaves are poorly adapted for this transfer. They may lack a completely developed waxy cuticle (Sutter and Langhans 1979), and/or the stomatal response to stress conditions may be sluggish, resulting in considerable water loss (Brainerd and Fuchigami 1982). Nevertheless, some plants, e.g., strawberry, readily adapt to greenhouse conditions. In contrast, leaves of tissue-cultured apples are particularly susceptible to injury when the plants are removed for transplanting, and can desiccate within a few minutes if not kept moist. This response of apple leaves before acclimatization is due to the stomates remaining open when subjected to stress (Brainerd and Fuchigami 1981, 1982).

The typical procedure for acclimatization is to remove the rooted plants from the culture container, wash the medium from the roots, and pot the plants in a growing mix appropriate for the crop. The potted plants are moved under a mist system or into an enclosure in which the humidity can be maintained above ambient. The misting or humidity is gradually reduced until the plants can be exposed to ambient temperature and humidity without desiccation or wilting, usually 5 to 14 days, depending on the crop. During acclimatization and sometimes for a few days during rooting, the light intensity is increased so that the plant can become autotrophic.

Uncapping the culture vessel for several days prior to removing the rooted plants and maintaining these opened containers in the growth room can speed and improve the acclimatization process for apples. If the opening on the culture vessel is too large, air exchange occurs too rapidly and the leaves desiccate. This problem can be overcome by using perforated plastic covers. With this method, leaves adapt to conditions of lower humidity in 2 or 3 days so that stomatal closure occurs in response to water stress without the lag found in unadapted leaves. Some desiccation of leaf margins may occur, but the leaves expand and their chlorophyll content increases. After 5 or 6 days the plants can be potted and placed in the greenhouse without using mist or a humidity chamber.

FACTORS AFFECTING RAPID CLONAL PROPAGATION

Juvenility. While it would be easier to initiate cultures from juvenile seedlings, such plants cannot be used as explant sources for plants of which clones are selected only in the mature phase. These clones must be established in culture from mature-phase plants, a task that has been accomplished for many crops. In culture, however, the explants of these mature-phase clones will later produce leaves characteristic of the seedling, or juvenile, stage (Zimmerman 1982b). These leaves seem to be produced only from leaf primordia initiated during culture and not from primordia preexisting on the explants at the time of culture establishment. Eventually, leaves characteristic of the mature phase, e.g., trifoliolate leaves of strawberry, can develop in vitro, but this is found most often at the rooting stage. More commonly, such mature-form leaves develop only after acclimatized plants are growing in the greenhouse or field.

This apparent reversion to the juvenile phase might also explain the increased rooting of shoot cuttings after a certain number of subcultures (Sriskandarajah, Mullins, and Nair 1982). Thus, cuttings of 'Jonathan' apple did not root at all when taken from the original plant, 8% rooted from the initial culture, 32% from the first subculture, 70% from the fifth, and 95% from the ninth and later subcultures. Similar results, but of lesser magnitude and requiring many more subcultures, were obtained with 'Delicious' (Sriskandarajah, Mullins, and Nair 1982) and have been noted for several other apple cultivars (R. H. Zimmerman, unpublished).

Nevertheless, apple trees propagated by these in vitro techniques do not show delayed flowering in the field, a response that would be anticipated if they had truly reverted to the juvenile phase in culture (Zimmerman 1981). In fact, trees of 'Jonathan' apple have flowered after one growing season in the field (18 months after removal from the culture tubes), and trees of 'Nugget,' 'Rome Beauty,' and 'Spuree Rome' have flowered and set fruit after two growing seasons in the field (about 28 months out of culture) (R. H. Zimmerman, unpublished). These results compare favorably to those obtained with conventionally propagated budded trees.

Latent bacterial contamination of cultures. A persistent problem in the culture of apples from meristem-tips or shoot tips has been the sudden appearance of bacteria in the medium around the base of the explants after some months of apparently contaminant-free growth of the cultures. Since these bacteria seem to emerge only from cut surfaces on the explant, one suspects that the organisms existed within or on the tissue of the original explant and were not eliminated by the procedures used to disinfest the original explant. The organism found most frequently in our cultures has been *Acinetobacter calcoaceticus,* which exists in soil and water and on human skin, but it is apparently non-pathogenic to apple. This organism has also been isolated from apple tissue cultures elsewhere in the United States. Although chlortetracycline limited the growth of the organism in vitro, the antibiotic did not eradicate the organism, and growth of the shoot cultures was inhibited.

A number of other bacteria, including *Bacillus pumilis,* have survived the usual surface-sterilization procedures and appear in cultures in the same manner as *Acinetobacter* (James and Thurbon 1979a; Constantine, Wiltshire, and Beddows 1980). Even when shoot tip explants are washed in running tap water for 30 to 90 minutes prior to disinfestation treatments (Jones, Pontikis, and Hopgood 1979), or meristem-tip explants less than 1 mm in height are dissected following a rigorous treatment with ethanol and hypochlorite, a sizeable percentage of the cultures will eventually show bacterial contamination. Tissue sections from in vitro cultured apples, screened monthly on yeast peptone agar with negative results for five consecutive times, still contained bacterial contaminants in 25% of the cases when screened a sixth time (James and Thurbon 1979a).

Tissue vitrification. Shoots of apple and many other crops grown in vitro sometimes produce succulent leaves and stems that appear waxy, water-soaked, or translucent, an appearance described as vitreous. The young leaves are malformed, often straplike and curled and may elongate much more than normal leaves (Debergh, Harbaoui, and Lemeur 1981; Zimmerman 1982b). Stem elongation is often inhibited and blades of older leaves may become partly or totally translucent. The vitreous leaves have no palisade layer, but only spongy mesophyll (Debergh, Harbaoui, and Lemeur 1981).

Apple cultures develop this condition more often and more rapidly when grown in liquid medium, but form of the medium had no effect on its development in globe artichoke (Debergh, Harbaoui, and Lemeur 1981). Cultures that become vitreous can sometimes be induced to revert to normal-appearing growth by appropriate treatment. With apple cultures, the vitreous condition can sometimes be overcome by transferring them from medium containing BA to one with 2iP or no cytokinin at all (Zimmerman 1982b). This treatment results in complete loss of shoot proliferation, but the cultures often grow out of the vitreous condition. The nonvitreous shoots can then be subcultured on medium with BA to renew shoot proliferation. Changes in salt mixtures and concentrations, as well as placing apple cultures in 4°C for one or more months, have resulted in cultures growing out of the vitreous condition in some cases.

With globe artichoke, increasing the agar concentration from 0.6% to 1.1% eliminated shoot vitrification but at the cost of halving the shoot proliferation rate (Debergh, Harbaoui, and Lemeur 1981). Additional sucrose or mannitol in the medium did not have the same effect on the water potential of the medium as did the agar, nor did increased concentration of either of these two substances reduce vitrification of the shoots. Increased agar concentration in the medium also eliminates vitrification in apple shoots in some cases but, again, with some reduction in shoot proliferation.

Since several apparently quite different treatments can reduce tissue vitrification in apple, the causes of this disorder seem to be complex. Devising a solution to this problem, especially one that does not exact too high a cost in terms of shoot proliferation, will require attention to a number of different factors simultaneously.

Differential response among clones. Different cultivars of apple respond differently to the same medium and environmental conditions during establishment, proliferation, and rooting in vitro. Differential sensitivity to the type and concentration of cytokinin and auxin and to the ratio between them are important, particularly in the first two stages. All apple cultivars are probably difficult to root when first established in culture. After they have been adapted to culture conditions and have been subcultured at least several times, some cultivars consistently root with ease while others remain difficult (Zimmerman and Broome 1981; Sriskandarajah, Mullins, and Nair 1982; R. H. Zimmerman, unpublished). Two separate lines of the apple rootstock M.9, differing only in the times at which the original stock plants were taken from cold storage, were found to differ significantly in rooting percentage in vitro (James and Thurbon 1981). Whether this difference relates to the difference in treatment of the stock plants or to inherent differences between individual plants of M.9 is not known. Similar differences among clones have been observed with blueberry and azalea in Stages I and II (R. H. Zimmerman, unpublished) and exist for many other crops as well.

Phenotypic stability of regenerated plants. Rapid clonal propagation is ultimately successful only if the plants produced are phenotypically identical to the original clones and are as stable genetically as plants propagated by other vegetative methods. Confirming the identity and stability of apples requires long-term field testing because data must be collected on vegetative and fruiting characteristics as well as on continuing field performance. Such testing has been done with strawberry (Swartz, Galletta, and Zimmerman 1981) and thornless blackberry (H. J. Swartz, G. J. Galletta, and R. H. Zimmerman, unpublished). These tissue culture–propagated plants were found to perform as well as, or better than, conventionally propagated ones. No variant plants of thornless blackberry, and very few of strawberry, were found in the field. Variant plants of in vitro propagated apples have not been found in the field, but the first test plantings have been in the orchard for less than 4 years so that fruiting and long-term yield data are not yet available (R. H. Zimmerman, unpublished).

CONCLUSIONS

The rapid propagation of selected plant clones is feasible using in vitro methods, and this fact is now being applied for commerical production of millions of plants annually. Most of these plants are produced by stimulating production of axillary shoots from meristem-tip or shoot-tip explants. Regeneration of plants following organogenesis from callus is used for some crops, but the possibility of producing variant plants may limit its wider use. Somatic embryogenesis, little used at present, has the potential for much greater exploitation in the future.

All the techniques used follow a sequence consisting of establishment of explants in vitro, proliferation of tissues or organs, regeneration of whole plants, and reestablishment of these plants in greenhouse or field conditions. Utilization of in vitro propagation techniques has been greatest by producers of ornamental plants, mainly herbaceous ones. As methods for woody plants have been developed, the range of crops being propagated has broadened, but new problems have arisen. These concern mainly the long-term performance of such plants in the field. Evaluation of in vitro–propagated plants requires many years, in some cases, with the result that the application of rapid clonal propagation to some of these crops may need to be delayed until adequate field performance testing has been completed.

LITERATURE CITED

Boxus, P., M. Quoirin, and J. M. Laine. 1977. *Large scale propagation of strawberry plants from tissue culture*. Pages 130–43 in J. Reinert and Y. P. S. Bajaj, eds., *Applied and Fundamental Aspects of Plant Cell, Tissue, and Organ Culture*. Springer-Verlag, Berlin-Heidelberg-New York.

Brainerd, K. E., and L. H. Fuchigami. 1981. *Acclimatization of aseptically cultured apple plants to low relative humidity*. J. Amer. Soc. Hort. Sci. 106:515–18.

———. 1982. *Stomatal functioning of in vitro and greenhouse apple leaves in darkness, mannitol, ABA, and CO_2*. J. Exp. Bot. 33:388–92.

Broome, O. C., and R. H. Zimmerman. 1978. *In vitro propagation of blackberry*. HortScience 13:151–53.

Constantine, D. R., S. Wiltshire, and C. Beddows. 1980. *Contamination of cultures*. Long Ashton Research Stn., Report 1979:74.

Debergh, P., Y. Harbaoui, and R. Lemeur. 1981. *Mass propagation of globe artichoke (Cynara scolymus): Evaluation of different hypotheses to overcome vitrification with special reference to water potential*. Physiol. Plant. 53:181–87.

Evans, D. A., and W. R. Sharp. 1982. *Application of tissue culture technology in the agricultural industry*. Pages 209–31 in D. T. Tomes, B. E. Ellis, P. M. Harney, K. J. Kasha, and R. L. Peterson, eds., *Application of Plant Cell and Tissue Culture to Agriculture and Industry*. University of Guelph, Ontario.

Hammerschlag, F. 1982a. *Factors influencing in vitro multiplication and rooting of the plum rootstock myrobalan (Prunus cerasifera Ehrh.).* J. Amer. Soc. Hort. Sci. 107:44–47.

———. 1982b. *Factors affecting establishment and growth of peach shoots in vitro.* HortScience 17:85–86.

Harney, P. M. 1982. *Tissue culture propagation of some herbaceous horticultural plants.* Pages 187–208 in D. T. Tomes, B. E. Ellis, P. M. Harney, K. J. Kasha, and R. L. Peterson, eds., *Application of Plant Cell and Tissue Culture to Agriculture and Industry.* University of Guelph, Ontario.

Holdgate, D. P. 1977. *Propagation of ornamentals by tissue culture.* Pages 18–43 in J. Reinert and Y. P. S. Bajaj, eds., *Applied and Fundamental Aspects of Plant Cell, Tissue, and Organ Culture.* Springer-Verlag, Berlin-Heidelberg-New York.

James, D. J. 1979. *The role of auxins and phloroglucinol in adventitious root formation in Rubus and Fragaria grown in vitro.* J. Hort. Sci. 54:273–77.

James, D. J., V. H. Knight, and I. J. Thurbon. 1980. *Micropropagation of red raspberry and the influence of phloroglucinol.* Scientia Hort. 12:313–19.

James, D. J., and I. J. Thurbon. 1979a. *Culture in vitro of M.9 apple.* Rep. East Malling Res. Stn. for 1978: 179–80.

———. 1979b. *Rapid in vitro rooting of the apple rootstock M.9.* J. Hort. Sci. 54:309–11.

———. 1981. *Shoot and root initiation in vitro in the apple rootstock M.9 and the promotive effects of phloroglucinol.* J. Hort. Sci. 56:15–20.

Jones, J. B. 1979. *Commercial use of tissue culture for the production of disease-free plants.* Pages 441–52 in W. R. Sharp, P. O. Larsen, E. F. Paddock, and V. Raghavan, eds., *Plant Cell and Tissue Culture.* Ohio State University Press, Columbus.

Jones, O. P. 1976. *Effect of phloridzin and phloroglucinol on apple shoots.* Nature (Lond.) 262:392–93.

———. 1979. *Propagation in vitro of apple trees and other woody fruit plants: Methods and applications.* Scientific Hort. 30:44–48.

Jones, O. P., C. A. Pontikis and M. E. Hopgood. 1979. *Propagation in vitro of five apple scion cultivars.* J. Hort. Sci. 54:155–58.

Lane, W. D. 1978. *Regeneration of apple plants from shoot meristem-tips.* Plant Sci. Lett. 13:281–85.

———. 1982. *Tissue culture and in vitro propagation of deciduous fruit and nut species.* Pages 163–86 in D. T. Tomes, B. E. Ellis, P. M. Harney, K. J. Kasha, and R. L. Peterson, eds., *Application of Plant Cell and Tissue Culture to Agriculture and Industry.* University of Guelph, Ontario.

Levy, L. W. 1981. *A large-scale application of tissue culture: The mass propagation of pyrethrum clones in Ecuador.* Environ. and Exp. Bot. 21:389–95.

Lloyd, G., and B. McCown. 1980. *Commercially feasible micropropagation of mountain laurel, Kalmia latifolia, by use of shoot-tip culture.* Proc. Intern. Plant Propagators' Soc. 30:421–27.

Mosella Ch., L., and J. J. Macheix. 1979. *Le microbouturage in vitro du Pecher (Prunus persica Batsch): Influence de certains composés phenoliques.* C. R. Acad. Sci. Paris, Ser. D 289:567–70.

Murashige, T. 1974. *Plant propagation through tissue cultures.* Ann. Rev. Plant Physiol. 25:135–66.

Murashige, T., and F. Skoog. 1962. *A revised medium for rapid growth and bioassays with tobacco tissue cultures.* Physiol. Plant. 15:473–97.

Nasir, F. R., and N. W. Miles. 1981. *Histological origin of EMLA 26 apple shoots generated during micropropagation.* HortScience 16:417 (Abstr.).

Quoirin, M., and P. Lepoivre. 1977. *Etude de milieux adaptés aux cultures in vitro de Prunus.* Acta Hort. 78:437–42.

Rosati, P., G. Marino, and C. Swierczewski. 1980. *In vitro propagation of Japanese plum (Prunus salicina Lindl. cv. Calita).* J. Amer. Soc. Hort. Sci. 105:126–29.

Sharp, W. R., M. R. Sondahl, L. S. Caldas, and S. B. Maraffa. 1980. *The physiology of in vitro asexual embryogenesis*. Hort. Rev. 2:268–310.

Smith, S. H., and W. A. Oglevee-O'Donovan. 1979. *Meristem-tip culture from virus-infected plant material and commercial implications*. Pages 453–60 in W. R. Sharp, P. O. Larsen, E. F. Paddock, and V. Raghavan, eds., *Plant Cell and Tissue Culture*. Ohio State University Press, Columbus.

Snir, I., and A. Erez. 1980. *In vitro propagation of Malling Merton apple rootstocks*. HortScience 15:597–98.

Sriskandarajah, S., M. G. Mullins, and Y. Nair. 1982. *Induction of adventitious rooting in vitro in difficult-to-propagate cultivars of apple*. Plant Sci. Lett. 24:1–9.

Sutter, E., and R. W. Langhans. 1979. *Epicuticular wax formation on carnation plantlets regenerated from shoot tip culture*. J. Amer. Soc. Hort. Sci. 104:493–96.

Swartz, H. J., G. J. Galletta and R. H. Zimmerman. 1981. *Field performance and phenotypic stability of tissue culture-propagated strawberries*. J. Amer. Soc. Hort. Sci. 106:667–73.

Thorpe, T. A. 1982. *Callus organization and de novo formation of shoots, roots and embryos in vitro*. Pages 115–138 in D. T. Tomes, B. E. Ellis, P. M. Harney, K. J. Kasha, and R. L. Peterson, eds., *Application of Plant Cell and Tissue Culture to Agriculture and Industry*. University of Guelph, Ontario.

Tisserat, B., E. B. Esan, and T. Murashige. 1979. *Somatic embryogenesis in angiosperms*. Hort. Reviews 1:1–78.

Vasil, I. K., and V. Vasil. 1980. *Clonal propagation*. Pages 145–73 in I. K. Vasil, ed., *Perspectives in Plant Cell and Tissue Culture*. Intern. Rev. of Cytology Suppl. 11A. Academic Press, New York.

Zimmerman, R. H. 1979. *The Laboratory of Micropropagation at Cesena, Italy*. Proc. Intern. Plant Propagators' Soc. 29:398–400.

———. 1981. *Micropropagation of fruit plants*. Acta Hort. 120:217–22.

———. 1982a. *Tissue culture*. Pages 124–35 in J. N. Moore and J. Janick, eds., *Methods in Fruit Breeding*. Purdue Univ. Press, West Lafayette, Indiana.

———. 1982b. *Apple tissue culture*. In D. A. Evans, W. R. Sharp, and P. V. Ammirato, eds., *Handbook of Plant Cell Culture,* vol. 2 Macmillan, New York (forthcoming).

Zimmerman, R. H., and O. C. Broome. 1980. *Apple cultivar micropropagation*. Pages 54–58 in *Proc. Conference on Nursery Production of Fruit Plants through Tissue Culture—Applications and Feasibility*. USDA-SEA-Agr. Res. Results ARR-NE-11, Beltsville, MD.

———. 1981. *Phloroglucinol and in vitro rooting of apple cultivar cuttings*. J. Amer. Soc. Hort. Sci. 106:648–52.

Index of Authors*

*Only the senior authors of literature cited and the pages upon which the full citation appears are listed. Authors' names in capital letters are contributors to chapters in these proceedings.

Index of Subjects